Using Graywater and Stormwater to Enhance Local Water Supplies

AN ASSESSMENT OF RISKS, COSTS, AND BENEFITS

Committee on the Beneficial Use of Graywater and Stormwater:
An Assessment of Risks, Costs, and Benefits

Water Science and Technology Board

Division on Earth and Life Studies

The National Academies of
SCIENCES · ENGINEERING · MEDICINE

THE NATIONAL ACADEMIES PRESS
Washington, DC
www.nap.edu

THE NATIONAL ACADEMIES PRESS **500 Fifth Street, NW** **Washington, DC 20001**

This activity was supported by the City of Madison, Wisconsin; Grant No. EP-C-09-003, TO #23 and EP-C-14-005, TO #9 from the U.S. Environmental Protection Agency; the Los Angeles Department of Water and Power; this material is based on work supported by the National Science Foundation under Grant No. CBET-1321776; the National Water Research Institute; the Water Environment Research Foundation; the Water Research Foundation Agreement No. 04521; the WateReuse Research Foundation; and with additional support from the National Academy of Sciences Arthur L. Day Fund, the National Academy of Sciences W.K. Kellogg Foundation Fund, and the National Academy of Sciences George and Cynthia Mitchell Endowment for Sustainability Science. Any opinions, findings, conclusions, or recommendations expressed in this publication do not necessarily reflect the views of any organization or agency that provided support for the project.

International Standard Book Number-13: 978-0-309-38835-1
International Standard Book Number-10: 0-309-38835-X
Digital Object Identifier: 10.17226/21866

Cover credit (*clockwise from top left*): Images courtesy of Water Replenishment District of Southern California; Innovative Water Solutions, LLC, Austin, Texas; and Mother Earth News.

Additional copies of this report are available for sale from the National Academies Press, 500 Fifth Street, NW, Keck 360, Washington, DC 20001; (800) 624-6242 or (202) 334-3313; Internet, http://www.nap.edu.

Printed in the United States of America

Suggested citation: National Academies of Sciences, Engineering, and Medicine. 2016. *Using Graywater and Stormwater to Enhance Local Water Supplies: An Assessment of Risks, Costs, and Benefits*. Washington, DC: The National Academies Press. doi: 10.17226/21866.

The National Academies of

SCIENCES • ENGINEERING • MEDICINE

The **National Academy of Sciences** was established in 1863 by an Act of Congress, signed by President Lincoln, as a private, nongovernmental institution to advise the nation on issues related to science and technology. Members are elected by their peers for outstanding contributions to research. Dr. Ralph J. Cicerone is president.

The **National Academy of Engineering** was established in 1964 under the charter of the National Academy of Sciences to bring the practices of engineering to advising the nation. Members are elected by their peers for extraordinary contributions to engineering. Dr. C. D. Mote, Jr., is president.

The **National Academy of Medicine** (formerly the Institute of Medicine) was established in 1970 under the charter of the National Academy of Sciences to advise the nation on medical and health issues. Members are elected by their peers for distinguished contributions to medicine and health. Dr. Victor J. Dzau is president.

The three Academies work together as the **National Academies of Sciences, Engineering, and Medicine** to provide independent, objective analysis and advice to the nation and conduct other activities to solve complex problems and inform public policy decisions. The Academies also encourage education and research, recognize outstanding contributions to knowledge, and increase public understanding in matters of science, engineering, and medicine.

Learn more about the National Academies of Sciences, Engineering, and Medicine at **www.national-academies.org**.

COMMITTEE ON THE BENEFICIAL USE OF GRAYWATER AND STORMWATER: AN ASSESSMENT OF RISKS, COSTS, AND BENEFITS

RICHARD G. LUTHY, Chair, Stanford University, California
RICHARD W. ATWATER, Southern California Water Committee, Studio City, California
GLEN T. DAIGGER, University of Michigan, Ann Arbor
JÖRG DREWES, Technische Universität München, Garching, Germany
BENJAMIN H. GRUMBLES, Maryland Department of the Environment, Baltimore
ARPAD HORVATH, University of California, Berkeley
ROBERT E. PITT, University of Alabama, Tuscaloosa
MARCUS M. QUIGLEY, OptiRTC, Boston, Massachusetts
ROBERT S. RAUCHER, Stratus Consulting/Abt Associates, Boulder, Colorado
SYBIL SHARVELLE, Colorado State University, Fort Collins
CLAIRE WELTY, University of Maryland Baltimore County
MARYLYNN V. YATES, University of California, Riverside

Academies Staff

STEPHANIE E. JOHNSON, Study Director, Water Science and Technology Board
MICHAEL J. STOEVER, Research Associate, Water Science and Technology Board

Preface

Much of the United States faces chronic or episodic water shortages. It is the topic of daily news in the West, where a historic 4-year drought has caused California to restrict the delivery of water to cities and farms. At the same time, the Midwest and Northeast have received drenching rains and heavier than normal snow. Against this backdrop—of not enough water or too much water—the National Academies of Sciences, Engineering, and Medicine's Water Science and Technology Board initiated a study on the beneficial use of stormwater and graywater. Graywater is a year-round source of water for nonpotable use, and use of urban stormwater can augment local water supplies, reduce demand for imported water, and lessen impacts from discharge.

As detailed in this report, increased attention to the use of stormwater and graywater has been driven by factors forcing change in the design and management of urban water supplies and infrastructure. Among the drivers are water scarcity in regions of the country facing water shortages and the impacts of climate change and population growth that exacerbate these shortages. In these places stormwater and graywater use may diversify the water supply portfolio, thereby achieving greater resiliency in the face of uncertain water deliveries. Furthermore, in many parts of the country—from the humid midcontinent to coastal cities—pollution control and discharges to impaired water bodies are driving changes in the ways that stormwater is managed, and stormwater capture and use can reduce pollution from urban runoff, including combined sewer overflows.

Stormwater and graywater use exemplify a growing trend of embracing sustainable urban water management and green design practices. The concept of a re-imagined urban water infrastructure—variously termed low-impact design, blue-green city, or water sensitive city—embraces sustainable practices in which metropolitan regions could serve as water supply catchments, provide ecosystem services, and prioritize livability, sustainability, and resilience. However, realizing this vision raises questions on exactly how graywater and urban stormwater should be captured, stored, and used. Because of the absence of ample documentation of costs, performance, and risks, many utilities are hesitant to integrate the practices into their long-term water resource plans beyond the simplest applications. Potential public health risks from microbial or chemical contamination associated with graywater or stormwater use raise concerns about safety, regulation, and management. To better address these challenges, the Academies formed a committee to study the risks, costs, and benefits of stormwater and graywater use to augment and conserve existing water supplies. Although there are challenges in advancing ever-more use of graywater and urban stormwater, this report documents the committee's finding that graywater and urban stormwater have substantial potential to contribute to local water supply needs while providing other benefits such as stormwater pollution reduction, water supply diversification, and increased local control of water supplies. Graywater and stormwater use could be an important part of a broader effort to reimagine urban water infrastructure to efficiently use water, energy, and financial resources while enhancing water supply reliability and resiliency and the livability of cities.

This study was supported with funding from the Environmental Protection Agency Office of Water and Office of Research and Development; National Science Foundation; Water Research Foundation; Water Environment Research Foundation; Los Angeles Department of Water and Power; WateReuse; City of Madison, Wisconsin; National Water Research Institute; and the National Academies' President's fund. We appreciate the sponsor liaisons, including Robert Bastian, Robert Goo, Christopher Kloss, John Whitler, and Andy Niknafs, for help with information gathering in support of the study and the many presenters to the committee for the helpful insights provided. The committee also appreciates the research assistance from Amy Streitwieser, Adam Schempp, Will Derwin, Jonathan Bradshaw, and Thomas Hendrickson.

The committee had the excellent fortune to be assisted by a dedicated and talented Academies staff, including Stephanie Johnson and Michael Stoever. I speak for the entire committee in expressing our profound respect and appreciation to Stephanie Johnson for her tireless effort and clear thinking. This report would not have been possible without her exceptional support and good humor.

I very much enjoyed working with the Academies' staff and the committee members. I am sure each of us learned more than we contributed, and we offer this report in hopes that it will advance our nation on a path toward more sustainable urban water futures.

Richard Luthy, *Chair*
Committee on the Beneficial Use of Graywater and Stormwater: An Assessment of Risks, Costs, and Benefits

Acknowledgment of Reviewers

This report has been reviewed in draft form by persons chosen for their diverse perspectives and technical expertise. The purpose of this independent review is to provide candid and critical comments that will assist the institution in making the published report as sound as possible and to ensure that it meets institutional standards of objectivity, evidence, and responsiveness to the study charge. The review comments and draft manuscript remain confidential to protect the integrity of the deliberative process. We wish to thank the following individuals for their review of this report:

Nicholas Ashbolt, University of Alberta
Michael Barrett, University of Texas
Peter Dillon, International Association of Hydrogeologists Commission on Management of Aquifer Recharge
Jerome B. Gilbert, Consulting Engineer
Peter Gleick, Pacific Institute for Studies in Development, Environment, and Security
Eberhard Morgenroth, Eawag
Sheila Olmstead, University of Texas
Kevin Price, Middle East Desalination Research Center
Karl Rockne, University of Illinois at Chicago
Larry A. Roesner, Colorado State University
Bahman Sheik, Independent Water Resources and Reuse Specialist
Eric W. Strecker, Geosyntec, Inc.
Rhodes Trussell, Trussell Technologies, Inc.
Wendy E. Wagner, University of Texas School of Law

Although the reviewers listed above have provided many constructive comments and suggestions, they were not asked to endorse the conclusions or recommendations, nor did they see the final draft of the report before its release. The review of this report was overseen by Kenneth W. Potter, University of Wisconsin, and Michael Kavanaugh, Geosyntec, Inc. They were responsible for making certain that an independent examination of this report was carried out in accordance with institutional procedures and that all review comments were carefully considered. Responsibility for the final content of this report rests entirely with the authoring committee and the institution.

Contents

APPENDIXES

Summary

Chronic and episodic water shortages are becoming common in many regions of the United States, and urban population growth in water scarce regions further compounds the challenges. In mid-2015, much of California faced an "exceptional drought" within an already moderate to severe drought throughout much of the western United States, but non-arid regions, such as the Southeast, have not been exempt from major water shortages. Increasingly, alternative water sources such as stormwater and graywater are being viewed as resources to supplement scarce water supplies, particularly in urban areas experiencing large population growth.

Stormwater runoff is the water from rainfall or snow that can be measured downstream in a pipe, culvert, or stream shortly after the precipitation event. For the purposes of this report, the term "stormwater" is used broadly to include runoff from rooftops, as well as other runoff from small to large source areas. Graywater is untreated wastewater that does not include water from the toilet or kitchen, and may include water from bathroom sinks, showers, bathtubs, clothes washers, and laundry sinks. Both can offer on-site alternative water supplies to a household or building, although stormwater can be captured and used at neighborhood and regional scales and graywater can be reused in neighborhoods and large multi-residential developments.

Stormwater and graywater can serve a range of nonpotable uses, including irrigation, toilet flushing, washing, and cooling, although treatment may be needed. Stormwater may also be used to recharge groundwater, which may ultimately be tapped for potable use. In addition to increasing of local water supply, harvesting stormwater has many potential benefits, including saving energy, preventing pollution, reducing the impacts of development on urban streams, and enhancing the livability of cities. Similarly, the reuse of graywater can enhance water supply reliability and extend the capacity of existing wastewater systems in growing cities.

Despite the benefits of using local alternative water sources to address water demands, many questions remain that have limited the broader application of graywater and stormwater capture and use. In particular, limited information is available on the costs, benefits, and risks of these projects, and beyond the simplest applications, many state and local public health agencies have not developed regulatory frameworks for full use of these local water resources. With funding support from the Environmental Protection Agency (EPA); National Science Foundation; Water Research Foundation; Water Environment Research Foundation; Los Angeles Department of Water and Power; WateReuse; City of Madison, Wisconsin; National Water Research Institute; and the National Academies of Sciences, Engineering, and Medicine's President's Fund, the Academies formed the Committee on the Beneficial Use of Graywater and Stormwater to analyze the risks, costs, and benefits on various uses of stormwater and graywater, as described in Box S-1. This study addresses technical, economic, regulatory, and social issues associated with graywater and stormwater capture and use across a range of uses and scales.

Graywater and stormwater capture and use can expand local water availability while providing additional financial, environmental, and social benefits, such as reduced water pollution and combined sewer overflow discharges (for stormwater), drought-resistant year-round water availability (for graywater), and diversification of water supplies. For stormwater, neighborhood- and regional-scale stormwater capture projects can contribute significantly to urban water supplies. In most cases, the technology is mature, and treatment can be provided to address contaminants to meet "fit-for-purpose" water quality objectives. However, broader implementation is hindered by the absence of risk-based guidelines for stormwater and graywater use across a range of applications, as well as water quality data (particularly for human pathogens) necessary to assess these risks.

BOX S-1 Statement of Task

An ad hoc committee will conduct a study and prepare a report that will analyze the risks, costs, and benefits of various beneficial uses of stormwater and graywater and approaches needed for its safe use. The study will address:

1. **Quantity and suitability.** How much stormwater capture and graywater reuse occurs in the United States and for what applications? What is the suitability—in terms of water quality and quantity—of captured stormwater and graywater to significantly increase in the United States, and where regionally would increases in these practices have the most benefit? How would significant increases in the beneficial use of stormwater and graywater affect water demand, downstream water availability, aquifer recharge, and ecological stream flows? What research should be pursued to understand these issues?

2. **Treatment and storage.** What are typical levels and methods of treatment and storage for stormwater capture and graywater reuse for various end uses? What types of treatment are available to address contaminants, odors, and pathogens, and how do these treatment methods compare in terms of cost and energy use? What research opportunities should be pursued to produce improved technologies and delivery and ensure adequate safeguards to protect public health and the environment?

3. **Risks.** What are the human health and environmental risks of using captured stormwater and graywater for various purposes? What existing state and regulatory frameworks address the beneficial use of stormwater and graywater, and how effective are they in assuring the safety and reliability of these practices? What lessons can be learned from experiences using captured stormwater and graywater both within and outside the United States that shed light on appropriate uses with varying levels of treatment? What local measures can be taken to reduce risk?

4. **Costs and benefits.** What are the costs and benefits of the beneficial use of stormwater and graywater (including nonmonetized costs and benefits, such as effects on water and energy conservation, environmental impacts, and wastewater infrastructure)? How do the economic costs and benefits generally compare with other supply alternatives? Can cost improvements be achieved through research?

5. **Implementation.** What are the legal and regulatory constraints on the use of captured stormwater and graywater? What are the policy implications regarding the potential increased use of stormwater and graywater as significant alternative sources of water for human consumption and use?

As part of its review, and to help benchmark U.S. standing worldwide, the committee will consider international experiences in onsite stormwater and graywater management, as it deems relevant.

There is no single best way to use graywater or stormwater to address local water needs because project drivers and objectives, legal and regulatory constraints, potential applications, local site and climatic conditions, source water availability, and project scales all vary widely. The report instead recommends clear objectives and provides a decision framework in Chapter 9 that can be used when considering the use of graywater or stormwater, with supporting information for each of the decision steps. Major report findings are highlighted below along with recommended research needs to improve support for decision making.

WATER AVAILABILITY

Potential potable water savings from graywater and stormwater will vary based on factors such as local climatic conditions, approaches, and scales. Chapter 3 details lessons learned regarding water saving potential largely based on an original scenario analysis in six U.S. locations. The scenarios considered medium-density residential development using graywater or stormwater for conservation irrigation of turfgrass, toilet flushing, or both.

Water savings from stormwater capture and use are dependent on tank size and the amount and timing of precipitation relative to water demand. Substantial potential household-scale water savings (24 to 28 percent) from the capture and use of roof runoff were calculated for scenario analyses in four of the six cities analyzed using one moderately sized (2,200-gallon [8,300-liter]) storage tank per house. These cities—Lincoln, Nebraska; Madison, Wisconsin; Birmingham, Alabama; and Newark, New Jersey (all located in the Midwest or East Coast)—have year-round rainfall closely matching irrigation demands. In contrast, the scenario analysis showed lower potential potable water savings for Los Angeles and Seattle (5 and 15 percent, respectively). In much of the arid West, the timing and intensity of rainfall limits the capacity of household-scale stormwater collection to reduce potable water use. Very small stormwater water storage volumes provide much lower water savings benefits (less than 2 percent in Los Angeles to up to 10 percent in Newark using two 35-gallon [130 liter] rain barrels per house, for example).

Neighborhood- and regional-scale stormwater capture projects can contribute significantly to urban water supplies. This is especially important for arid climates in which stormwater can be stored in aquifers for use during drought or the dry season. Based on 1995-1999 data for Los Angeles, average stormwater runoff from medium-density residential developments, if captured and stored, would be roughly sufficient to meet indoor residential water needs in those areas.

Graywater reuse offers the potential for substantial potable water savings and could provide a reliable source of water for arid regions. Based on the committee's scenario analyses, graywater reuse in Los Angeles and Seattle provides greater potential potable water savings than does household-scale stormwater capture, because graywater provides a steady water source during summer months with little or no rainfall. Additionally, the analyses showed that graywater can more effectively meet toilet flushing demand compared to stormwater in all cities analyzed. Graywater use for toilet flushing has been demonstrated to achieve potable water savings as theoretically expected without impacting water availability to downstream users, but water savings associated with graywater irrigation at the household scale have not been demonstrated with confidence. Little is known about the impact of installing on-site nonpotable water systems on human water use behavior, which points to the need to study behavioral responses to conservation measures.

Beneficial use of graywater is typically more appropriate for residential and multi-residential applications than commercial application. Most commercial facilities do not generate enough graywater to justify use for toilet flushing or irrigation. Even offices that have on-site showers are not likely to generate enough graywater to meet end-use demands (toilet or irrigation). Some commercial applications for which graywater use may be appropriate include fitness facilities, hotels, and laundromats.

If water conservation is the primary objective for stormwater and graywater investments, then strategies that reduce outdoor water use should first be examined. In arid regions, potential potable water savings for residential and multi-residential use of stormwater and/or graywater are significant, but small relative to today's outdoor water demand. Although use of graywater or roof runoff for toilet flushing can reduce indoor demand by up to 24 percent, the committee's scenario analysis estimated potential water savings of only 13 percent with graywater use in the Los Angeles area (and significantly less for stormwater capture, even using large tanks). Significantly reducing or eliminating irrigation demand, for example through the use of xeriscaping, would provide much larger reductions in water demand in arid regions. In these circumstances, graywater could be used to supply irrigation water to meet specific small irrigation needs. Otherwise, graywater and stormwater may help facilitate the continued use of landscaping that is not sustainable in the long term and inappropriate for local climate conditions.

WATER QUALITY

Understanding the potential applications of graywater and stormwater as on-site water supplies and associated treatment needs requires a clear understanding of source water quality.

Pathogens and organic matter in graywater impact opportunities for beneficial uses without treatment. Human pathogens are likely to occur in graywater, although the specific types and concentrations vary substantially among sources and their occurrence and fate are not yet well understood. Organic matter is present in high enough concentration in graywater to enhance microbial growth, thus limiting the potential uses of graywater without disinfection. Sodium, chloride, boron, and other chemicals can impact the quality of graywater for irrigation uses. Best management practices exist for source control of microbial and chemical constituents, and such practices can be implemented at the household scale to reduce concentrations of these constituents in graywater.

Stormwater quality is highly variable over space and time and might contain elevated levels of microorganisms, metals, organic chemicals, and sediments, potentially necessitating treatment to facilitate various beneficial uses. Stormwater quality is a direct function of land use, source area, catchment size, and climatic and seasonal factors. Existing data suggest that most stormwater contains elevated levels of organic matter, suspended sediment, and indicator bacteria. Metals are also commonly found in urban stormwater runoff and may pose concerns for some beneficial uses, including irrigation and surface reservoirs or wetland features. Despite the enormous spatial and temporal variability of stormwater quality, the treatment systems required for achieving end uses may be relatively consistent over a wide variety of catchments. Land uses, contributing areas, and collection materials can be selected that minimize contaminants of concern to optimize stormwater quality and minimize treatment requirements for the intended use.

Little is known regarding the occurrence of human pathogens and organic chemicals in stormwater, and additional research is needed to characterize their occurrence and fate. Studies on the presence of microorganisms in stormwater have consistently reported high concentrations of fecal indicator microorganisms across different source areas. In the few studies that have analyzed for pathogenic microorganisms in stormwater, they have generally been detected, at least in some samples. However, more work is needed to characterize their occurrence and fate, particularly for roof runoff systems where the beneficial use of untreated stormwater is common and raises concerns for uses with the potential for human exposure. More research is also needed to characterize the occurrence of organic chemicals in stormwater and their fate during various uses.

HUMAN HEALTH AND ENVIRONMENTAL RISKS

Although no documented reports of adverse human health effects from the beneficial use of stormwater or graywater have been identified, additional examination of risk is necessary to support safe and appropriate design and implementation of stormwater and graywater use systems.

Risk assessment provides a means to determine "fit-for-purpose" water quality criteria or treatment needs based on human exposures. Risk from graywater or stormwater is a factor of chemical or microbial concentrations and exposure (typically, the amount of water ingested). Thus, unlike drinking water criteria, which are established based on 2 liters of water consumed per day, criteria for applications with minimal human exposures might allow for much higher concentrations of contaminants in graywater or stormwater and still result in acceptably low health risks. Risk assessment tools provide a ready means for developing such criteria for many chemicals and microbes for which drinking water criteria exist. As nonpotable on-site use of graywater and stormwater becomes more common, additional public health risk communication efforts would be beneficial to help the public understand risk-based treatment objectives and appropriate safeguards.

Considering the low exposures in most nonpotable graywater and stormwater applications, pathogens represent the most significant acute risks. Available risk assessments and the committee's risk calculations using limited, observed pathogen data and various possible exposure scenarios suggest that disinfection is necessary for many uses of graywater, including spray irrigation, food crop irrigation, and toilet flushing, to protect human health. Subsurface landscape irrigation (including drip systems covered by landscape) with graywater does not pose significant risk, if best practices are followed, because human exposure is minimized. These findings are consistent with most regulatory guidance, although the risk of surface drip irrigation (without landscape cover) at the household scale remains unresolved. Limited data on pathogens in roof runoff suggest that treatment may also be needed, even for low levels of human exposure, such as toilet flushing, although more research on pathogens in roof runoff is needed. Chemicals become of concern in groundwater infiltration projects, where drinking water supplies could be impacted.

Extremely limited data are available on the pathogen content in graywater and roof runoff, which precludes a full assessment of microbial risks. Most water quality monitoring assesses microbial indicator data, and microbial risk assessments are conducted using assumed relationships between the concentrations of indicator microorganisms and pathogenic microorganisms. Consistent relationships between surrogates and contaminants have not been established for graywater or stormwater. This is a particular concern for roof runoff, which may include microbial indicator organisms from the waste of animals that do not transmit human pathogens.

Enhanced infiltration of stormwater for groundwater recharge poses risks of groundwater contamination and necessitates careful design to minimize those risks. The risk of groundwater contamination from stormwater recharge is related to the contaminants present, any pretreatment processes installed, the capacity for the subsurface soil and engineered media used in the infiltration basin to remove them, and the proximity to groundwater used as a drinking water supply. Dry wells, which directly inject water into the subsurface, and surface infiltration through sandy soils do not effectively attenuate chemical contaminants.

Environmental impacts from the outdoor use of graywater and stormwater generally appear low, but risks depend upon several factors, including water quality, application rates, and plant or animal species exposed. Effects of irrigation on plant and soil health can occur from salts, boron, and metals, but source control practices and appropriate irrigation rates can reduce these impacts. If not controlled at the source, then long-term build-up of boron or salt can pose risk to plant and soil health, depending on soil and climatic conditions. Constructed stormwater ponds and wetlands typically contain elevated contaminant levels sufficient to impair reproduction among some aquatic species, often leading to a habitat dominated by pollution-tolerant organisms. Such ecological affects may be acceptable, considering the overall environmental benefits provided by such features, including reduced pollution to other surface waters, but the ecological objectives of such projects are often unclear, hindering efforts to limit ecological risks through improved management and design.

STATE OF PRACTICE AND SYSTEM DESIGN

The report also outlines the state of practice for graywater and stormwater system designs at household, neighborhood, and regional scales and treatment that may be used to meet specific quality objectives.

Graywater irrigation at the household scale can be achieved with simple systems that require little energy and maintenance. These simple systems, such as the laundry-to-landscape system or systems that include storage, coarse filtration, and pumps, typically do not include organic matter removal or disinfection, and risk is managed through a series of best management practices. Neighborhood-scale systems typically provide disinfection where access-control is not feasible, which creates more system complexity and requires more energy.

Graywater reuse for toilet flushing requires plumbing components and treatment systems that are most appropriate in multi-residential buildings or neighborhoods. Graywater systems for toilet flushing require dual plumbing with a connection to potable water and backflow preventers that require annual inspection. Treatment systems for toilet flushing should include disinfection to reduce risk and prevent bacterial growth, and existing technologies are available. Even the simplest treatment systems require periodic maintenance that can be a burden at the household level, although such maintenance is more easily managed by contractors or on-site staff at the neighborhood/multi-residential scale. For broader adoption of graywater for toilet flushing at the household scale, treatment systems are needed that are low maintenance and include process automation and control to ensure safe use at a reasonable cost.

Many state graywater treatment standards for toilet flushing are not risk-based or fit-for-purpose. Standards vary widely across states, resulting in an inconsistency in treatment systems that can be applied, and several are based on standards unrelated to residential or multi-residential toilet flushing. Many standards for toilet flushing may be unnecessarily strict in terms of organic content and turbidity removal, resulting in requirements for technologies that are costly, energy-intensive, and require frequent maintenance. Additional research is needed to determine appropriate design standards for dissolved organic carbon and turbidity that prevent aesthetic and maintenance issues while allowing proper function of disinfection systems when using graywater for toilet flushing.

New developments and future urban planning provide opportunities for rethinking the conveyance and use of various water and waste streams for maximum cost, energy, and water savings. Separation of graywater results in blackwater that is more concentrated in solids and organic matter than conventional domestic wastewater and may be amenable to methane biogas production. These systems can also be integrated with urine separation including nutrient capture. Thus, graywater reuse can be a key element of energy-efficient urban water and resource management systems that not only minimize net water abstraction from the environment but also achieve a high level of energy and nutrient recovery.

The state of practice and development of cost-effective and safe stormwater capture systems for roof runoff are hindered by the lack of data on human pathogens and the risk associated with various uses. Design and treatment standards are generally well accepted for nonpotable use of runoff collected from land surfaces, and no treatment other than coarse solids removal is needed for subsurface irrigation where human exposures are minimal. For beneficial uses of roof runoff with low to moderate exposures, additional pathogen data and risk analyses are needed to establish a consistent state of practice for on-site stormwater use. Technologies are mature and can be readily adapted for various scales and uses.

Operations and maintenance of household and neighborhood graywater and stormwater use systems is not well guided or monitored. All systems that capture graywater and stormwater for beneficial use require routine maintenance. For systems where disinfection is not required (e.g., subsurface irrigation), failure to conduct needed maintenance poses operational concerns but does not pose a significant risk for human health or environmental quality. However, for systems with disinfection processes to protect human health (i.e., systems for toilet flushing), ongoing maintenance is critical. Although many states require that installed systems meet certain water quality targets, ongoing monitoring is not required. More guidance is needed to ensure safe operations of graywater and stormwater treatment systems at household and neighborhood scales. Because frequent routine water quality analyses are expensive and impractical even at the neighborhood scale, system operational performance standards and online monitoring of surrogate parameters (e.g., residual chlorine, suspended solids, or turbidity) should be considered.

Stormwater infiltration for aquifer recharge is commonly practiced, but designs and regulations in the United States may not be adequately protective of groundwater quality. Design for large-scale stormwater infiltration projects are still emerging. For many locations, the design and performance standards for stormwater infiltration have been developed to address surface water regulatory drivers rather than the protection of groundwater quality. Of particular concern is the infiltration of organic contaminants and salts from highly urbanized areas into water supply aquifers, although human pathogens may also be of concern depending on the infiltration site characteristics. Thoughtful planning, source area selection, source control, and mechanisms to integrate treatment into the watershed could improve efficiency of these systems and reduce the amount of treatment required. Treatment systems, such as engineered wetlands and filter media, may also be needed for regional-scale systems where source control is challenging.

COSTS AND BENEFITS

It is important to recognize the full suite of benefits—as well as the full costs—of graywater and stormwater projects, although it may be empirically challenging to do so. Some of these benefits are financial and can be readily estimated and portrayed in monetary terms, such as the value of water savings or the avoided cost of obtaining water from an alternative supply. In addition, important social and environmental benefits may apply but may be difficult to quantify or monetize. Costs for graywater and stormwater projects are highly dependent on scale, system design, and plumbing requirements, and generally are better understood than the benefits, yet there is a lack of well-documented and complete cost information for many of the possible applications. The following findings are based on limited available cost data and some example analyses of potential water savings based on the committee's scenario analysis.

Simple household-scale graywater reuse or roof runoff capture systems can offer reasonable financial payback periods under certain water use scenarios and appropriate climate conditions. For example, considering the committee's scenario analysis of potential water savings in medium-density residential development, simple laundry-to-landscape graywater systems can offer payback periods as low as 2.5-6 years (not accounting for the cost of labor), with the shortest payback periods in the Southwest and central United States. These estimates assume graywater for irrigation actually offsets potable use—an assumption that remains to be demonstrated. Longer payback periods were estimated for rain barrels (5-26 years) and cisterns (14 to more than 50 years, not accounting for labor) used for conservation irrigation. The longer payback periods reflect locations where distinct wet and dry seasons do not coordinate well with irrigation demands, as in the arid Southwest. The cost of installation (whether by contracting with a paid professional or valuing homeowner-provided labor) greatly extends the payback period, as do water uses in which additional plumbing and treatment are required.

Economies of scale are evident for large stormwater and graywater use projects. Several regional stormwater capture and recharge projects in Southern California, for example, can pay back large dividends by avoiding the cost of expensive imported water in addition to other social and environmental benefits. Based on available unit cost data, stormwater alternatives designed to recharge groundwater at neighborhood and regional scales tend to be much less expensive than on-site or neighborhood tank capture. Published cost data from larger-scale graywater projects is extremely limited, but some efficiencies of scale would be associated with graywater toilet flushing systems in large, new multi-residential developments (particularly compared to smaller retrofits). Additional incentives may be possible if such investments defer water and wastewater infrastructure expansion in densely populated urban areas.

Depending on the stormwater or graywater system design, energy savings are possible compared with conventional water supplies, but data for a sound assessment are lacking. Conventional water systems in the United States are reported to provide water to customers at an energy cost of between less than 1 kWh/m^3 to as much as 5 kWh/m^3, depending mostly on pumping costs for conveying the water from the source to the water treatment plant. Rooftop stormwater capture systems have been reported in a limited number of studies to have a greater energy demand (median is 1.4 kWh/m^3) in practice than in theoretical studies (0.2 kWh/m^3), but many potential variables (e.g., scale, pumping, treatment, material inputs) will drastically affect the life-cycle energy demands of these systems, and the effects of these variables in practice remain poorly understood.

LEGAL AND REGULATORY ISSUES

As technologies and strategies continue to advance, graywater and stormwater use is being incorporated into law in a variety of respects at the federal, state, and local levels. However, as is often the case with innovative technologies, the law has not evolved quickly enough to keep up with the technology and its use. Several legal and regulatory constraints remain that hinder the capacity for graywater and stormwater to significantly expand the nation's water supplies.

In most western states, acquisition of water rights is a requirement for large-scale stormwater capture and use projects, and water rights may limit widespread implementation of smaller-scale stormwater and graywater projects for consumptive uses. Unless water rights can be acquired or legislative solutions developed, opportunities for large-scale stormwater capture projects to expand existing water supplies would largely be limited to coastal regions with no downstream users or to non-consumptive uses (e.g., toilet flushing). Several states (e.g., California, Kansas, Oregon, Utah, and Washington) have established regulations that allow small-scale roof runoff capture projects to proceed without water rights permits, and only one state (Colorado) has strict limits on stormwater capture and use out of concern for water rights impacts. The right to stormwater and graywater use in most prior-appropriation states has not been firmly resolved through judicial decisions, leaving an unclear outlook for projects that have not acquired water rights, because they could be vulnerable to legal challenges. New scientific analyses of the impacts to return flows of various on-site water uses in different regions would help clarify these concerns, but additional legal research and guidance could better facilitate the use of on-site water supplies, considering potential legal challenges.

There is substantial variation in on-site graywater and stormwater regulations at the state level with respect to design and water quality for household-scale projects, which leads to varying exposures and risk. As one example, there is lack of consistency among states on whether outdoor graywater use is limited to subsurface irrigation. At least three states allow drip irrigation without landscape cover, which could lead to higher pathogen exposures. In addition, states vary on their regulation of untreated graywater irrigation of food crops. Whether such exposures would lead to unacceptable risks at various scales has not been definitively resolved, but higher risks are likely with increased exposures. Regulations affecting large-scale graywater and stormwater use where public access is not controlled tend to include conservative public health protection measures, such as disinfection.

The lack of authoritative, risk-based guidelines for the design and potential applications of graywater and stormwater in the United States is a major impediment to their expanded use. The wide variability in existing regulations and absence of federal guidance leaves stakeholders and local decision makers uncertain about the safety of these practices and the appropriate level of treatment necessary for particular uses. Development of rigorous, risk-based guidelines for graywater and stormwater across a range of possible uses and exposures could improve safety, build public confidence in the practices, reduce expenditures on unnecessary treatment, and assist communities that lack an existing regulatory framework for on-site water supplies. Such guidelines could be developed by the Environmental Protection Agency (EPA), a collaboration of states, or a collaboration of U.S. water organizations working with the EPA. This guidance could then serve as a basis for developing standards of practice for on-site nonpotable water use. Oversight and enforcement of water quality standards for applications with significant exposures is also important but challenging, and local enforcement agencies would benefit from additional guidance on appropriate, cost-effective maintenance, monitoring, and reporting strategies.

BOX S-2 Summary of Research Needs to Enhance the Safe and Reliable Use of Graywater and Stormwater and Conserve Water, Energy, Environmental, and Financial Resources

Risk and water quality

1. Assess the occurrence and fate of pathogens in graywater and stormwater
2. Assess the occurrence and fate of chemical contaminants in stormwater
3. Understand the implications of enhanced water conservation on graywater quality and use
4. Develop risk-based water quality guidance for various uses that could serve as a basis to develop standards of practice
5. Develop monitoring technology and strategies to assure compliance with water quality criteria

Treatment technology

6. Develop treatment systems to meet tailored (fit-for-purpose) water quality objectives across a range of scales
7. Understand the long-term performance and reliability of graywater and stormwater treatment systems (from small to large scales)

Infrastructure

8. Envision opportunities for water- and energy-conserving infrastructure designs in new construction and demonstrate their performance
9. Identify strategies to retrofit existing infrastructure for enhanced beneficial use of stormwater

Social science and decision analysis

10. Understand behavioral impacts on overall water use in the context of graywater and stormwater projects
11. Collect performance data (including cost, energy, water savings, water quality, and other benefits) in support of integrated water supply management, decision making, and refinement of decision tools

Policy and regulatory issues

12. Identify incentives and various regulatory strategies that have proven effective in the implementation of stormwater or graywater systems to conserve water supplies

RESEARCH NEEDS

Information is generally available to support water management decision making for simple, household-scale graywater and stormwater systems with minimal human exposures, but additional research would enhance decision making for larger systems or those with significant exposures. Key uncertainties affect the capacity to make fully informed decisions on appropriate and cost-effective designs, particularly for larger or more complex graywater or stormwater beneficial use systems, including

- Fit-for-purpose water quality objectives that are protective of public health;
- The occurrence and fate of pathogens in stormwater and graywater;
- Costs and benefits for neighborhood- and regional-scale systems, including nonmonetized benefits, such as water pollution control and community amenities;
- Energy implications of on-site alternative water supplies; and
- Long-term system performance and maintenance needs.

A summary of research needs to enhance decision making and ensure the safe and reliable use of graywater and stormwater to reduce water demand is provided in Box S-2.

1

Introduction

Many parts of the United States face chronic or episodic water shortages. In the Colorado River Basin and California, recent multi-year droughts have resulted in reservoirs at near record low levels (Figure 1-1), forcing state drought declarations and decreased water allocations for many users. Climate change is anticipated to further impact water supplies by altering the timing and amounts of precipitation, increasing evapotranspiration, and altering snowmelt and the timing of runoff in the western states (Barnett and Pennell, 2004). Also, longer-term droughts are expected to intensify in the southwest, the Great Plains, and the southeast (Melillo et al., 2014). Meanwhile, population growth in the more water scarce regions of the United States (Figure 1-2) compounds the issue, placing additional strain on water supplies and infrastructure. For example, the states of California, Nevada, Arizona, Texas, and Florida saw their populations increase between 85 percent and 400 percent between 1970 and 2009, while the overall population of the United States increased by less than 50 percent during that same timespan (NRC, 2012a). The current population in the United States (321 million) is expected to grow by 30 percent by 2060, mostly in cities (Colby and Ortman, 2014).

To help alleviate these water shortage problems, alternative water sources such as stormwater and graywater are increasingly being viewed as resources to supplement scarce water supplies rather than as waste or nuisance water. Harvesting stormwater has many potential benefits including water conservation, energy savings, and reduced impacts of urban development on the environment. Even in the more humid areas of the United States, stormwater capture and use are growing in popularity as a means to enhance water supply and reduce nutrient loads to receiving waters. Similarly, the reuse of graywater for residential and building landscape irrigation or other nonpotable uses can reduce the year-round demand on public water supplies treated to drinking water standards (hereafter called potable water).

Stormwater and graywater use are two options among many in a diverse water supply portfolio, including conservation, desalination, managed aquifer recharge, and wastewater reuse (see NRC, 2008a, 2009a, 2012a). Conservation and water use efficiency are generally the best ways to address water supply problems on a broad scale. In some water-challenged areas, numerous water conservation initiatives have already been implemented, including installing

FIGURE 1-1 The impacts of the recent drought in California are exemplified in Folsom Lake, which was at 97 percent capacity in 2011 and at 17 percent capacity in January 2014. SOURCE: California Department of Water Resources.

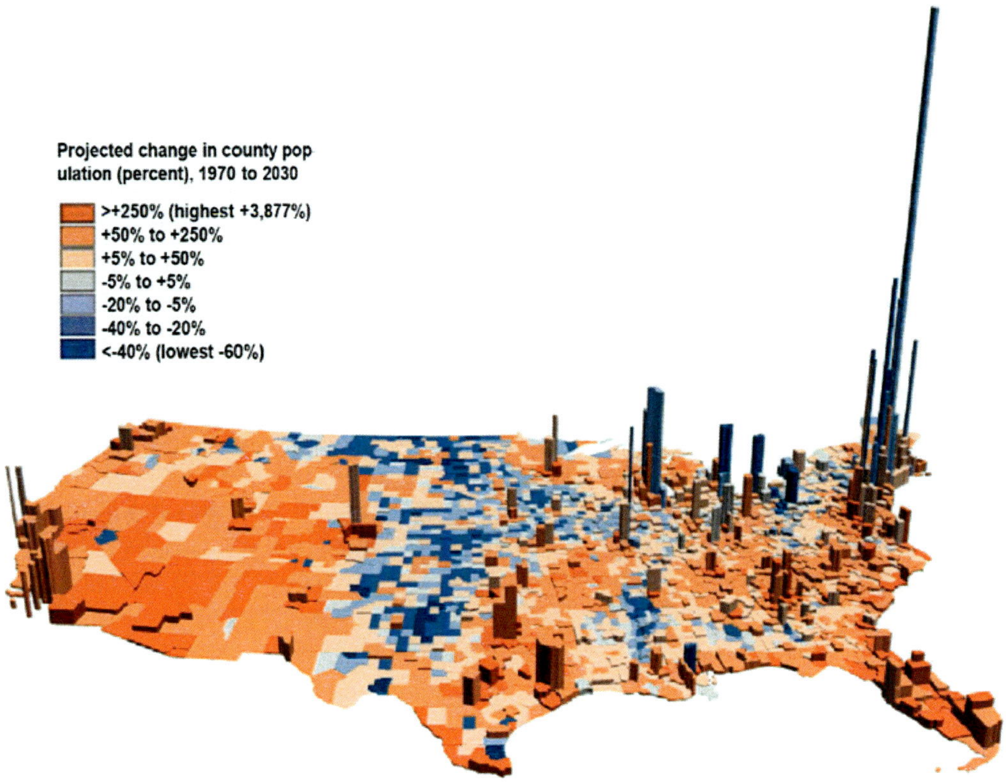

FIGURE 1-2 County-level population trends in the United States between 1970 and 2030. Each block on the map illustrates one county in the United States. The height of each block is proportional to that county's population density in the year 2000, and so the volume of the block is proportional to the county's total population. The color of each block shows the county's projected change in population between 1970 and 2030, with shades of orange denoting increases and blue denoting decreases. SOURCE: National Assessment Synthesis Team (2001).

low-flow devices in buildings and incentivizing less-water-intensive landscaping, but water utility managers need additional strategies to address current and future challenges of water supply reliability and water demand. For example, an analysis of the water supply portfolio for the City of Los Angeles shows that continued conservation efforts over the next 20 years (based on utility projections) may accommodate future population growth but are not likely to significantly reduce the present demand for imported water (Luthy and Sedlak, 2015).

Brackish and seawater desalination and wastewater reuse have received substantial attention as means to augment public water supplies (NRC, 2008a, 2012a), but less information is available on the beneficial use of graywater and stormwater. The process through which graywater and urban stormwater are captured, stored, and used has mostly developed in an ad hoc manner (Grebel et al., 2013). Because of the absence of ample documentation of costs, performance, and risks, many utilities are hesitant to integrate the practices into their long-term water resource plans. Potential public health risks from microbial or chemical contamination associated with graywater or stormwater use also raise concerns and subsequent debate over the appropriate regulatory framework to protect public health without adding excessive cost and permitting burdens to these projects (EBMUD, 2009). To better address these challenges, the National Academies of Sciences, Engineering, and Medicine (Academies) formed a committee to conduct a study on the risks, costs, and benefits of stormwater and graywater use to augment and conserve existing water supplies.

DEFINITION OF STORMWATER RUNOFF AND GRAYWATER

Stormwater runoff is the water from rainfall or snow that can be measured downstream in a pipe, culvert, or stream shortly after the precipitation event (NRC, 2009a). What constitutes "shortly" depends on the size of the watershed and the efficiency of the drainage system. From a practical perspective, stormwater runoff is water that may be in an engineered feature, running over the ground surface, or seeping into the shallow subsurface and soon reemerges as seeps.

Stormwater runoff is distinct from deeper percolation of precipitation that moves slowly through the ground and sustains the base flow of streams and rivers and recharges groundwater. Urbanization results in an increase in the amount of land covered with impervious surfaces, resulting in a greater percentage of precipitation appearing as stormwater runoff.

Rainfall that is captured directly from rooftops and stored on site in barrels or cisterns is frequently called rainwater harvesting. For the purposes of this report, the term "stormwater" is used broadly to include runoff captured directly from rooftops, and the term "roof runoff" is used when flows from the ground surface are not included.

Graywater is the wastewater produced from bathroom sinks, showers, bathtubs, clothes washers, and laundry sinks and is derived from residential buildings or commercial establishments. Graywater is mainly a byproduct of washing and does not include toilet water (sometimes called "blackwater"). This report and most recent scientific papers also exclude water from kitchen sinks and dishwashers from the definition of graywater (Sharvelle et al., 2013), because kitchen water contains high levels of organic matter and solids along with foodborne pathogens (Eriksson et al., 2002), necessitating more extensive treatment (with unit processes similar to wastewater reuse). Nevertheless, some states regulations (e.g., Wash. Chap. 246-274) include water from kitchen sinks and dishwashers in the definition of graywater. Graywater, as defined in this report, accounts for about one-half of a typical indoor home wastewater flow (Sheikh, 2010). The definition of graywater could conceivably include condensate from air conditioning units, which represents an additional on-site water resource, although condensate is not typically included in the definition of graywater.

Stormwater and graywater can be captured and used at various scales using engineered conveyance, treatment, and storage systems of varying complexity. For stormwater, this may include the household, neighborhood (or multi-residential building), or regional scales. Accordingly, the annual capture can vary from several hundred gallons for rain barrels to millions of gallons in large subsurface tanks to billions of gallons for large, regional surface reservoirs or groundwater infiltration systems. Graywater systems also can be applied at varying scales, from the individual residence to the multi-residential building to a large residential development with a semi-centralized graywater treatment system serving as many as 10,000 people.

CURRENT DRIVERS

Many different drivers exist for local water capture and use, and these drivers vary by region as to their relative importance. Current interest in the beneficial use of stormwater and graywater is driven by water scarcity in regions of the country that experience chronic or episodic droughts. Additional drivers are flood control, pollution prevention, nutrient management, reduced hydromodification, and energy savings associated with locally sourced water supplies. Although water scarcity, flood control, and pollution prevention are primary drivers depending on local conditions, other co-benefits such as green space, community amenities, and public education may be equally important to decision making. The relative importance of each of these drivers determines the graywater and stormwater strategies that might be appropriate for a particular site or region.

Scarcity

Many areas in the West, Southwest, and Southeast face water shortages. Chronic water shortages exist from California and the desert Sun Belt to the Colorado Front Range and the Great Plains. A 14-year drought has lowered the water levels in the Colorado River basin to historic levels, forcing communities in the southwest that rely on water from the Colorado River to look for alternative supplies. In northern California, communities that have traditionally taken groundwater from the Carmel River basin (Monterey County) are under state mandate to preserve water for the river ecosystem, withdraw less, and find other supplies (MPWMD, 2014). In California, 2013 was the driest calendar year since the Gold Rush when record keeping began, and by the winter of 2015, the state had the lowest snowpack in recorded history. California has been under a drought state of emergency since 2014 (Governor of the State of California, 2014) with substantial water use and water delivery restrictions (CA DWR, 2014).

Such problems, however, are not confined to the West. The southeastern United States experiences extended periods of low rainfall as a normal component of the climate system (Figure 1-3), resulting in conflicts between Georgia and Florida over water supply for the city of Atlanta versus water releases to the Chattahoochee River and Apalachicola Bay (NRC, 2009b). Water shortages and the over-pumping of groundwater resulted in Tampa becoming the first large U.S. city to substantially augment its water supply via seawater desalination (Tampa Bay Water, 2008).

Population growth and redistribution to water scarce regions has exacerbated these challenges. Figure 1-2 shows high levels of projected population growth in the Southwest and Southeast in areas already facing water challenges.

Climate change may add further stresses to water scarce regions that are already exceeding the limits of imported surface water supplies and sustainable yields from groundwater basins to meet the needs of urban users. These conditions

FIGURE 1-3 Lake Lanier reached record lows in 2007 during an extended drought. The lake is the water supply to 5 million people in greater Atlanta and periodic droughts continue to threaten this resource. SOURCE: Courtesy of Bill Kinsland; http://www.srh.noaa.gov/images/ffc/lanier12108.jpg.

are leading communities to conserve existing potable water supplies and seek out alternative sources, such as stormwater and graywater (see Box 1-1).

Water Supply Reliability and Diversification

The beneficial use of stormwater and graywater also provides ways to augment and diversify local water supplies and reduce reliance on imported water supplies. In Los Angeles, for example, 88 percent of the current water supply is imported, and the city seeks to diversify its water portfolio and increase the use of local water supply sources, such as stormwater (Box 1-2). In 2010, Los Angeles announced plans to meet at least 4 percent of its water supply through new stormwater capture systems by 2035 (Figure 1-4; LADWP, 2010), although this could grow to become even larger as a result of recent stormwater capture planning (see Box 1-2). More aggressive timelines for reducing dependence on imported water are presented in the 2015 sustainability plan for Los Angeles (City of Los Angeles, 2015). California's State Water Resources Control Board has ambitious goals for increasing stormwater capture and use by an additional 500,000 AF/yr (620 million m^3/yr) by 2020 and by 1 million AF/yr (1.2 billion m^3/yr) by 2030 to reduce the state's reliance on imported water (SWRCB, 2013). For comparison, the Metropolitan Water District of Southern California (MWDSC, 2010) estimated that as of 2007 approximately 470,000 AF/yr (580 million m^3/yr) of stormwater was being captured in the coastal plain of southern California, with significantly less urban stormwater captured in northern California (e.g., the San Francisco Bay Area). The Pacific Institute concluded that additional urban stormwater capture in southern California and the San Francisco Bay Area could potentially increase water supplies by 420,000 to 630,000 AF/yr (520 to 780 million m^3/yr) (Pacific Institute and NRDC, 2014).

Graywater systems offer a reliable, year-round source of water to irrigate landscaping or flush toilets that can help conserve existing water supply sources. This reliability offers a major benefit in areas that face frequent outdoor water use restrictions during times of drought.

Pollution Prevention

Urban stormwater contains a number of contaminants and is a major source of nonpoint pollution to surface waters for chemicals and pathogens (EPA, 1994). Chemical contaminants include those derived from paving materials, automobile tires, and urban biocides (Grebel et al., 2013). Stormwater runoff also contributes substantial loads of nitrogen, phosphorus, and sediment, which can cause algal blooms, low dissolved oxygen, and reduced water clarity and significantly impact aquatic life in inland water bodies and coastal estuaries. Nutrient discharges from urban and agricultural runoff, wastewater discharges, and air pollution have created a "dead zone" with low oxygen in the Chesapeake Bay (Figure 1-5), which has motivated a multi-state pact to reduce pollution loads.[1]

---

[1] Urban stormwater is estimated to contribute 8 percent of the total nitrogen loads and 15 percent of the phosphorus loads to the Chesapeake Bay. See http://stat.chesapeakebay.net.

BOX 1-1 Arizona Prison

A graywater reuse system that uses shower and hand-washing water to flush toilets was installed during construction of the Eloy Detention Center located in Arizona, which houses up to 6,492 inmates (Figure 1-1-1). Drivers for the installation of the system included water conservation, a desire for environmentally friendly practices in the facility design, and cost savings associated with a reduced hydraulic load to the on-site septic tank, which facilitated the use of a smaller septic system.

The graywater is treated by filtration and chlorination, with the goal of no detection of fecal coliform bacteria. Water samples are monitored and reported to the Arizona Department of Environmental Quality on a weekly basis. The system has been in compliance for non-detectable fecal coliform bacteria since the permit was issued in 2008. Because this system was the first large-scale application of graywater use for toilet flushing in Arizona and the state lacks a standard regulation that addresses large-scale (i.e., commercial or multi-residential) graywater use for toilet flushing, the permitting process was lengthy. Since the system has been in operation, water savings of 20 gallons per day (gpd; 80 liters per day [lpd]) per inmate have been observed (or approximately 130,000 gpd [145 AF/yr or 179,000 m^3/yr] at full prison capacity).

SOURCES: C. Graf, Arizona Department of Environmental Quality, personal communication, 2014; T. Valentine, Valentine Engineers, personal communication, 2014.

FIGURE 1-1-1 The Eloy Detention Center, located in Arizona, reuses graywater for toilet flushing to stretch its water supplies. SOURCE: http://www.law.arizona.edu/clinics/Immigration_Law_Clinic/deportation_defense.cfm.

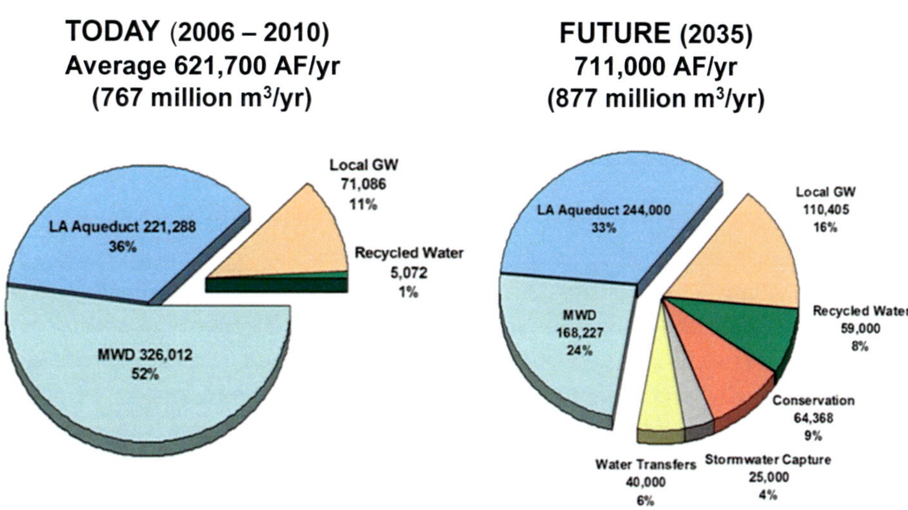

FIGURE 1-4 Current plans for diversifying Los Angeles' water supplies by 2035 through increased use of local sources, including stormwater capture. Existing stormwater capture through spreading operations, about 27,000 AF/yr (33 million m^3/yr), is included within "local groundwater" in the left-hand chart. Ongoing stormwater capture planning could further increase this percentage. SOURCE: LADWP, Water Management Group.

BOX 1-2 Stormwater Capture in Los Angeles

The 2013 update of the Greater Los Angeles County Region Integrated Regional Water Management Plan (GLAC IRWM, 2013) describes a 20-year horizon for meeting the Los Angeles region's water supply needs. In 2010, the Los Angeles County region's stormwater capture and direct use was 1,000 AF/yr (1.2 million m^3/yr) and stormwater capture for aquifer recharge totaled 196,000 AF/yr (242 million m^3/yr). In the next 20 years, Los Angeles County aims to increase stormwater capture and direct use by 26,000 AF/yr (32 million m^3/yr) and increase stormwater recharge by 75,000 AF/yr (93 million m^3/yr) (GLAC IRWM, 2013). Thus, in the Greater Los Angeles region, additional stormwater capture for water supply is expected to increase by more than 50 percent over the next 20 years.

Presently in the City of Los Angeles, 29,000 AF/yr (36 million m^3/yr) of stormwater are recharged on average through spreading grounds; incidental stormwater infiltration accounts for an additional 35,000 AF/yr (43 million m^3/yr) in recharge to the city's aquifers (see Figure 1-2-1), but existing challenges from aquifer overpumping and groundwater contamination prevent all of this recharge from being used to meet current water demands. In 2010, the City of Los Angeles set a target of capturing an additional 25,000 AF/yr (31 million m^3/yr) of stormwater by 2035 (LADWP, 2010). However, additional planning is under way through the Stormwater Capture Master Plan that is likely to expand this goal. Los Angeles Department of Water and Power has identified additional stormwater capture strategies that could contribute between 115,000 (142 million m^3/yr) and 194,000 AF/yr (239 million m^3/yr) to the region's water supply (LADWP, 2015). Figure 1-2-1 shows potential stormwater capture by 2099 according to an aggressive and a conservative path, with the aggressive path potentially tripling the amount of stormwater currently captured. These goals are aspirational and need technical and financial feasibility assessment. The extent to which this stormwater capture would address future water demand depends on managing existing groundwater contamination, assumptions about land use and groundwater recharge and recovery sustainable yield analysis, financing, and other factors. Nonetheless, these analyses demonstrate that stormwater capture has the potential to contribute significantly to Los Angeles' water supply and reduce the need for imported water.

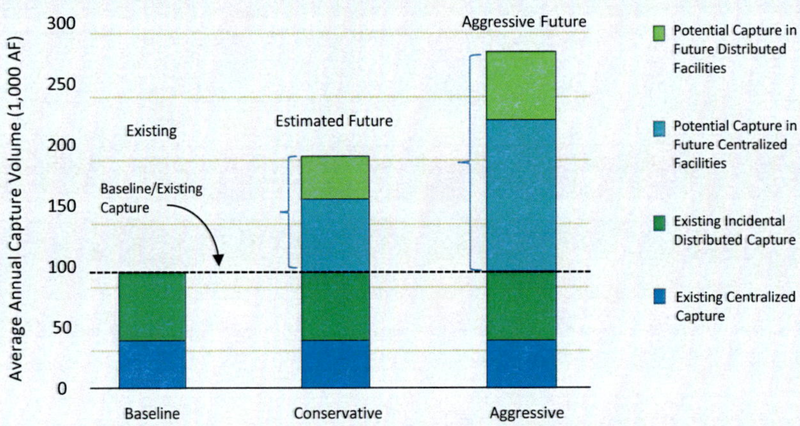

FIGURE 1-2-1 The City of Los Angeles' Stormwater Capture Master Plan is considering scenarios that potentially could nearly triple the amount of stormwater captured with aggressive action by 2099. SOURCE: LADWP (2015).

In numerous U.S. cities, particularly in the Midwest and Northeast, stormwater and wastewater infrastructure were constructed together (termed combined sewer systems), such that stormwater runoff drains into sewers and passes through the wastewater treatment plant (Figure 1-6). In areas with advanced wastewater treatment, such construction can be a benefit under low to normal hydrologic conditions, because the treatment plant can remove nutrients and sediment from both stormwater and wastewater. However, combined sewer systems were constructed with overflows that would prevent the wastewater treatment plant from being overloaded after storm events. Under extreme wet weather conditions or when blockages or mechanical failures occur, combined sewer overflows (CSOs) discharge untreated wastewater, polluting the surface waters with pathogens, organic matter, and nutrients. Improved stormwater capture and use provide a means to reduce CSOs and the associated pollution loads.

In coastal regions, stormwater may be a significant contributor to pathogens that pollute recreational waters. Stormwater runoff is the most frequently identified source of beach closings and advisory days, and the U.S. Environmental Protection Agency (EPA) estimates that more than 10 trillion gallons (38 trillion liters) of untreated stormwater make their way into our surface waters each year (EPA, 2004a). In 2012

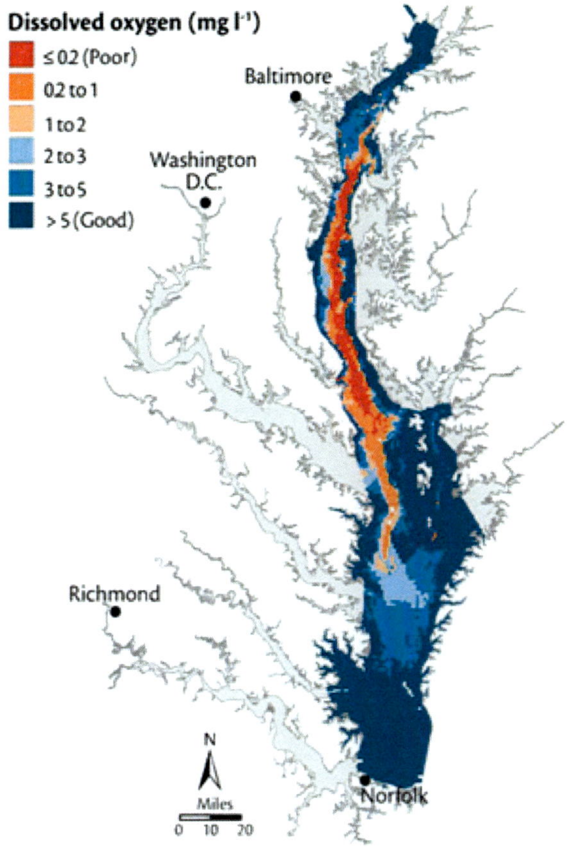

FIGURE 1-5 Dissolved oxygen conditions in August 2009 in the Chesapeake Bay. Dissolved oxygen levels below 2 mg/L are considered hypoxic, but impacts to biota have been observed at levels below 4 mg/L (Buchheister et al., 2013). SOURCE: http://www.cbf.org/about-the-bay/maps/pollution/dead-zones.

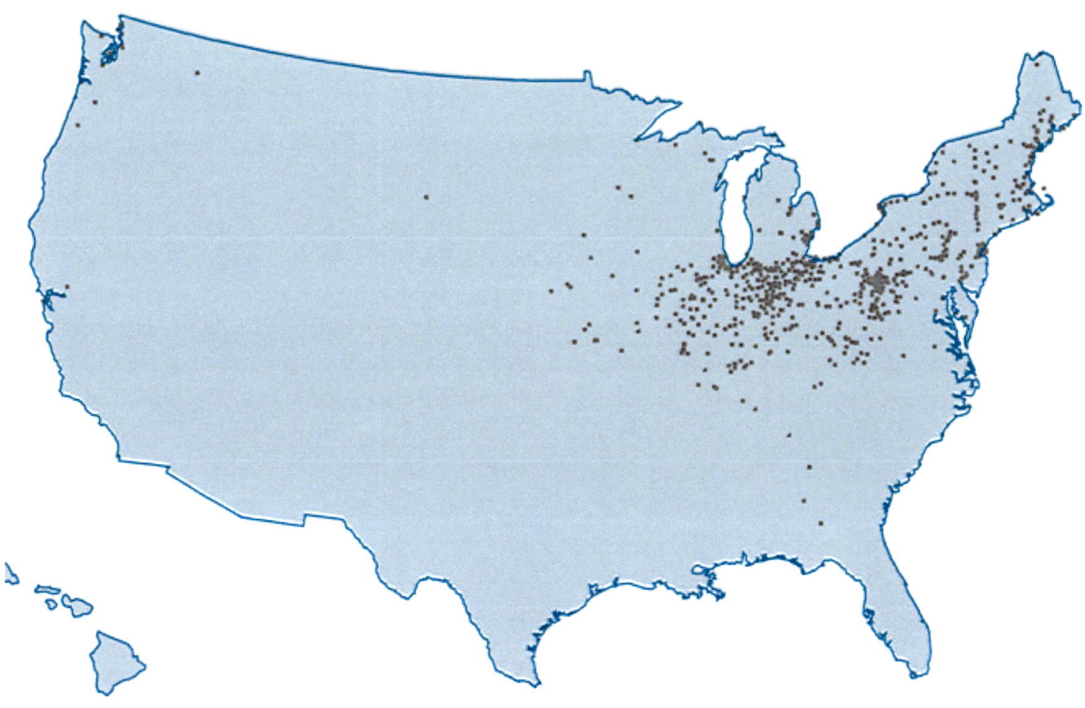

FIGURE 1-6 Combined sewer overflows exist in 772 cities, predominantly in the Northeast and Midwest. To avoid costly infrastructure replacement to meet water quality standards, these cities are increasingly turning to intensive stormwater management strategies, including stormwater capture and use. SOURCE: EPA (2004a).

there were more than 20,000 beach closing and advisory days, with more than 80 percent caused by bacteria levels in recreational waters that exceeded public health standards (Dorfman and Haren, 2013).

Discharges to impaired (i.e., degraded) water bodies are typically governed by "total maximum daily load" (TMDL) requirements, which outline the maximum amount of a pollutant that a water body may receive and still meet water quality standards (with a factor of safety). TMDLs have been approved for pathogens, nutrients, mercury, other metals, and sediment, among others.[2] Stormwater management efforts are often driven by TMDLs (discussed in more detail in Chapter 7), and meeting TMDLs can be an expensive and uncertain proposition. However, stormwater capture and use provides a strategy to reduce pollution while providing additional water supply benefits (see Box 1-3).

Hydromodification and Flood Management

Stormwater runoff in urban areas is characterized by increased volumes of runoff and more intense peak flows compared to the more natural state. This change in runoff regime, called "hydromodification," is caused by land use and altered landscapes in the watershed that destabilize streambeds and impair stream condition and function (Goodman and Austin, 2011). As watersheds urbanize and are covered with impervious surfaces, runoff is conveyed directly to streams via the conventional storm drain system. Infiltration into soil is reduced and overland flow increases. As a result, the magnitude and duration of flows entering receiving streams increase, which contributes to more erosive energy within the channel. Unless managed, hydromodification can cause channel erosion, unstable stream banks, altered base flow, change in bed material composition, and biological impacts to stream systems (Figure 1-7; OEHHA and SWRCB, 2009; Paul and Meyer, 2001).

Hydromodification management in new developments seeks to mimic natural hydrologic conditions by retaining stormwater and subsequently releasing it to match pre-development flow volumes, durations, and frequencies. The theory is that if the pre-development distribution of in-stream flows is maintained over a broad range of critical flow rates for long periods, then the baseline capacity to transport sediment, a proxy for the geomorphic condition, will be maintained as well. Stormwater infiltration or capture alters the runoff hydrograph of a site through the changes to the timing of discharges, and stormwater capture also reduces the overall volume. Evaluation of onsite stormwater use for hydromodification management typically involves sizing such strategies based on continuous simulation of both the pre-development and post-development conditions and incorporating demand for onsite use and iterative design of the facility until flow duration control is achieved. Hydromodification management can provide a flood control benefit by holding up and releasing water more slowly to waterways.

Since the early 1900s, water suppliers and flood control agencies in the southwestern United States have been capturing floodwaters behind dams and/or diverting stormwater into large-scale spreading basins to replenish groundwater basins and manage flood risk. These facilities can be combined with flow control (e.g., constructed wetlands) to benefit hydromodification control strategies (Santa Clara Valley, 2005). The development of multi-purpose flood control and stormwater capture facilities to enhance percolation of stormwater (or historically called flood water) supplies along the river channel or alongside the river banks into recharge percolation ponds has developed into a more sophisticated water resources management strategy in recent decades (see Box 1-4). Los Angeles County's Department of Public Works alone operates 27 spreading basins to enhance local water supplies.[3]

Energy Savings and Greenhouse Gas Reductions

Beneficial use of stormwater and graywater can save energy compared to conventional sources under certain conditions. In the western and southwestern United States, there is a mismatch between population centers and areas of precipitation, and massive systems have been constructed to convey water over long distances to urban areas. Many of the conventional water supplies are derived from surface water that is pumped long distances, with significant energy and infrastructure costs. A study commissioned by the California State Water Resources Control Board concluded that capturing 1 acre-foot (1,200 m^3) of southern California stormwater and storing it in the ground saves roughly a metric ton of carbon dioxide (CO_2) compared to imported water supplies (Spencer, 2013). Although energy savings from stormwater and graywater are highly variable (see Chapter 6), it may be feasible to reduce energy use and greenhouse gas emissions through the increased use of stormwater or graywater, particularly when these local water sources require minimal treatment or pumping.

Environmental Stewardship

The implementation of stormwater and graywater beneficial use projects may also be driven by a sense of environmental stewardship, even in the absence of specific water conservation or pollution prevention goals. Individuals, businesses,

[2] See http://water.epa.gov/lawsregs/lawsguidance/cwa/tmdl.

[3] See http://dpw.lacounty.gov/wrd/SpreadingGround.

BOX 1-3 Stormwater Use in the Twin Cities Brings Multiple Benefits

In the land of 10,000 lakes, stormwater use is receiving attention for the multiple benefits it can provide. In 2011, the Metropolitan Council for the Twin Cities region of Minneapolis/St. Paul issued a guide to stormwater capture and use with the dual goals of reducing potable water demands that would otherwise require costly treatment plant upgrades and decreasing the pollution loads associated with stormwater discharges (Metropolitan Council, 2011). Notable projects in the Twin Cities are the Saint Anthony Village water reuse facility and the Target Field Rainwater harvesting system.

At Saint Anthony Village stormwater is collected from 15 acres (6 hectares [ha]) along with filter backwash from the city's water treatment plant. The water is stored in a half million-gallon reservoir below an open stormwater pond (Figure 1-3-1) and is used to irrigate a 20-acre (8 ha) park and city hall campus. The capital costs for the project were \$1.5 million, with operating costs of about \$3,000/yr. The total annualized cost (amortized capital outlay plus annual operation and maintenance costs), assuming a 5 percent interest cost over a 30-year lifetime, amounts to approximately \$100,000 per year. The project reduces stormwater discharges by more than 4.6 million gallons (17 million liters) per year and saves \$16,000/yr in potable water charges and more than \$15,000/yr in avoided wastewater disposal fees. Thus, the net costs amount to about \$69,000 per year. The project has reduced total suspended solids loading by 95 percent and phosphorus loads by 77 percent (Metropolitan Council, 2011). Other non-monetary benefits are beautifying the urban landscape through the reflection pool and fountains, and creating community awareness of public water management issues through signage and tours.

At the Leadership in Engineering and Environmental Design (LEED) silver-certified Target Field—home of the Minnesota Twins in downtown Minneapolis—a runoff recycling system saves approximately 2 million gallons (7.6 million liters) per year. The system includes a 200,000-gallon (800,000-liter) underground cistern located below the playing field and a treatment system comprised of filtration, ultraviolet disinfection, and chlorination. The water is used for field irrigation and wash-down of the lower grandstands. Capital costs are estimated between \$150,000 and \$500,000, and operating and maintenance costs are \$50,000/yr including energy and winterization. Besides the water savings, the Target Field project provides public education regarding stormwater treatment and reuse (Metropolitan Council, 2011). The total annualized cost (amortized capital outlay plus annual operation and maintenance costs), based on the midpoint of the capital cost estimates and assuming a 5 percent interest cost over a 30-year lifetime, amounts to approximately \$70,000 per year. The potable water savings yield a cost savings benefit of approximately \$8,000 per year (assuming local potable water supplies costs of \$4 per 1,000 gallons). The net costs per year thus amount to approximately \$62,000 per year, which may be weighed against the non-monetized benefits associated with public recognition of environmental stewardship (through LEED certification), public education, stormwater-related impacts and loadings avoided, and other values generated by the project.

FIGURE 1-3-1 A half million-gallon stormwater storage reservoir at Saint Anthony Village lies below an attractive stormwater pond with fountains. The system reduces the use of water treated to drinking water standards for irrigation and reduces phosphorus discharges to surface water. SOURCE: Metropolitan Council (2011).

FIGURE 1-7 Urbanization has increased stormwater runoff in Paint Branch, in College Park, Maryland. The resulting hydromodification causes more erosion, deepening of urban streams, and unstable channels compared to the pre-development state. SOURCE: http://www.anacostiaws.org/news/blog/tags/12.

and municipalities may be driven toward "green" practices out of a motivation to be good stewards to the earth. Individuals and communities may be motivated by the positive emotional return from making investments in green infrastructure, and businesses and municipalities may aim to enhance their public image through such initiatives.

There is a growing trend of environmental practices that embrace the concepts of sustainable urban water management and "green design" (Allen et al., 2010; WERF, 2009). For example, low-impact development projects, including green roofs and enhanced stormwater capture and infiltration, help manage the quantity and quality of stormwater runoff and better mimic the undeveloped landscape. These projects can reduce pollution, provide habitat, and contribute to the creation of a greener, more aesthetically pleasing city. Once considered pioneering, these practices now are widely implemented (e.g., Prince George's County, Maryland, 1999) and are required in some cities for development or redevelopment (e.g., San Francisco [SFPUC, 2009]).

The U.S. Green Building Council's trademark Leadership in Energy and Environmental Design (LEED) certification program recognizes environmental practices in building design.[4] The LEED program credits graywater irrigation to reduce water consumption and wastewater discharges, rainwater capture systems, pervious pavements, and on-site infiltration to reduce stormwater runoff.[5] Innovative water management strategies, such as on-site use of graywater or stormwater to reduce indoor and outdoor water demand and stormwater discharge, can earn builders up to one-half of the points required for basic LEED certification,[6] which can be a motivator even in areas where water scarcity is not a primary driver (Box 1-3).

Likewise, the Sustainable Sites Initiative is a voluntary effort to transform land development and landscaping in ways that offset impacts and use less energy and water.[7]

Extending the Capacity of Existing Infrastructure

America's urban water infrastructure was largely developed in the middle of the twentieth century, a time of inexpensive energy, smaller urban populations, and less appreciation of environmental impacts such as damage to aquatic habitat and consequences of greenhouse gas emissions. Massive dams, aqueducts, and pipelines were built to supply water to metropolitan areas such as the Colorado Front Range, the San Francisco Bay Area, the Los Angeles basin, Phoenix, and Dallas. Many of these water systems were characterized by a linear pattern of taking, treating, and discharging, using capital- and energy-intensive technologies with high costs for maintenance and operation (Daigger, 2009). In many urban areas today, this water infrastructure is reaching the end of its design life (ASCE, 2009).

[4] See http://www.usgbc.org/leed#why.

[5] See http://www.leeduser.com/credit/NC-2009/SSc6.1 and http://www.leeduser.com/credit/NC-2009/WEp1?usgbc=1.

[6] See http//www.usgbc.org/credits.

[7] See http://www.sustainablesites.org.

BOX 1-4 Re-Operation of Impoundments to Capture and Recharge Stormwater in California

In several locations in California, existing impoundments are managed to capture and enhance the recharge of stormwater. On the Santa Ana River, two U.S. Army Corps of Engineers (Corps) dams (Seven Oaks and Prado Dams) for decades have been operated to store stormwater on a temporary basis to allow for percolation and recharge into the groundwater aquifers within the Santa Ana River watershed. Although the primary purposes of these impoundments are for flood control, efforts are ongoing to optimize the dam operations to increase the recharge of Santa Ana River stormwater flows. Over the past two decades, about 100,000 AF/yr (120 million m^3/yr) of stormwater have been recharged along the Santa Ana River based on the operations of the Seven Oaks and Prado Dams (Santa Ana River Watermaster, 2014). In addition, with improved forecasting of storm events the Corps is working with the local water districts to improve the coordinated operation of these facilities to increase the effective capture and recharge of stormwater.

The Los Angeles County Flood Control District has also been working to manage its impoundments along two Corps dams (Santa Fe and Whittier Narrows) within the San Gabriel River watershed to increase infiltration downstream. Between 90 and 95 percent of the precipitation in average years above Whittier Narrows Dam is conveyed to a network of facilities in a coordinated manner to recharge the Main San Gabriel and Central groundwater basins (on average about 150,000 acre-feet per year [190 million m^3/yr] of stormwater recharge is re-operated and recharged; Figure 1-4-1) (C. Stone, LA County Flood Control District, personal communication, 2014).[a]

[a] See www.dpw.lacounty.gov/wrd/index/cfm.

FIGURE 1-4-1 Rio Hondo Spreading Grounds, which receives controlled releases from San Gabriel Canyon, Santa Fe, and Whittier Narrows dams for groundwater recharge. SOURCE: http://www.wrd.org/engineering/groundwater-replenishment-spreading-grounds.php.

Alternative water supplies, such as stormwater and graywater, can provide a means to prolong the life of existing infrastructure and avoid or postpone costly upgrades to centralized infrastructure. For example, New York City launched a plan to address the city's pollution problems from combined sewer overflows that included $2.4 billion in spending on green infrastructure, including incentives for graywater use and stormwater capture. If realized over a 20-year period, then it is estimated to save the city $1.4 billion compared to the conventional infrastructure approach (NYSDEC, 2013). At the household scale, graywater reuse may prolong the operating life of septic systems.[8]

[8] See http://extension.missouri.edu/p/EQM104F.

In dense urban areas, population growth and rising real estate prices often spur even denser use and taller buildings. If water use patterns remain the same, then this would require that additional water supply capacity be provided, along with a commensurate increase in wastewater collection and conveyance capacity. Because water supply and wastewater collection facilities are generally located in existing streets and adjacent rights-of-way, constructing these facilities is quite costly both monetarily and in terms of the traffic and business disruption caused by construction in existing roadways. Engineers in Tokyo have found that localized wastewater or graywater reuse for nonpotable purposes can be quite cost-effective because it can eliminate the need to construct new water supply and wastewater collection facilities to serve the more dense areas. Implementing nonpotable water reuse can reduce the net water supply requirement and wastewater production volume by about one-half, meaning that even if the total demand is doubled, existing water supply and wastewater collection infrastructure can serve the new facility. Tokyo requires graywater reuse in all new buildings larger than 32,000-54,000 ft^2 (3,000-5,000 m^2) and in existing buildings larger than 320,000 ft^2 (30,000 m^2) or with the capacity to reuse 26,000 gpd (100 m^3/d) (Ogoshi et al., 2001; CSBE, 2003).

Financial Incentives and Business Opportunities

Builders and developers may also implement stormwater capture and graywater reuse systems to take advantage of financial incentives or even to permit development in some water scarce regions. For example, the California legislature passed two bills in 2003 (SB 221 and SB 610) to advance water supply planning for growing communities. These laws require future water reliability assessments for all development projects subject to the California Environmental Quality Act and written verification by the water agency serving that project prior to approval of the project. The result of these requirements has been that new urban development in California has to go through a process of ensuring that new development has a reliable supply. Developers are thus motivated to have a minimal impact on local/regional supply reliability and typically incorporate state-of-art water use efficiency and conservation practices for indoor and outdoor water use, including on-site stormwater capture with low-water-use landscapes. Some cities also offer tax incentives or zoning allowances to LEED-certified buildings.

Increasingly, regulated municipal separate storm sewer system operators are looking for effective strategies to reduce the cost of compliance. To this end, Washington, DC, was the first entity to establish a true stormwater credit trading system with the intent of using a market-based approach to improve the efficiency of implementation of stormwater controls. The District of Columbia Stormwater Retention Credit Trading Program[9] allows private and public developers the ability to both buy and sell "stormwater retention credits" (SRCs). Properties can generate SRCs by building, operating, and maintaining green infrastructure that reduces stormwater runoff, including stormwater capture and use systems. Owners can sell their SRCs in an open market to developers or other private or public entities who can use them to meet regulatory requirements for retaining stormwater within the District. The first SRC trades occurred in the fall 2014 at $2.27/gallon per year ($0.60/liter per year) for a total value of about $25,000 (Brian VanWye, DDOE, personal communication, 2015).

[9] See http://green.dc.gov/src.

Balancing Multiple, Sometimes Conflicting Drivers

The many drivers discussed here can lead to a wide variety of different water management strategies, depending on which drivers are given the highest priority. A stormwater capture system that maximizes pollution prevention could be designed quite differently from one that maximizes capture and use, although systems can be designed to optimize multiple drivers. Likewise, a household graywater system could be designed in concert with native plantings to maximize water conservation, while a similar graywater system to provide a supplemental low-cost, reliable water supply for new nonnative plantings in an arid climate could actually increase overall water use.

Because urban water systems typically provide social and environmental benefits in addition to financial benefits and costs, it is important to fully account for the broad array of benefits and costs that may be associated with a graywater or stormwater beneficial use project, in the context of overall objectives (see Chapters 7 and 9). Multi-criteria decision analysis or broadly defined benefit-cost analysis are two important tools that may be useful for evaluating future management strategies that create such a broad spectrum of valuable outcomes. These decision support methodologies may include both monetized and non-monetized objectives such as water reliability and resiliency, locally sourced water and reduced imports, energy savings and conservation, financial incentives and cost sharing, environmental outcomes, community acceptance, and support of nongovernmental organizations (see Chapter 9).

ASSESSMENT OF CURRENT STORMWATER AND GRAYWATER USE TRENDS

The combined drivers of water scarcity, pollution prevention, infrastructure replacement costs, and concerns about energy and the environment have expanded efforts to capture and use stormwater and graywater.

Stormwater

Historically stormwater management meant flood control, but since passage of new sections in the Clean Water Act in 1987, attention has focused on control of pollutants from runoff under stormwater programs. Today, stormwater management can mean many different things, including controlling pollution and improving urban waterways, improving aquatic habitat, creating green space, and recharging local groundwater. Urban stormwater control measures are also a vital part of managing flooding and drainage in a city (NRC, 2009a).

There has been a recent increase in the use of stormwater practices that recharge groundwater. In wetter climates infiltration can raise groundwater levels, increase base flows, and sustain wetlands and lakes. Stormwater infiltration has been practiced in scattered locations for a long time. On Long Island, New York, infiltration basins were built in the 1930s to reduce the need for a storm sewer system. In Los Angeles, managed aquifer recharge with stormwater has been practiced since 1938, when the Rio Hondo and San Gabriel Coastal Spreading Grounds were opened by the Los Angeles County Flood Control District (Johnson, 2011; see Box 1-4). Within the area served by the Metropolitan Water District in the greater Los Angeles area, an annual average of about 477,000 AF/yr (588 million m^3/yr) of stormwater runoff is captured for recharge (MWD IRP Technical Work Group, 2009). In the 1980s, Maryland took the lead on the East Coast in developing statewide infiltration practices, and the number of states embracing low-impact development infiltration practices has increased (NRC, 2009a). These facilities, and low-impact development systems, were originally managed for fast infiltration. Today there is interest in understanding how these infiltration systems can provide water treatment in addition to improved aesthetics and new habitats that enhance community acceptance.

Evidence shows increasing trends and interest for stormwater use spreading to other parts of the country. A number of states are viewing stormwater as a resource for development of additional, local water supplies and reductions in demands on an aquifer. With increasing population and greater extent of impervious surfaces in urban areas, the Texas Water Development Board is assessing stormwater runoff to augment water supplies. The Board's report (Alan Plummer Associates, 2010) goes beyond small-scale rainwater harvesting and provides guidance on intermediate-scale stormwater harvesting to capture larger amounts of overland flow. Even temperate regions that normally experience adequate rainfall are exploring stormwater use (see Box 1-3). The Minnesota Pollution Control Agency began a pilot program for beneficial use of stormwater in industrial processes that otherwise would use groundwater (MPCA, 2012).

National data on the use of urban stormwater for water supply are not currently available. Rather, the literature describes many examples of systems at different scales, ranging from household to regional scales, and in different U.S. climatic zones, from the humid East Coast to the dry Southwest. Evidence of increased interest in stormwater harvesting is illustrated by signature projects in the past dozen years designed to comply with LEED certification, help reduce stormwater discharges to combined sewer systems, or develop sustainable urban environments in water-stressed cities. As part of its planning, the San Francisco Public Utilities Commission (SFPUC, 2013) documented numerous neighborhood-scale case studies such as the Solaire in New York City, where stormwater is collected in a 10,000 gallon (38,000 liters) tank for irrigation, and the Olympic Village in Sydney, Australia, where stormwater runoff from the 1,580-acre (640 ha) Olympic Park is harvested and reused for irrigation, washing, and other uses. Singapore harvests stormwater on a large-scale for its drinking water supply (PUB, 2015). Within the service district of the Metropolitan Water District of southern California there are 34 stormwater projects anticipated for completion between 2009 and 2020 that could increase regional stormwater capture by about 50,000 AF/yr (60 million m^3/yr) (MWD IRP Technical Work Group, 2009), and even more new projects will be developed as part of the Los Angeles Stormwater Master Plan (see Box 1-2).

Graywater

In recent years, interest in graywater reuse has greatly increased. Documenting this trend are state laws that promote graywater reuse. As of 2013, 20 states allow some form of graywater reuse, and Arizona and California are considered leaders in promoting graywater reuse (Sharvelle et al., 2013). Arizona allows graywater reuse without a permit for systems less than 400 gpd (1,500 lpd), while California allows household laundry-to-landscape systems without a permit as long as they follow specific design guidelines (EBMUD, 2009; California Plumbing Code Ch. 16A, Sec. 108.4.1). The heightened interest in graywater reuse is also documented by an increase in the number of national conferences and workshops on graywater reuse and ways to promote the practice, such as an EPA workshop in Atlanta in 2010 on graywater practices and regulations (EPA, 2010). Professional societies have sponsored studies on graywater practices and effects (e.g., Sheikh, 2010) and offer official policy statements on water use efficiency including appropriate on-site graywater reuse (e.g., Olson, 2014).

Interest in safe and effective graywater use has prompted some limited surveys on the practice. A 1999 survey by the Soap and Detergent Association reported that 7 percent of U.S. households reuse graywater, mainly for irrigation, with the largest concentration residing in the West and Southwest (NPD Group, 1999). A 2000 survey in Pima County, Arizona, found about 8 percent of respondents employed some form of graywater reuse but this was highly variable depending on water district and ranged from 2 to 25 percent (Little, 2000).

FUTURE PERSPECTIVES

Recent trends suggest that beneficial use of graywater and stormwater are small but increasing parts of the nation's water supply portfolio. To meet future water demands amidst challenges from aging infrastructure, population growth and redistribution, and climate change, an array of water supply and conservation alternatives and innovative water management strategies will be necessary. The nature of this re-invention of urban water management is difficult to predict, but the nation's water infrastructure will likely look quite different in 50 years than it does today. Future water infrastructure designs could create more cost-effective opportunities for the use of local or on-site water sources, such as stormwater and graywater. Currently, few buildings contain dual plumbing to take advantage of nonpotable water use, but as buildings are redeveloped in the future, major changes in water distribution and use become possible. Thus, consideration of the potential for graywater and stormwater to augment the nation's water supply should not be limited by current infrastructure constraints. An example of a recent, rapid change in water infrastructure can be found in adoption of low-flow toilets (see Box 1-5).

Drivers of water management evolve over time with changing infrastructure, water availability, prices, and societal and individual values. Today's infrastructure investments may shape water management practices for decades and should ideally support future water management priorities rather than maintaining outdated or inefficient practices that are anticipated to decline over time. For example, early water reuse efforts in the Southwest provided dual delivery of nonpotable reclaimed wastewater for landscape irrigation in parks and highway medians; today, such infrastructure may be seen as wasteful, when native landscaping strategies can significantly reduce irrigation demand. Thus, graywater and stormwater infrastructure investment decisions ideally include the anticipation of the role of alternative water supplies in the future, rather than simply solving today's water management challenges.

Further advances in water use efficiency and conservation could impact the demand for alternative water sources, such as stormwater and graywater, as well as the supply of graywater available for reuse. Graywater relies upon the reuse of laundry, sink, and shower water, and additional improvements in water efficiency in these applications, such as further advances in the water efficiency of washing machines, would impact the amount and quality of graywater available for reuse. Developments in source separation (separating urine from solid waste) could reduce water use, increase the cost-effectiveness of water reuse, and allow the recovery of energy and nutrients to provide important sources of revenue in the future (Daigger, 2009; Guest et al., 2009). Similarly, climate change could increase the intensity of precipitation, which in some regions could result in less capture and/or recharge of stormwater with existing systems. Thus, the potential contributions of graywater and stormwater to the nation's water supply will evolve over time based on demand and supply and evolving requirements for urban water, wastewater, and stormwater utilities in the United States (Hering et al., 2013).

In the United States, despite steady population growth, total water use has declined since its peak in 1980 (Figure 1-8) because of improvements in industrial and domestic water efficiency, conservation, and recent declines in the use of once-through thermoelectric cooling. Even public water supply use, which supplies domestic, commercial, and industrial uses, declined 5 percent between 2005 and 2010 after increasing steadily between 1950 and 2005, although these declines may be related to the economic recession (Figure 1-9). Per capita domestic use declined at an even steeper rate, from 100 gpcd (380 lpcd) in 2005 to 89 gpcd (340 lpcd) in 2010 (Maupin et al., 2014). However, domestic water use varies widely across the country. At the city level, an analysis of data from 2005 to 2010 for 21 U.S. cities found median domestic

BOX 1-5 Rapid Water Infrastructure Changes: Low-Flow Toilets

Low-flow toilets were introduced in 1988 when California was in the midst of a significant drought. In 1988 water districts in southern California began conducting pilot studies with Swedish low-flow toilets that used about 1.6 gallons (6.1 liters) per flush. At the time, American manufacturers made toilets that used 3 or more gallons (11 liters) per flush. Based on these pilot studies, programs to install these low-flow toilets began in 1990. The programs were so successful that by 1992 U.S. manufacturers began making low-flow toilets, and the U.S. Congress enacted legislation that changed the standard for toilets to the 1.6 gallons per flush. By 1993 approximately 1 million toilets in Los Angeles County had been replaced through rebates and financing by the local water districts. Together with high-efficiency washing machines, low-flow toilets have been a major contributor to declines in indoor water use observed nationally over the past 15 years (DeOreo et al., 2016; Mayer et al., 1999). The efficiency standard for toilets in California was further reduced to 1.28 gallons (4.85 liters) per flush in April 2015 (CEC, 2015).

water use ranged from 43 to 177 gpcd (160 to 670 lpcd) with some correlation with a city's precipitation and temperature (Kenny and Juracek, 2012). In drier climates outdoor residential water use is typically much greater than indoor water use. For example, the Water Research Foundation (Coomes et al., 2010) reported three times or more residential water use in Dallas and Phoenix compared to Seattle; similarly, a U.S. Geological Survey (USGS) state-by-state survey of public water supplies showed a threefold difference in residential water use from 55 gpcd (210 lpcd) in Maine to 168 gpcd (636 lpcd) in Idaho (Maupin et al., 2014). By contrast, in Australia residential water use is less at about 39 gpcd (150 lpcd) in Melbourne on the wetter east coast (Gan and Redhead, 2013; Melbourne Water, 2013/2014) and about 76 gpcd (290 lpcd) in arid Perth on the west coast with about 39 percent for irrigation (The Water Corporation, 2010).

At a household level, substantial improvements in outdoor water use efficiency are possible through the use of native landscaping, but once high-efficiency appliances and plumbing fixtures are installed, indoor water use is unlikely to see substantial additional declines without the use/reuse of on-site sources such as stormwater or graywater. Meanwhile, population growth and redistribution will continue to increase urban water demand.

Domestic water conservation and trends in water use in agricultural and industrial sectors will influence regional water availability and the benefits of graywater and stormwater use in the future. Today water professionals and urban designers understand that more efficient use of water and resources is possible. The "Cities of the Future" initiative by the International Water Association seeks to integrate water and city planning much more closely and support innovation and strategic thinking (Daigger, 2011). So-called water-centric urban design can lead to both better use of resources and an enhanced urban environment. In the same fashion, the concept of the "water sensitive city" embraces sustainable urban water planning and management practices in which cities serve as water supply catchments, provide ecosystem services and prioritize livability, sustainability and resilience (Ferguson et al., 2013). These concepts are examples of where the future could be going and the role that graywater and stormwater could play.

COMMITTEE CHARGE

Despite several drivers supporting increased use of local alternative water supplies to address water demands, many questions remain that have limited the broader application of graywater and stormwater capture and use. In particular, limited information has been available on the costs, benefits, and risks of these projects, and beyond the simplest applications, many state and local public health agencies have not developed regulatory frameworks for full use of these local water resources. With the timeliness of a severe drought in the West and periodic water shortages elsewhere and funding support from the EPA Office of Water, EPA Office of Research and Development, EPA Region 9, National Science Foundation, Water Research, Water Environment Research

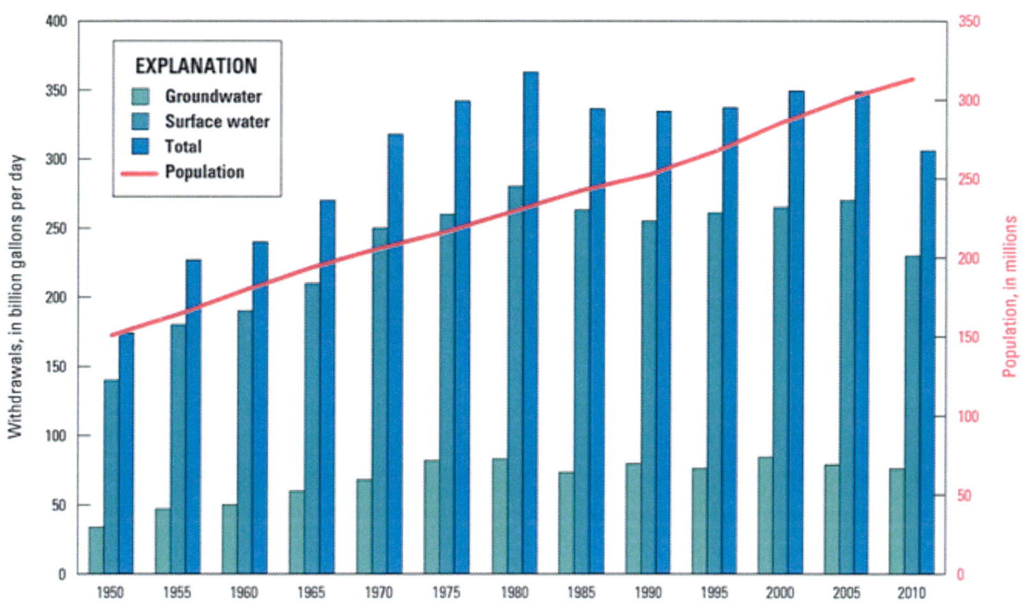

FIGURE 1-8 Trends in U.S. freshwater withdrawals and population growth. SOURCE: Maupin et al. (2014).

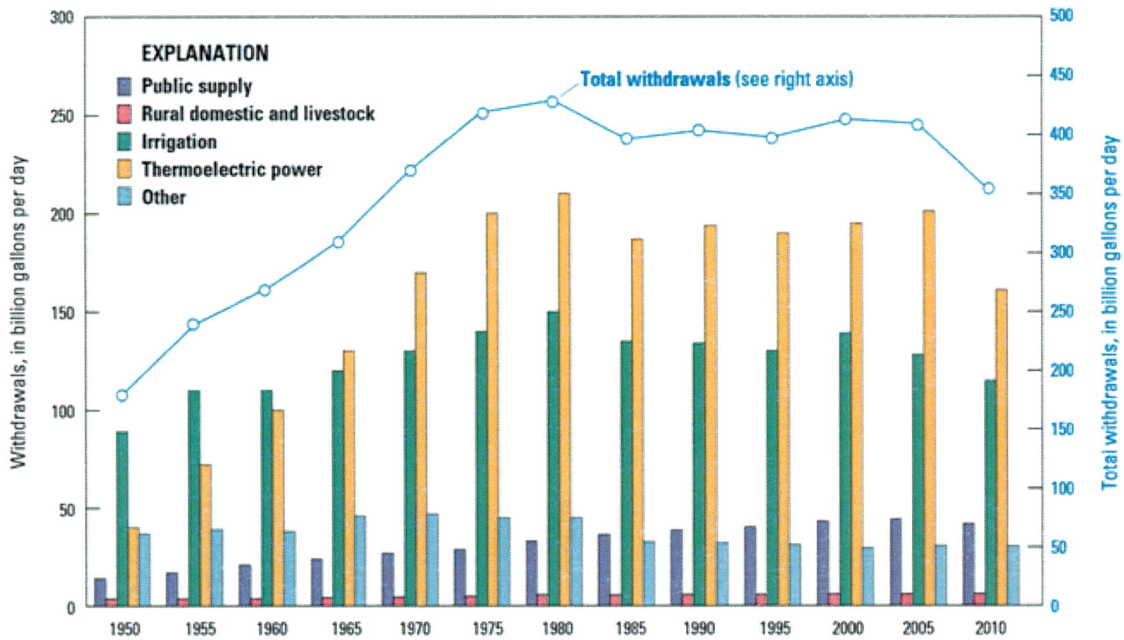

FIGURE 1-9 Trends in total U.S. freshwater withdrawals by category of water use. Public supply includes domestic, commercial, and industrial uses. Uses may be consumptive or nonconsumptive. SOURCE: Maupin et al. (2014).

Foundation, Los Angeles Department of Water and Power, WateReuse Foundation, City of Madison, Wisconsin, National Water Research Institute, and the Academies' President's fund, the Academies formed a Committee on the Beneficial Use of Graywater and Stormwater. This study builds on previous work of the Academies to assess the augmentation of urban water supplies by desalination (NRC, 2008a) and reuse of municipal wastewater (NRC, 2012a), with a new focus on two on-site water sources—graywater and stormwater. The goals of the committee are to be forward looking and to conduct a study and prepare a report on the risks, costs, and benefits of various uses of stormwater and graywater and the approaches needed for their safe use. The committee's work considers multiple scales for these approaches—from the household scale to multi-residential or neighborhood scales to large municipal systems. The study will address both technology and policy questions:

1. **Quantity and suitability.** How much stormwater capture and graywater reuse occurs in the United States and for what applications? What is the suitability—in terms of water quality and quantity—of captured stormwater and graywater to significantly increase in the United States, and where regionally would increases in these practices have the most benefit? How would significant increases in the beneficial use of stormwater and graywater affect water demand, downstream water availability, aquifer recharge, and ecological stream flows? What research should be pursued to understand these issues?
2. **Treatment and storage.** What are typical levels and methods of treatment and storage for stormwater capture and graywater reuse for various end uses? What types of treatment are available to address contaminants, odors, and pathogens, and how do these treatment methods compare in terms of cost and energy use? What research opportunities should be pursued to produce improved technologies and delivery and ensure adequate safeguards to protect public health and the environment?
3. **Risks.** What are the human health and environmental risks of using captured stormwater and graywater for various purposes? What existing state and regulatory frameworks address the beneficial use of stormwater and graywater, and how effective are they in assuring the safety and reliability of these practices? What lessons can be learned from experiences using captured stormwater and graywater both within and outside the United States that shed light on appropriate uses with varying levels of treatment? What local measures can be taken to reduce risk?
4. **Costs and benefits.** What are the costs and benefits of the beneficial use of stormwater and graywater (including non-monetized costs and benefits, such as ef-

fects on water and energy conservation, environmental impacts, and wastewater infrastructure)? How do the economic costs and benefits generally compare with other supply alternatives? Can cost improvements be achieved through research?

5. **Implementation.** What are the legal and regulatory constraints on the use of captured stormwater and graywater? What are the policy implications regarding the potential increased use of stormwater and graywater as significant alternative sources of water for human consumption and use?

The committee's report and its conclusions and recommendations are based on a review of relevant technical literature, briefings, discussions, and field trips at its six meetings, and the experience and knowledge of the committee members in their fields of expertise. The committee received briefings from a range of experts, including water utilities, practitioners, government and public health officials, nongovernmental organizations (including public interest and industry groups), and academics.

The report focuses on nonpotable uses of graywater and stormwater, such as irrigation and toilet flushing (see Chapter 2), which can be met with little or no additional treatment, and recharge of groundwater that eventually may be used for drinking water supplies. The committee did not examine technologies for direct on-site use of graywater or stormwater for drinking water, because of the associated unique safety issues and treatment requirements and the unlikely potential for expanding such uses in the United States in ways that significantly augment existing water supplies. The committee recognizes that roof runoff is used to meet potable needs in many rural areas in the United States and around the world, and other reports are available that recommend specific capture and treatment practices (e.g., Macomber, 2010; TWDB, 2005).[10] In light of this report's focus on the potential water supply contributions of graywater and stormwater projects, the report's stormwater components focus on stormwater capture and infiltration efforts that have intentional or inadvertent water supply benefits. A full discussion of stormwater management strategies, including those designed primarily for water quality benefits, are described in NRC (2009a).

[10] See also http://www.who.int/water_sanitation_health/gdwqrevision/rainwater.pdf.

OUTLINE OF THE REPORT

Following this introduction, the statement of task is addressed in eight subsequent chapters of this report:

- Chapter 2 describes potential uses of graywater and stormwater and specific water quality constraints on these uses.
- Chapter 3 describes the potential water savings provided by stormwater or graywater, and provides the results of an original scenario analysis in six different locations of the country.
- Chapter 4 describes what is known about the water qualities of various sources of stormwater and graywater before treatment.
- Chapter 5 describes approaches for characterizing the risk of on-site nonpotable uses of graywater and stormwater and summarizes the research to characterize these risks.
- Chapter 6 examines the state of the technology of graywater and stormwater system components.
- Chapter 7 analyzes the costs and benefits (including both financial and non-monetized costs and benefits) of stormwater and graywater projects.
- Chapter 8 outlines the legal and regulatory controls on graywater and stormwater projects and identifies the largest impediments to expanding the use of such sources.
- Chapter 9 synthesizes the major report findings, as they are relevant to the perspective of a local decision maker, and presents major steps to consider in the development of such projects.
- Chapter 10 summarizes major research needs to enhance the implementation of graywater and stormwater to meet current and future water demands.

2

Beneficial Use Options for Graywater and Stormwater

Stormwater and graywater can provide on-site or local sources of water for an array of uses. As discussed in Chapter 1, the practice of capturing graywater and urban stormwater for safe use has many potential benefits, such as reducing the impact of urban development on water quality and stream flow and contributing toward water conservation objectives. This chapter provides an overview of the suitability of graywater and stormwater for various beneficial applications.

GRAYWATER AND STORMWATER USE ACROSS MULTIPLE SCALES

Graywater and stormwater can be captured and used to meet water demands or to conserve conventional water supplies across a range of scales. The scale of the application usually determines which beneficial uses are reasonable, the feasibility of treatment (considering maintenance requirements and cost), and appropriate sizing of cost-effective storage. Both graywater and stormwater use can occur at the household (or small building) scale, at the neighborhood/multi-residential scale, or at the regional scale (Figure 2-1). The committee's definitions of these scales are provided in Table 2-1.

At the household (or small building) scale, storage capacities and treatment for captured stormwater and graywater are typically limited because of cost. At this scale, irrigation with untreated graywater is most common and provides a small but steady year-round source of irrigation water. Larger storage and treatment systems at the household scale are possible, but they are expensive and require periodic maintenance. Stormwater capture tanks at the residential scale also tend to be small (rain barrels or cisterns) and generally provide supplemental water for irrigation and outdoor uses near times of rainfall. Although rooftop capture at the household scale can reduce some domestic water use and reduce stormwater discharges, it does not represent a reliable long-term source of water, especially in arid regions. For residential stormwater harvesting, most of the attention is given to roof runoff, because its quality is typically better than other source areas, and it is elevated above the likely use areas, easily captured, and therefore hydraulically simpler to utilize.

At the neighborhood or multi-residential scales (including large commercial buildings and institutions), more complex stormwater and graywater system designs with storage and treatment become feasible. The added treatment expands the array of applications available, particularly those in which human contact and inadvertent ingestion is possible. For stormwater, larger storage extends the capacity to address water supply needs during periods without rain through the use of retention basins or subsurface tanks. Construction of infiltration facilities resulting in managed groundwater recharge is feasible at a neighborhood scale, where blended runoff from several source areas is utilized.

At the regional scale, stormwater capture and infiltration systems can be developed in some areas to collect runoff from multiple neighborhoods or a large drainage basin resulting in managed aquifer recharge for later recovery. Regional stormwater capture systems might also offer opportunities to establish water features in public parks for aesthetics and recreational purposes in addition to water storage. Stormwater runoff collected at a regional scale comes from a wide variety of land uses resulting in a water quality that requires a higher level of treatment prior to use. Regional-scale graywater capture and beneficial reuse have been adopted very recently in large new developments (see Box 2-1). Because of the infrastructure requirements to separate graywater from blackwater at larger scales, regional graywater use might not always be economically feasible, particularly in existing buildings, and instead, reuse of municipal wastewater effluents can be implemented at large scales to make use of this locally available water supply (NRC, 2012a).

Currently, captured stormwater and graywater are used primarily for nonpotable applications, mainly irrigation (i.e., lawn, trees, xeriscape) and selected urban applications (e.g., toilet flushing, air conditioning, car washing). In addition, stormwater can be used for wildlife habitat maintenance and recreational uses and to recharge groundwater, which eventually may serve potable uses or maintain stream base flow for aquatic habitat. Common and specific uses for stormwater and graywater, their typical scale, and some limitations of these applications are summarized in Table 2-2. Although it is feasible to treat stormwater or graywater for on-site potable use (Ahmed et al., 2010), such applications are uncommon in the United States, except in remote areas with no reliable groundwater supplies where rainwater may be harvested from roofs for household supplies.[1] Given the complexity and unique safety issues associated with on-site potable use of graywater and stormwater at the household and neighborhood scales and the unlikely potential for expanding such uses in the United States to maintain and expand existing water supplies, the committee did not examine issues surrounding on-site potable use in this report.

[1] For example, see http://health.hawaii.gov/sdwb/raincatchment.

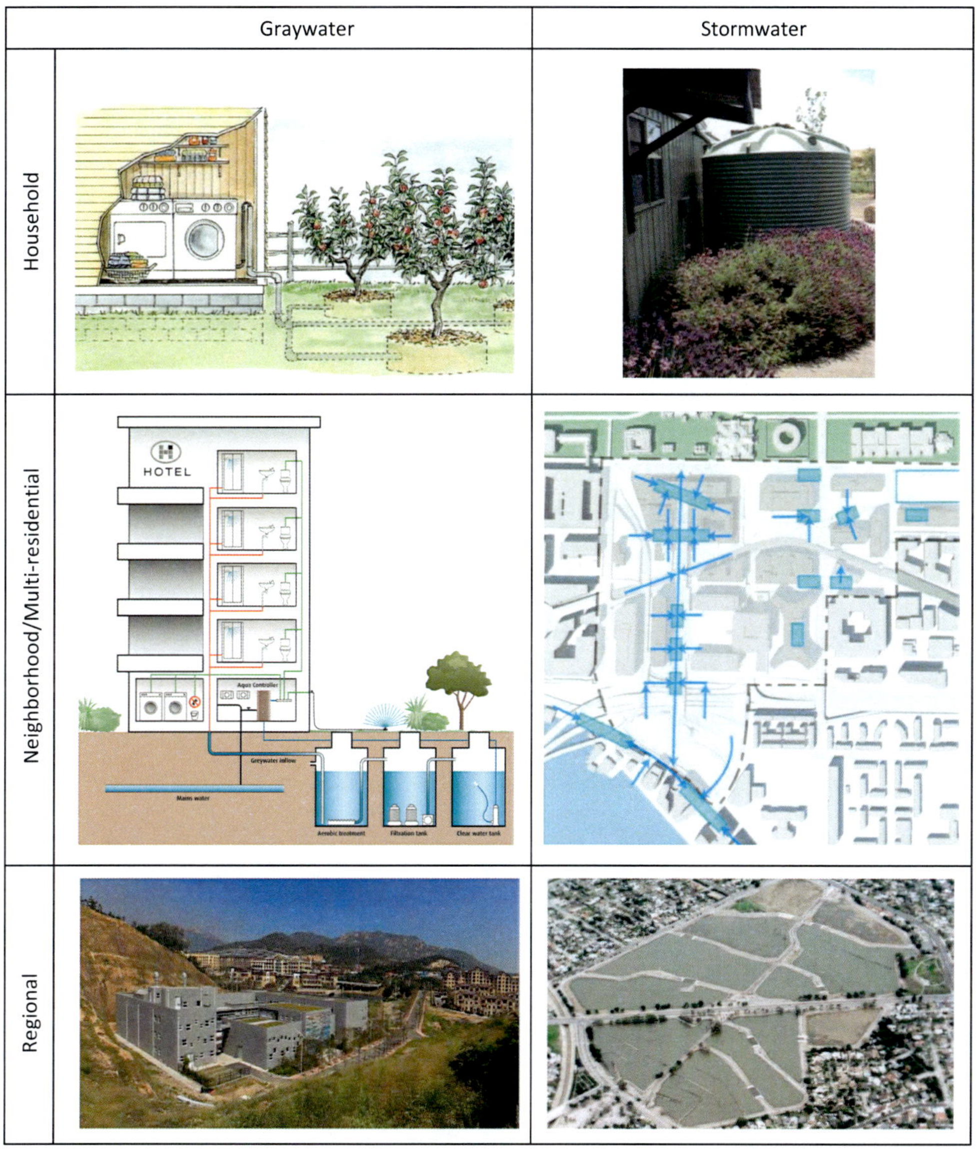

FIGURE 2-1 Illustrations of graywater and stormwater beneficial uses across scales. For graywater, the scale of use can range from simple laundry-to-landscape systems, to multi-residential capture and treatment for use in toilet flushing, to large new communities with centralized graywater capture and treatment (see Box 2-1). For stormwater, the scale of use can range from household capture if roof runoff in rainbarrels or cisterns, to neighborhood stormwater capture facilities, to large regional capture and infiltration systems. SOURCES: Allen and Woelfle-Erskine 2011; Robert Pitt Waterscan 2014; Diane Sullivan, National Capital Planning Commission, personal communication, 2013; http://www.semizentral.de; http://dpw.lacounty.gov/adm/sustainability/ProjectSP.aspx?id=69.

TABLE 2-1 Scales of Beneficial Use of Graywater and Stormwater, as Defined by the Committee

	Household[a]	Neighborhood[b]	Regional
Approximate number of people served	< 10	10-8,000	>8,000
Number of land uses	1	≤3[c]	>3[b]
Approximate area [d]	< 1 acre	1 acre-1 mi^2	>1 mi^2

[a]The parameters for household scale were identified as those likely to fall under the lowest tier of typical state tiered regulatory systems, which are commonly limited to 400 gpd (1,500 lpd) or less (see Chapter 8).
[b]Multi-residential units fall under the neighborhood scale.
[c]At neighborhood scales, land uses are typically selected to minimize sources of contaminants, such as residential, commercial, or institutional. At regional scales, many urban land uses are typically included within the project footprint and may include industrial uses and high traffic roadways.
[d]1 acre = 0.4 hr; 1 mi^2 = 260 hr.

BOX 2-1 The Semicentral Resource Recovery Center, Qingdao, China

In most fast-growing urban areas, the existing infrastructure is challenged by the high demand for drinking water and energy as well as rapidly increasing amounts of wastewater and solid waste. One possible approach to address these challenges is the so-called "semicentralized" supply and treatment system. This innovative approach was developed by the Technical University of Darmstadt, Germany, for applications in fast-growing urban areas, such as those in China. It focuses on the integrative assessment of the different material and energy flows, in particular water, wastewater, and waste. The treatment of waste, graywater, and blackwater takes place in a semicentralized Resource Recovery Center (RRC) within or close to the residential area, and the water is reused for nonpotable uses. Depending on site-specific conditions, 30-100 percent of wastewater can be reused, resulting in a significantly lower amount of wastewater to be discharged into water bodies. The other main advantage is internal energy recovery enabled by the increased biogas production through co-digesting biowaste with waste-activated sludge.

The first RRC opened in April 2014 in Qingdao Shiyuan, China, serving a total of 12,000 population equivalents (Figure 2-1). Its service area consists of residential areas, a large administration center with guest houses, and two hotel complexes. Graywater from showers, hand washbasins, and washing machines is collected and transported separately from the blackwater to the RRC (Figure 2-1-1). Graywater treatment consists of mechanical pre-treatment, biological treatment (elimination of organic carbon compounds), and disinfection. To meet the strict quality standards for service water and irrigation water, a membrane bioreactor (MBR) with subsequent chlorine disinfection is employed. The RRC will provide service water for toilet flushing via a separate distribution system. In addition, service water is used for street cleaning. Blackwater from kitchens and toilets is also conveyed to the RRC and treated separately from graywater (Figure 2-2-1).

The third important material flow within the RRC in Qingdao is food waste from restaurants and canteens. Following pre-treatment, the food waste is mixed with waste-activated sludge and thermophilically co-digested. The digestate is used as biosolids in landscaping, utilizing phosphorus from the wastewater as a fertilizer. The biogas derived from anaerobic treatment and a combined heat and power unit generates electricity and heat, resulting in a self-sufficient operation of the RRC.

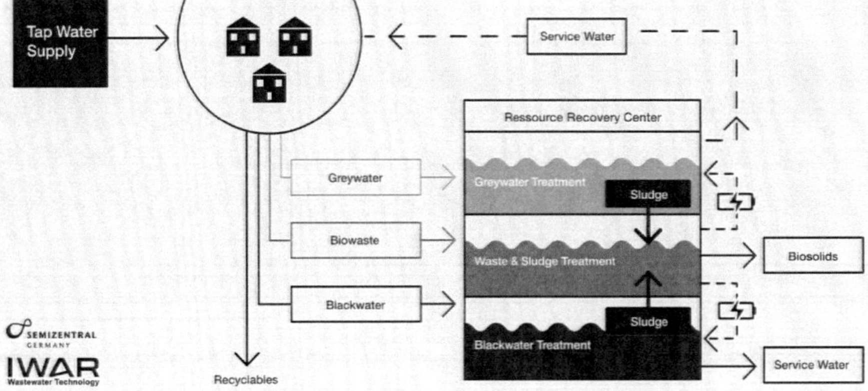

FIGURE 2-1-1 Flow schematic of the semicentralized Resource Recovery Center in Qingdao, China. SOURCE: J. Tolksdorf, S. Bieker, P. Cornel, Technische Universität Darmstadt, Germany, http://www.semizentral.de.

TABLE 2-2 Current Uses of Graywater and Stormwater in the United States to Enhance Water Supplies and Relevant Scales

		Graywater Application			Stormwater Application			
Category of Use	**Specific Types of Use**	**Household/ Small building**	**Neighborhood/ Multi-residential**	**Regional**	**Household/ Small building**	**Neighborhood/ Institutions**	**Regional**	**Limitations**
Urban Applications	Toilet and urinal flushing	X	X	X	X	X		• Dual distribution system costs • Building-level dual plumbing may be required • Greater burden on cross-connection control
	Heating and air conditioning makeup water or evaporative cooling				X	X		
	Fire fighting					X		
	Laundries					X		
	Vehicle washing	X	X		X	X		
	Street cleaning		X	X		X		
	Decorative fountains/other ornamental water features				X	X		
Landscape irrigation	Lawns, flowerbeds	X	X	X	X	X		• Dual distribution system costs • Uneven seasonal demand • High total dissolved solids (TDS) and sodium adsorption ratio (SAR) of graywater water can adversely affect plant health
	Parks, golf courses					X		
	Playgrounds/schools					X		
	Agriculture				X	X		
	Greenways					X		
Wildlife habitats and recreational uses	Wetlands		X	X		X	X	• Dual distribution system costs • Nutrient removal may be required to prevent algal growth • Potential ecological impacts depending on graywater quality and sensitivity of species
	Ornamental or recreational water body					X	X	
Large-scale water supply augmentation	Surface impoundments					X	X	• Treatment for contaminant removal may be needed • Recharge requires specific hydrogeological conditions • Potential for water quality degradation in subsurface
	Groundwater recharge					X	X	

NOTES: Other uses are possible but not commonly applied. Certain uses require a high level of treatment. Scales are defined in Table 2-1.

COMMON APPLICATIONS TO CONSERVE CONVENTIONAL WATER SUPPLIES

Graywater and stormwater are being used in many places in the United States and worldwide in ways that conserve conventional water supplies. Common graywater and stormwater uses are illustrated in Table 2-2, but other uses could be considered in the future for each water source. As noted in Chapter 1, there are many drivers behind graywater and stormwater use, and water supply may be a primary objective or only a peripheral benefit supplementing the primary drivers for the project. These applications are discussed in the following section, along with water quality concerns for particular end uses.

Urban Applications

An array of on-site uses for stormwater and graywater exist for urban applications, including toilet and urinal flushing, wash water, air conditioner chiller water, firefighting, commercial laundries, vehicle washing, street cleaning, decorative fountains, and other water features. As an alternative supply source for these nonpotable uses, stormwater requires treatment mainly to address particulate matter and pathogens for applications with significant potential human exposures (see also Chapter 5). Graywater requires treatment prior to most uses other than subsurface irrigation to minimize human health risks from microbial contaminants and to provide a stable water quality because microbial growth occurs when untreated graywater is stored (see Chapter 5). Treatment systems (discussed in Chapter 6) can vary substantially in complexity and the degree of treatment achieved.

Toilet Flushing

A popular beneficial application that reduces consumption of conventional water supplies is the use of captured stormwater or graywater for toilet flushing. Toilet flushing makes up approximately 24 percent of domestic water use (equivalent to about 14 gallons [53 liters] per person per day); thus, the savings potential for potable water is significant (Reichel et al., 2011; DeOreo et al., 2016). Water use for toilet flushing varies with population, toilet design (i.e., low-flow versus standard toilets), and demand fluctuations. For example, weekend versus weekday water use patterns are different, and seasonal variations occur because of varying workforce numbers or student populations at schools or in cities.

Using a nonpotable water supply for toilet flushing requires a plumbing system separate from the potable water system (known as dual plumbing). Dual plumbing is most cost-effective to install in new construction, although some buildings have been retrofitted to accommodate stormwater and graywater for toilet flushing. Treatment systems are typically used, with the primary objective of controlling microbial growth and human health risks from pathogens (discussed further in Chapters 5 and 6). Given the treatment system maintenance requirements, the use of on-site water sources for toilet flushing is more common in multi-residential or office buildings (see Box 2-2) rather than individual households. Treatment systems for single residences are commercially available and have been successfully installed and operated.

Outdoor Water Features

Stormwater can be used to provide water for outdoor fountains or other ornamental water features (Box 1-3). In Germany, there are several examples of stormwater beneficial uses where aesthetic values were important in the stormwater management system. Dreiseitl (1998) states, "[S]tormwater is a valuable resource and opportunity to provide an aesthetic experience for the city dweller while furthering environmental awareness and citizen interest and involvement." For example, in Berlin at the Potsdamer Plaza, roof runoff is captured in large underground tanks. Some of the runoff is treated for toilet flushing and irrigation use, and the rest flows into a 3.8-acre (1.4 ha) artificial lake in the center of the developed area (Figure 2-2), reducing stormwater discharges into local rivers. The Potsdamer Plaza project also has numerous fountains (Dreiseitl, 1998). Pitt et al. (2011) describe several other cases where stormwater has been used for aesthetic purposes. At the Cincinnati Zoo, harvested stormwater is used as makeup water for moats surrounding animal enclosures and for flowing water features (artificial streams) through the zoo grounds (Box 2-3). Water treatment typically focuses on removal of human pathogens and possibly particulate matter to minimize clogging and other maintenance issues depending on the storage and conveyance systems used.

BOX 2-2 Graywater Use for Toilet Flushing at a Colorado State University Residence Hall

At Aspen Residence Hall on the campus of Colorado State University, shower and handwash water from 14 rooms (up to 28 residents) is collected and subsequently used for toilet flushing. The drivers for project installation were the university's desire to save water and green building initiatives (including LEED certification). The water savings are estimated to be 5.4 gallons (20 liters) per person per day.

Graywater is treated by coarse filtration (particle exclusion > 1 mm) and disinfection (Figure 2-2-1). Online chlorine monitoring is integrated in the system to ensure safety in case of malfunction of the disinfection system. If the chlorine residual drops below a set point (originally 1.0 mg/L), then supply is automatically switched to municipal water.

The capital cost of the system totaled \$66,000 with operation and maintenance (O&M) costs of \$5,000 per year. Implementation issues involved costly plumbing modifications necessary to meet state plumbing board requirements and weekly cleaning for the chlorine measurement probe. Odor problems occurred when the chlorine residual approached 1 mg/L, and the low set point for chlorine was increased to 1.75 mg/L to address this concern.

FIGURE 2-2-1 Storage and treatment system for the graywater toilet flushing system at Aspen Residential Hall, Colorado State University. SOURCE: Courtesy of Sybil Sharvelle.

Washing

Stormwater and graywater can be used for washing vehicles, equipment, and paved surfaces. Graywater reuse is commonly practiced by commercial car washes, which often reuse their wash water to substantially reduce overall water use. In Washington, DC, a number of fire stations are equipped with stormwater capture systems, and the harvested stormwater is available for daily fire engine spray washing and refill of fire truck day tanks. Although the DC stormwater capture systems are primarily designed to reduce runoff flowing to the combined sewer during wet weather, the captured stormwater is also available onsite to reduce potable water use. Typical vehicle wash water is filtered and disinfected at the time of use to reduce risk associated with any human pathogens that might be present.

Firefighting

Water for firefighting can be supplemented with stormwater runoff generated on site or within the immediate neighborhood, provided there is sufficient storage. Wet detention ponds can provide much needed reliability of access to firefighting water during times of natural disasters. As an example, during re-building of Veteran's Administration hospitals in Los Angeles after the Northridge earthquake, stormwater detention ponds were planned that would hold sufficient additional water for firefighting needs. If stormwater flow is insufficient to maintain the water level in the ponds to the level needed to meet firefighting needs, then the detention pond would be supplemented with makeup water. Firefighting water volume needs for specific building types and sizes can be calculated based on the International Fire Code (ICC, 2012). As an example, a 50,000 ft^2 building made from heavy timber with noncombustible external materials would require a fire flow of 4,000 gpm (15,000 L/min) for a duration of 4 hours. Therefore, a total volume of about 960,000 gallons (3.6 million liters) must be available for on-site storage of firefighting water. In most cases, sufficient treatment is provided in the wet detention pond, although additional disinfection may be required if fecal indicator bacteria levels exceed use requirements.

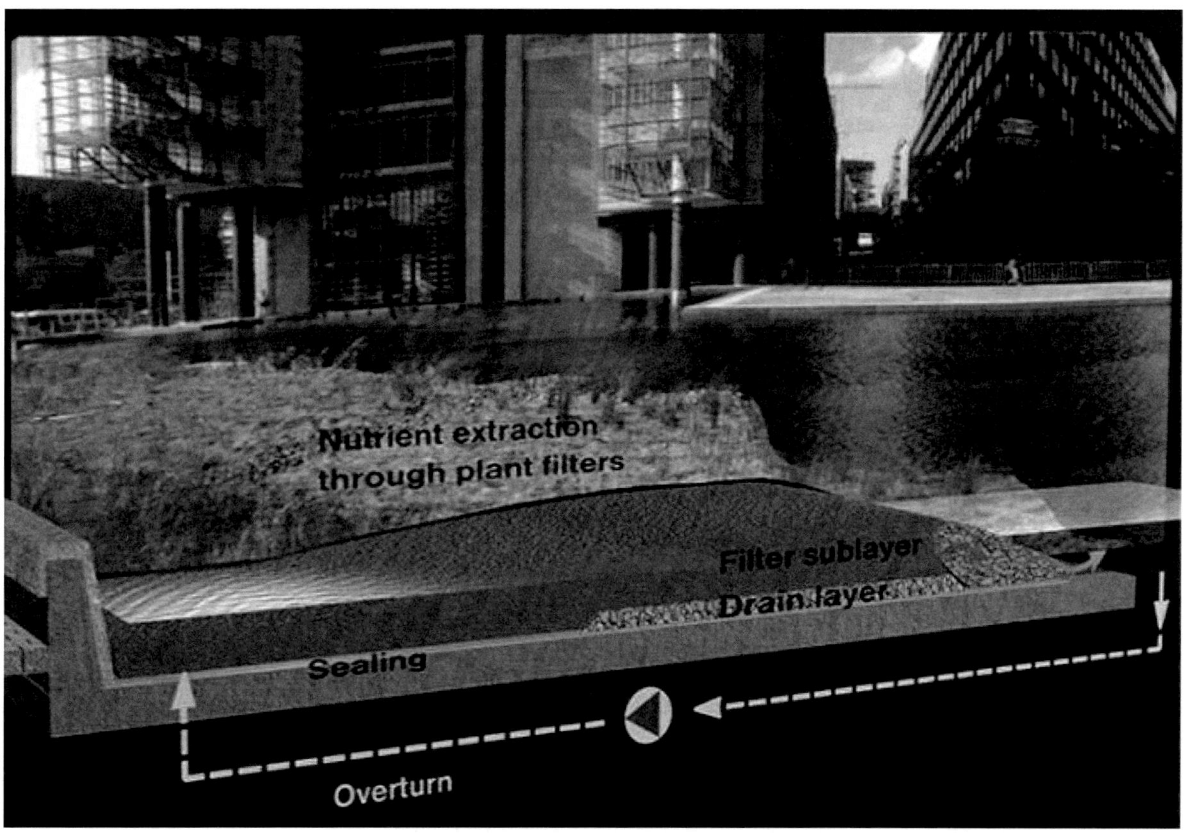

FIGURE 2-2 Beneficial use of stormwater at the Potsdamer Plaza, Berlin, Germany. SOURCE: http://www.dreiseitl.com/index.php?id=82&lang=en. Reprinted with permission; copyright 2016, Ramboll Studio Dreiseitl.

BOX 2-3 Stormwater Use at the Cincinnati Zoo

The Cincinnati Zoo's stormwater management objective is to have no site runoff from the new African Savannah exhibit area for all rains up to the 50-year storm event (Talebi and Pitt, 2013). To meet this objective, the Cincinnati Zoo has implemented a number of stormwater management strategies, including permeable pavers and a large stormwater capture and storage system in its new African Savannah exhibit (Figure 2-3-1). Roof runoff along with other runoff from parking areas and trails is redirected from the existing combined sewer toward the stormwater capture system.

Sand filtration is used for pre-treatment prior to storage in the 410,000-gallon (1.6 million-liter) subsurface storage tanks. Water is treated with ultrafiltration prior to use. The collected water is used for exhibit water (such as the polar bear pool), on-site irrigation of the horticulture gardens, and replenishment of the zoo's other outdoor water features (Figure 2-3-2; Warren, 2013). The total annual water savings from this installation is about 13 million gallons (49 million liters). The total project costs (including the new exhibit areas plus the stormwater management efforts) are estimated to be about $26 million.[a]

[a] See http://www.youtube.com/watch?v=6SJLraXsxys.

FIGURE 2-3-1 Construction of subsurface stormwater storage at the Cincinnati Zoo (top). SOURCE: http://www.youtube.com/watch?v=6SJLraXsxys.

FIGURE 2-3-2 An ornamental water feature at the Cincinnati Zoo that is replenished by a stormwater capture system. SOURCE: Robert Pitt.

Heating, Ventilation, and Air Conditioning (HVAC)

Increasingly HVAC systems are being integrated into on-site water use. Captured and treated stormwater can be used as makeup water in evaporative cooling systems. Cooling water has been estimated as 15 percent of total water use for commercial and industrial land uses (Gleick et al., 2003). Barclays Center in Brooklyn, New York, utilizes retained stormwater as a water supply source for the cooling towers.

Air conditioning condensate is also a reliable source of onsite water supply, although the water can be high in heavy metals (e.g., copper, lead). Some on-site stormwater capture systems include collection of condensate for landscape irrigation or other nonpotable uses. This approach has been used at a range of scales from small household air conditioning systems to major commercial properties. The rate of condensate generation in HVAC systems typically correlates well with seasonal demands for irrigation water.

Landscape Irrigation

Stormwater and graywater are commonly used for landscape irrigation at both household and neighborhood scales. Three basic irrigation methods are as follows:

- *Subsurface irrigation*, in which graywater or stormwater is supplied through drip systems either through buried distribution pipes below the ground surface or beneath a thick layer of mulch or through direct drainage into mulch basins;
- *Surface irrigation*, in which water is supplied through a drip system without a thick mulch cover; and
- *Spray irrigation*, in which graywater or stormwater is supplied through spray nozzles (i.e., sprinklers).

Irrigation systems that use graywater and stormwater can be very simple (see Chapter 6) and have been widely accepted within the regulatory community, although state laws may restrict surface or spray irrigation and the irrigation of food-crops (see Chapter 8). Important chemicals of concern in graywater for irrigation uses are sodium, chloride, and boron. In stormwater, deicing salts (particularly, sodium chloride) also pose a concern because they can negatively impact soil quality and plant health. These contaminants are typically managed through source control.

Graywater, at the household scale, is most commonly used for landscape irrigation via simple laundry-to-landscape systems, which provide subsurface irrigation of untreated graywater (Box 2-4). Larger graywater systems in offices and other commercial buildings may also be used for landscape irrigation, although large-scale graywater supplies are commonly disinfected prior to use to reduce possible human health risks, unless exposures are carefully controlled (see Chapter 5 for a more detailed discussion of risk and exposure).

Many cities and water utilities have implemented rain barrel rebate programs to encourage household stormwater capture and outdoor irrigation use with the hope of reducing adverse environmental impacts of stormwater. Small household irrigation systems do not typically provide treatment, but treatment systems can be added to facilitate longer-term storage. Similarly, neighborhood-scale stormwater systems have captured area runoff for irrigation of parks, playgrounds, school lawns, and greenways. For example, the National Park Service recently completed a major project designed to capture local stormwater for irrigation of the National Mall in Washington, DC, for the purpose of reducing potable water use (Box 2-5). Large storage facilities can be designed to extend the use of stormwater for landscape irrigation, in particular through periods with low and no precipitation, at greater project cost.

Wildlife Habitats and Recreational Uses

The capture and use of stormwater can enhance and in some cases create aquatic habitat. Possible uses include neighborhood-scale wetlands to absorb nutrients and filter sediment from stormwater or stormwater retention ponds. The use of stormwater for habitat creation is a potentially important application of stormwater, especially in rapidly growing regions with limited availability of surface water. Wetlands, such as the Rory M. Shaw Wetlands Park (Box 2-6), provide recreational features, community amenities, educational opportunities, and urban habitat. As such, stormwater-created wetlands are gaining support of environmental groups and stakeholder communities. However, inherently variable stormwater flows complicate the design of systems intended to enhance or create aquatic habitats and may necessitate access to other water sources during extended dry periods. Wet ponds are generally better at providing wildlife habitat and require more space than do dry ponds (Barnes and Adams, 1998). The Rio Salado Environmental Restoration Project established through a partnership between the City of Phoenix and the U.S. Army Corps of Engineers serves as an example to restore native wetlands and riparian habitats with recreational and educational uses at a regional scale using a 5-mile section of the Salt River (DeSemple, 2006).

Stormwater wetlands provide habitat services when designed properly (NCSU, 2011; Duffield, 1986). Because urban stormwater wetlands may accumulate contaminants from the urban landscape, these effects should be monitored and mitigated if possible. Sparling et al. (2004) investigated

BOX 2-4 Long Beach, California, Laundry to Landscape Program

In 2011, the City of Long Beach, California, created a pilot program called the Laundry-to-Landscape Backyard Irrigation Program to explore water conservation through graywater systems. The pilot program was motivated by the possibility of reducing potable water use by an estimated 14 to 40 percent. This savings would represent a fundamental shift in the city's water demand and contribute to the its reputation as a leader in conservation and sustainability (City of Long Beach, 2011)

The program tested the use of graywater from home washing machines for subsurface irrigation of trees, shrubs, and gardens (Figure 2-4-1), anticipating potable water savings of about 15 gallons (57 liters) per person per day. The pilot program selected 33 single-family homes by lottery throughout the city for free installation. Long Beach is believed to be the first community in southern California to municipally fund a graywater program. The systems included a valve to divert washing machine water directly to mulch basins in the garden for landscape irrigation, or to the sewer if necessary if bleach or harsh detergents were used in the wash cycle. Installations were performed in 2012 and 2013 at an average cost of $1,248 for materials and labor per household.

Follow-up interviews for systems installed for at least 6 months found a high level of satisfaction with the systems. The main problem reported was unequal distribution of water to discharge points located in mulch basins, which was overcome by a simple adjustment of valves. Although residents reported heightened awareness of water usage, only 7 of 33 homes showed decreased water use. On average potable water usage increased about 700 gallons (3,000 liters) per month based on monthly metering (City of Long Beach, 2011). Though limited by the small number of homes, this finding ran counter to the expected water savings. Reasons for this increase in potable water use compared to pre-project conditions remain unknown but could include a lower than average winter rainfall or added landscaping because of a perceived more abundant water supply, or the perception that more water use was not wasteful (J. Gallop, City of Long Beach, personal communication, 2014). Research needs to better understand behavioral impacts on water use are discussed in Chapter 10.

FIGURE 2-3-1 The City of Long Beach piloted a laundry-to-landscape graywater program at 33 homes to assess costs, homeowner satisfaction, and water savings. Laundry water is conveyed directly from a clothes washer via buried pipes to drip irrigation systems. SOURCE: Photo courtesy of Jason Gallup, Office of Sustainability, City of Long Beach.

BOX 2-5 Stormwater Capture for Turf Irrigation on the National Mall, Washington, DC

As part of a major turf restoration project on the National Mall in Washington, DC, the National Park Service is installing a 1-million gallon (3.8 million liters) stormwater capture and treatment system to supply water for irrigation. In Phase 1 of the project, completed in December 2012, new turf, engineered soil, and a new irrigation and drainage system were installed along with two 250,000-gallon (950,000-liter) subsurface cisterns to capture stormwater runoff from the lawns and walkways (see Figures 2-5-1 and 2-5-2). Prior to use, the stormwater is treated using filtration and ultraviolet disinfection. The remaining 500,000-gallon (1.9 million-liter) storage is anticipated to be installed by 2016. The stormwater capture and treatment system will serve as the primary water source for irrigation, reducing potable water demand and reducing harmful stormwater discharges. The capital cost for all four cisterns is estimated to be $1.75 million, which does not include the costs of the treatment system or the piping and drainage components. Full operating costs (including electrical costs for pumping and the treatment system) have yet to be determined, although maintenance costs are estimated at $35,000 per year. The project is expected to provide approximately 68 percent of the 11 million gallons (42 million liters) of irrigation water needed annually, while improving area surface water quality.

SOURCE: M. Stachowicz, NPS, personal communication, 2014, 2015.

FIGURE 2-5-1 Phase 1 of the National Mall turf restoration project rehabilitated eight lawn panels between 3rd and 7th Street, NW, and installed 500,000 gallons (2 million liters) of stormwater storage capacity. SOURCE: M. Stachowicz, NPS, personal communication, 2014.

FIGURE 2-5-2 Installation of a 250,000-gallon (950,000-liter) cistern beneath the National Mall in Washington, DC. The captured stormwater is used to irrigate the turf along the National Mall. SOURCE: M. Stachowicz, NPS, personal communication, 2014.

BOX 2-6 Neighborhood-scale Stormwater Capture in Sun Valley-Los Angeles, California

When completed in 2020, the Rory M. Shaw Wetlands Park located in the Sun Valley neighborhood of Los Angeles will convert a former quarry and inert material landfill into a stormwater capture facility and multipurpose park. The 46-acre (19-ha) site will provide recreational space, reduce flooding, treat stormwater in naturalistic wetlands, and recharge local groundwater. A storm drain system will collect runoff from a 929-acre (376-ha) drainage area, and the detention ponds and wetlands will capture and treat stormwater. The treated stormwater will be pumped to adjacent infiltration basins in Sun Valley Park. The amount of water infiltrated is estimated to be 900 AF/yr (1 million m^3/yr) (Hagekhalil, et al., 2014). Total project cost is about $52 million with $28 million for purchase of the property (LACFCD, 2012).

The central feature of the project is the conversion of a defunct landfill and blighted landscape into 46 acres of green space and recreation for a community that is currently underserved for recreational opportunities. This project illustrates how co-benefits and public support may be achieved with neighborhood-scale stormwater capture. By working with local groups the project generated greater effectiveness in community engagement and links between decision makers and the people they serve. One such group, TreePeople Inc., an environmental nonprofit, partnered on this project with Los Angeles Department of Water and Power and the Los Angeles Bureau of Sanitation to promote sustainable solutions and encourage communities to take personal responsibility for their urban environment. The Sun Valley Watershed Stakeholders Group promoted features such as trails, tennis and basketball courts, a tot lot, and picnic and play areas, while addressing chronic stormwater flooding problems, retaining water in the watershed, and reducing stormwater pollution.

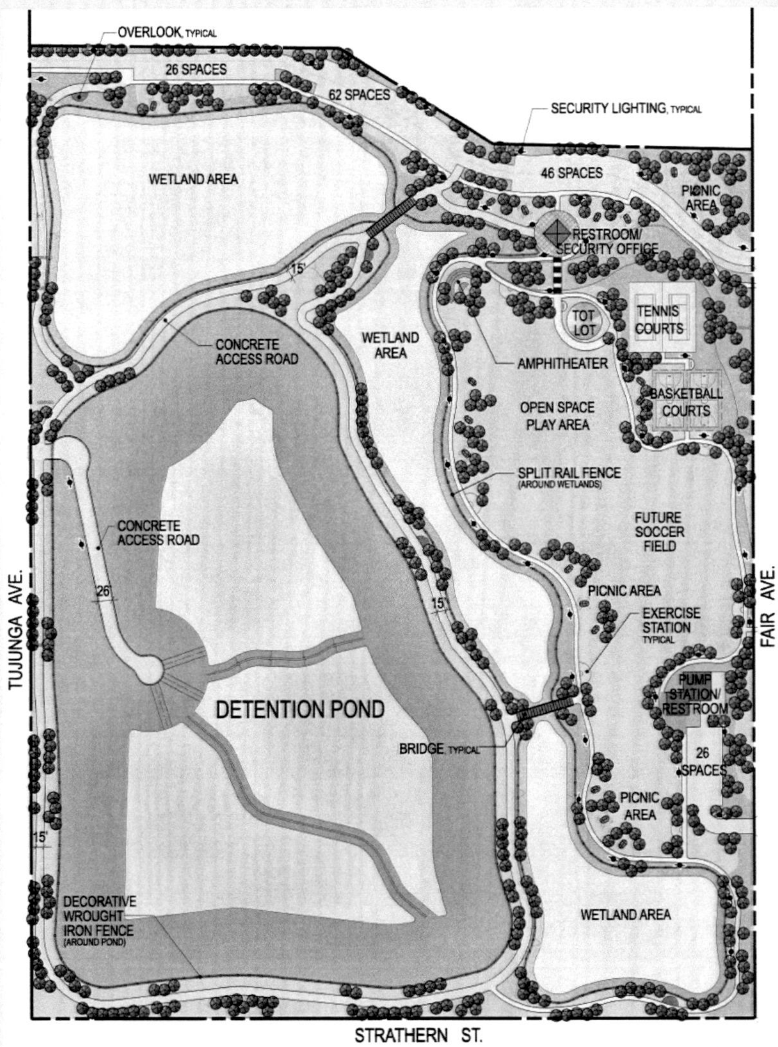

FIGURE 2-6-1 The Rory M. Shaw Wetlands Park project in the Sun Valley neighborhood of Los Angeles will augment local groundwater supplies in a multipurpose facility. A former 46-acre quarry will provide open space and recreational opportunities along with stormwater capture, treatment, and recharge. SOURCE: LACFCD (2012).

13 stormwater wetlands in the Maryland suburbs near Washington, DC. The accumulation of zinc in hatchlings of red-winged blackbirds inhabiting wetlands in industrial areas suggested nestling stress and impairment. Overall nestling success compared favorably with national averages. Although the zinc concentrations were elevated, the authors concluded that the benefit of stormwater habitat provided by stormwater wetlands might outweigh the negative impacts of contaminant accumulation provided by the stormwater habitat because of the scarcity of such habitats in urban areas. Controlling sedimentation rates and periodic dredging of sediments would prevent urban stormwater wetlands from accumulating toxic chemicals (Sparling et al., 2004).

Surface Impoundments and Groundwater Recharge

On-site capture and infiltration of urban stormwater is often used to mitigate stormwater runoff and contaminant loading to surface waters and reduce flows to combined sewer systems. Under appropriate hydrogeological conditions and project design, stormwater infiltration can be used to recharge local aquifers and thereby expand groundwater supplies. Several projects exist in southern California to recharge stormwater (see Box 1-4 and Figure 2-3), and more are in the planning stage (Box 1-1). In some areas underlain by low permeability rock or sediment (e.g., clay), infiltration does not reach the deeper aquifers that provide water supply (Figure 2-4). However, in areas with appropriate hydrogeology, stormwater recharge and storage basins can be designed at neighborhood or regional scales in engineered infiltration facilities to recharge local aquifers.

Groundwater infiltration projects capture surface runoff from relatively large and diverse source areas, and the stormwater is likely to contain a broad array of contaminants, such as petrochemicals, urban pesticides, and flame retardants (discussed in more detail in Chapter 4) (Eriksson et al., 2007). If these chemicals are not removed by soil-aquifer treatment upon infiltration, then additional water quality treatment may be necessary to prevent groundwater contamination. Suspended sediment also needs to be removed prior to infiltration to prevent clogging of the infiltration basins.

Surface water impoundments may also capture stormwater runoff for water supply and can sometimes be operated to optimize groundwater recharge (Box 1-4). Singapore harvests and treats stormwater on a large-scale for its public drinking water supply (PUB, 2015). Referred to as one of the "four taps," stormwater is harvested from a network of drains and canals and stored in reservoirs—including river estuaries dammed to hold runoff—as part of a diversified water supply portfolio for water self-sufficiency.

The major benefits from use of regional solutions for captured stormwater are that larger infiltration or storage sites can

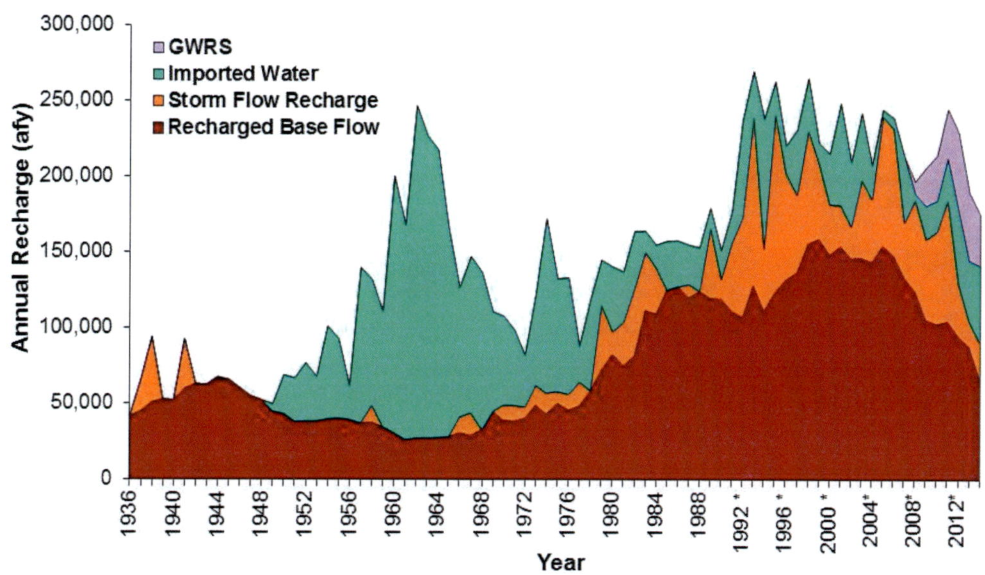

FIGURE 2-3 Managed recharge of stormwater and other water sources by the Orange County Water District has increased substantially over the past 30 years in the Santa Ana River basin. The figure shows recharge of surface water base flow, managed stormwater recharge, managed recharge of imported water, and reclaimed water from the Orange County Groundwater Replenishment System (GWRS). NOTE: From 1936 to 1990, the year shown reflects the water year (October to September). From 1991 to 2014, the year reflects the fiscal year (July to June). SOURCE: A. Hutchison, OCWD, personal communication, 2015.

be designed, operated, and maintained by water supply agencies to maximize stormwater capture and provide treatment as needed to remove hazardous contaminants. Regional facilities can also bring an efficiency of scale (see Chapter 6).

SUMMARY

A large number of potential beneficial use options exist for graywater and stormwater depending on scale and treatment provided. Stormwater capture projects can serve a range of uses from irrigation to toilet flushing and HVAC cooling water to large-scale groundwater recharge. Stormwater treatment needs will vary with land use and catchment size, potential human exposures, and whether captured stormwater will contribute to urban water supplies. Traditional applications of graywater have focused on household-level nonpotable uses, including landscape irrigation and toilet flushing. However, neighborhood- and regional-scale projects are now emerging worldwide that integrate graywater reuse into the design of fast-growing urban centers and facilitate decentralized resource recovery in large new developments.

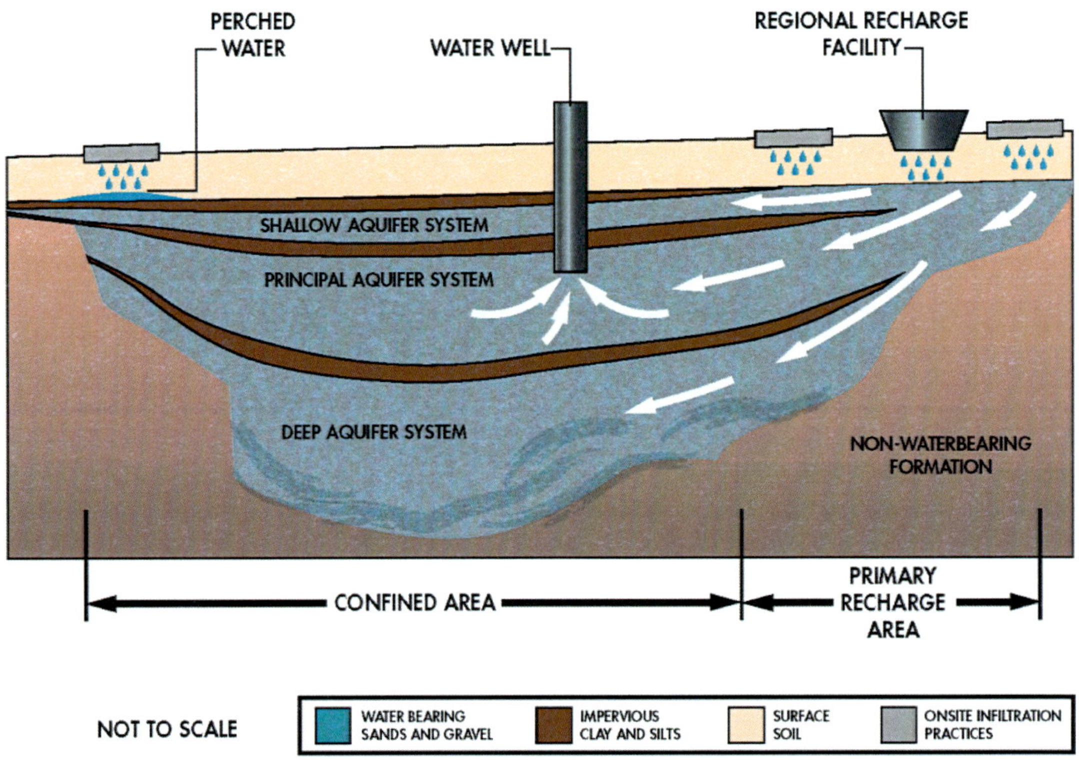

FIGURE 2-4 Stormwater infiltration effects on groundwater supplies using regional recharge facilities. This illustrates the favorable location of recharge facilities in places with permeable soil to recharge aquifer systems. SOURCE: SCWC (2012).

3

Quantities Available for Beneficial Use and Potential Impacts on Water Demand

Graywater and stormwater can be used in many applications to reduce the reliance on conventional water supply sources (Chapter 2), but the availability of that water varies in most cases locally, regionally, and by source area and land use. This chapter presents an approach that can be used to examine the quantities of graywater and stormwater available for beneficial uses and the potential impacts on water demand for various applications. The results from a medium-density residential scenario at six U.S. locations are discussed.

GRAYWATER

Graywater can be reused in the residential setting, at household, multi-residential, and even regional scales, and in commercial and industrial settings. Each offers different quantities available for reuse and different potential impacts on water demand.

Household and Multi-residential Use

Indoor water use has been determined to be largely consistent throughout the United States and among varying socioeconomic groups (Mayer and DeOreo, 1999). Thus, residential graywater production does not vary regionally, and graywater can be considered a more consistently available source of water than stormwater in arid regions. A study by DeOreo et al. (2016)[1] revealed that indoor water use in nine North American cities was on average 138 gallons per household per day (gphd) or 59 gallons per capita per day (gpcd; 522 liters per household per day [lphd] and 220 liters per capita per day [lpcd]). Based on these data, graywater[2] comprises approximately 45 percent of household wastewater (Figure 3-1).

In a typical household, the graywater produced (63 gphd or 26 gpcd [240 lphd or 98 lpcd]) exceeds the demand for toilet flushing (33 gphd or 14 gpcd [120 lphd and 53 lpcd]; Figure 3-1), enabling use of graywater to meet the total toilet flushing demand in residential settings, with excess graywater available for irrigation or other nonpotable uses (DeOreo et al., 2016). Use of graywater for toilet flushing alone would reduce indoor water demand on average by 24 percent. For comparison, a study on 10 homes using graywater to flush toilets showed consistent reductions in water use of 6 gpcd (23 lpcd; City of Guelph, 2012). This is lower than the previous reported savings of 14 gpcd above, but 8 of the 10 homes were using low-flow toilets (1.2 gallons [4.5 liters] per flush). Case studies report savings from toilet flushing ranging from 5.4 gpcd [20 lpcd] in a dormitory setting (Box 2-2) and 20 gpd [76 lpd] per inmate in a prison setting (Box 1-1).

Indoor water use does not vary substantially between single residence and multi-residential units when reported on a per-capita basis (Mayer and DeOreo, 1999). Thus, water savings reported in this chapter also apply to multi-residential units and regional-scale applications. For example, if graywater were used for all of the toilet flushing needs in a multi-residential unit with 100 housing units, then assuming 33 gphd water use for toilet flushing and no change in water use behavior as the result of graywater reuse, a total of 3,300 gallons (12,000 liters) water per day could be saved in the building.

Water savings achieved by using graywater instead of potable water for irrigation will depend on irrigation demand. If all residential graywater is used for irrigation to replace prior watering using potable supplies, then potable water demand can be reduced by 63 gphd (240 lphd), the av-

[1] When this report was written and then approved by the National Academies of Sciences, Engineering, and Medicine, the committee referenced a research paper, DeOreo and Mayer (in press), which was subsequently published as DeOreo et al. (2016). The per capita water use data reported in the final paper differed from the preliminary data provided to the committee (M. Hodgins, WaterRF, personal communication, 2013)—the same household water use data were reported but the final analysis reported fewer persons per household, resulting in higher per capita water use values (for details, see Box 3-1). The newer data did not change the committee's findings, conclusions, or recommendations; however, the text and Figure 3-4 were updated with the newer data.

[2] As defined in Chapter 1, graywater includes water from bathroom sinks, showers, bathtubs, clothes washers, and laundry sinks but does not include water from toilets, kitchen sinks, or dishwashers.

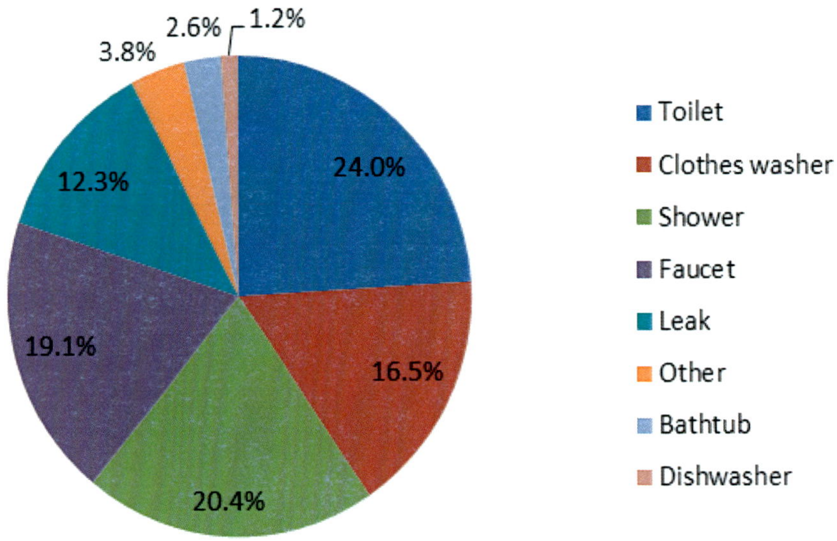

FIGURE 3-1 Average allocation of indoor household water use (138 gphd) in nine North American cities. SOURCE: Based on data from DeOreo et al. (2016).

erage graywater available. This is typically not enough water to meet the entire outdoor residential demand in most areas, depending on the vegetation being irrigated. Native vegetation requires minimal additional irrigation, while turfgrass in arid regions could demand substantial irrigation volumes. In areas where irrigation is seasonal, the noted reduction in irrigation demand resulting from graywater use will only occur during irrigation months. Thus, water demand reduction associated with seasonal graywater irrigation will be lower than 63 gphd over the course of the year.

A common practice at the household scale is to use laundry water to irrigate landscape vegetation, particularly because systems can be installed in existing households without major plumbing retrofits (SFPUC, 2012; see Box 2-4). Laundry water makes up about 36 percent of household graywater on average, or 9.6 gpcd (DeOreo et al., 2016). Thus, savings of approximately 23 gphd (87 lphd) can be achieved when all of the laundry water is used for landscape irrigation, compared to 63 gphd when all graywater is used. These savings can only be realized if irrigation rates do not increase with installation of the graywater system. Laundry-to-landscape initiatives in Long Beach and San Francisco, California (Box 2-4) documented more households with water demand increases than decreases after installation of graywater systems. In Long Beach, only 7 of the 33 homes showed a reduction in water use, and program-wide an average household increase of 700 gallons (2,600 liters) per month was observed (City of Long Beach, 2013). In San Francisco, 29 sites had increased water use compared to 27 sites with reduced water use (P. Kehoe, SFPUC, personal communication, 2014), likely from new landscape installations, variability in the data, or changes in behavior during the study period. More research is needed to evaluate the quantity of water savings actually achieved when graywater is used for irrigation and whether there are behavioral changes that result in increased use of water when graywater systems are installed (see Chapter 10).

Published Assessments of Graywater Impact on Water Demand

A modeling study evaluated potential water demand reduction from graywater use with varying irrigation demand (Reichel et al., 2011). This study calculated a 5 to 6 percent reduction in municipal water demand when 50 percent of the population adopted graywater reuse at the residential level (10 percent use for toilet flushing, 30 percent use for irrigation, 10 percent use for combined toilet flushing and irrigation). The study cities were Fort Collins, Colorado; Orlando, Florida; Philadelphia, Pennsylvania; San Diego, California; and Seattle, Washington. Although total per capita water demand reduction was greater in areas with year-round irrigation (Orlando and San Diego), the higher irrigation demand in those regions resulted in similar percentage reductions as those cities with seasonal or limited reductions. Although average water savings of 63 gphd in water demand are possible through graywater reuse, the largest savings are only likely to be achieved in new development or redevelopment areas. Retrofitting existing homes to either collect graywater or use treated graywater for toilet flushing is a large barrier for adoption. However, new homes can easily be built with dual plumbing systems for graywater collection and use at a low marginal cost (see also Chapters 6 and 7).

Committee's Assessment of Graywater Impact on Water Demand

To examine potential potable water savings from graywater use, the committee developed scenarios of wide-scale adoption of graywater use in six U.S. cities (Los Angeles, California; Seattle, Washington; Lincoln, Nebraska; Madison, Wisconsin; Birmingham, Alabama; and Newark, New Jersey), representing a range of climatic conditions and geographical distribution. These cities were also selected because comparable data were available on land use characteristics for all six locations that would support parallel analyses of potential water savings from stormwater or graywater use. The committee conducted an original analysis to examine changes in potable water demand for a hypothetical 100-acre (40-ha), medium-density, residential community (with 12 persons per acre) in each of the six cities. The analysis considered the following four water demand and graywater use scenarios detailed in Box 3-1:

- **Base scenario:** Potable water is used for all indoor and outdoor household water needs. Outdoor irrigation demand was calculated as that necessary to meet the minimum evapotranspiration deficit for turfgrass (the most common household irrigated vegetation), which varies by location (see Figure 3-2).
- **Scenario 1 (irrigation):** Graywater and supplemental potable water as needed is used to irrigate turfgrass to barely meet the evapotranspiration deficit (see Box 3-1). Indoor potable water use is unchanged.
- **Scenario 2 (toilet flushing):** Water demand for toilet flushing is entirely met by graywater. Potable demand for outdoor irrigation is unchanged from the base scenario.
- **Scenario 3 (irrigation and toilet flushing):** Toilet flushing demands are entirely met by graywater, and remaining graywater is available to address irrigation demand, as needed.

Graywater production was estimated considering typical wastewater generation from showers, baths, laundry, and one-third of faucets (excludes kitchen sink faucets) based on preliminary data from the DeOreo et al. (2016) survey of indoor water use (M. Hodgins, WaterRF, personal communication, 2013). In each scenario, it was assumed that 100 percent of the residential population in a 100-acre, medium-

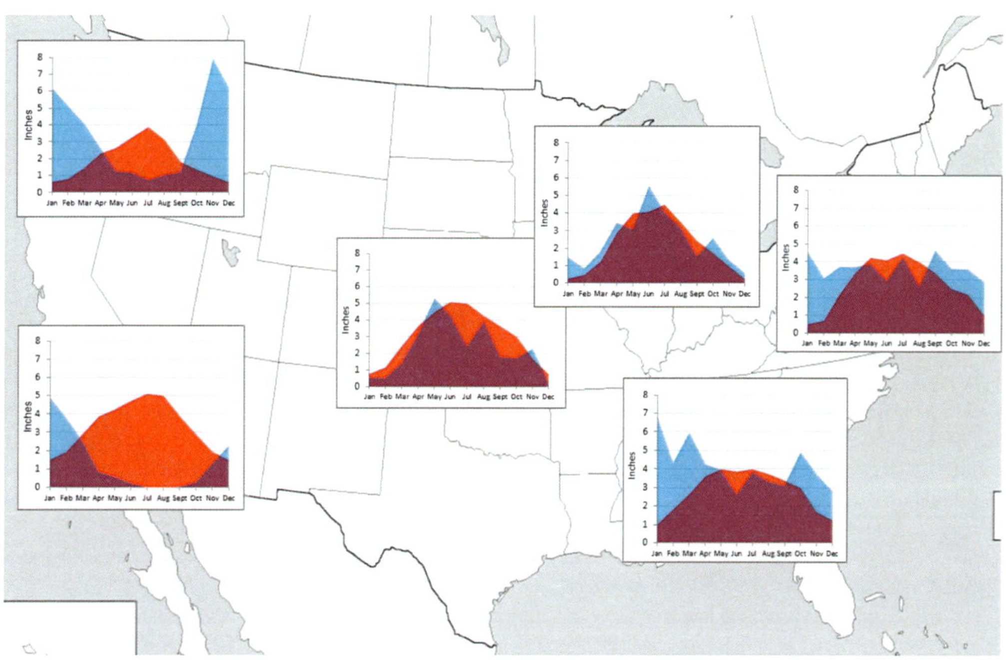

FIGURE 3-2 Minimum monthly irrigation demands (bright red) that are not met by rainfall, shown in six locations based on monthly evapotranspiration (ET) (red/purple) for standard turfgrass versus average monthly rainfall from 1995 to 1999 (blue/purple). Data sets to develop the figure, including reference ET rates, are described in Appendix A. Monthly precipitation averages as shown here may underestimate irrigation demand if precipitation events are clustered over several days rather than spread evenly across a month.

BOX 3-1 Nonpotable Water Demand Scenarios: Irrigation and Toilet Flushing

To assess the potential for stormwater and graywater to meet water needs and provide potable water savings, the committee considered four water demand and use scenarios in six cities.

Base scenario. Potable water is used for all indoor and outdoor household water needs, and all analyses are based on a 100-acre (40-ha), medium-density, residential development with 12 people per acre (40 people per hectare). Indoor water use does not vary for different climatic conditions and was therefore assumed to be consistent for all of the cities examined. An indoor water use rate of 46 gpcd (170 lpcd or 20.3 Mgal/yr per 100 acres) was used based on preliminary updated data on indoor water use in nine North American cities (M. Hodgins, WaterRF, personal communication, 2013). The water use data reflects the same household-level water use data as that reported in DeOreo et al. (2016), but the preliminary data used by the committee assumed 3 persons per household, while DeOreo et al. (2016) ultimately reported fewer persons per household and a nonlinear relationship between household size and water use. This difference resulted in lower per capital values for indoor water use, total graywater production, laundry, and toilet flushing (46, 21, 7.6, and 11 gpcd, respectively) used in this analysis compared to the data reported in DeOreo et al. (2016) (58.6, 26, 9.6, and 14 gpcd, respectively). Outdoor irrigation demand was calculated according to conservation irrigation requirements (irrigation only to meet the minimum evapotranspiration deficit) for turfgrass, the most common household irrigated vegetation. The number of houses and percentage of landscaped area in each of the six locations were determined from regional data (see Table A-2 in Appendix A).

Monthly long-term averaged evapotranspiration rates for the candidate cities were obtained from the literature and adjusted for turfgrass conditions (see Pitt et al., 2011, and Appendix A). Average monthly rainfall values based on 1995-1999 rainfall records (1996-1999 for Lincoln, due to missing data) were then subtracted from the monthly evapotranspiration values to calculate typical monthly irrigation demands for each location. Irrigation demands only occur in months when the evapotranspiration requirements are greater than the recorded average monthly rainfall. As is shown in Figure 3-2, monthly rainfall (blue) and evaporation deficit (red) patterns vary greatly in different parts of the country. The largest evapotranspiration-driven deficits occur during the mid-summer months in arid areas when irrigation water is usually applied to make up the deficit. Irrigation requirements tend to be much lower during the winter (especially in those areas with dormant growing conditions).

Daily irrigation rates for each month were applied at a constant rate for each month reflecting the prior calculated monthly irrigation demands. The daily use rate is the same for the same month in different years, an assumption made necessary by the lack of monthly evapotranspiration data at each site for the time periods analyzed and use of monthly long-term averaged evapotranspiration data. Table 3-1-1 summarizes average monthly conservation irrigation demands for conventional warm season turfgrass for residential areas for these six geographical areas based on 1995-1999 rainfalls (1996-1999 for Lincoln). These 4- to 5-year rainfall records were not significantly different from the complete several-decade rain records, although there are some apparent differences (see Appendix A and Box 3-2). The water volumes necessary to barely meet conservation irrigation demands were then calculated, considering the average landscaped portion of the medium-density residential land use for each location. In the base scenario, the irrigation volume is provided by potable water supply.

Scenario 1: Irrigation. Using the irrigation demand data described under the base scenario, the committee examined the potential for graywater and stormwater to address irrigation needs with indoor potable water use unchanged. Irrigation of site vegetation is one of the most direct and popular beneficial uses of graywater and stormwater. Irrigation with nonpotable on-site water can reduce potable water use, as long as the landscaped area requiring irrigation is not increased once a graywater or stormwater capture system is installed or the amount of irrigation water applied is not significantly increased with the use of the newly available water. As with the base scenario, the committee's analysis assumes that irrigation is applied to barely satisfy the evapotranspiration deficits (conservation irrigation) for turfgrass, which is usually recommended when minimizing water use. However, it should be recognized that most homeowners use much more water than this minimum amount. The same analyses could also be conducted with different landscaping choices. Careful planting (such as xeriscaping with native plants) could result in minimal (or no) irrigation requirements, which would likely result in even greater reductions of domestic water used for irrigation compared to the beneficial uses of graywater or stormwater for irrigation.

For the graywater analysis, potable water demand for irrigation for each city was calculated by subtracting monthly graywater generation (21 gpcd x 1,200 people) from the monthly irrigation demand over the 100-acre, medium-density, residential area calculated in the base scenario. It was assumed that a storage tank was installed with sufficient volume to store the daily graywater generated, and thus that all graywater generated was available for use. No change occurred in indoor water use, and the total potable water use was summed over the year for each location (Figure 3-2).

The ability of stormwater to reduce potable water use depends on the size of the tank storage facilities and by location (dry periods and available rain) compared to the irrigation demand. A model was used to conduct multi-year, water-balance analyses to assess the availability of roof runoff to meet daily irrigation demands, described in more detail later in this chapter and in Appendix A.

Scenario 2: Toilet flushing. Under scenario 2, 11 gpcd graywater or stormwater is used to meet toilet flushing water demand. Potable demand for outdoor irrigation is unchanged from the base scenario. The typical household produces more graywater per day than is needed to meet toilet flushing demand, and therefore total indoor potable water demand was reduced by 11 gpcd multiplied by the population (1,200) within the 100-acre area for each location. The ability for stormwater to meet toilet flushing demand is dependent on storage facilities and the frequency and amount of rainfall, and was calculated based on continuous model simulations.

Scenario 3: Irrigation and toilet flushing. For graywater, the analysis assumed that toilet demand (11 gpcd) was entirely met by graywater and the remaining 10 gpcd graywater was applied toward the monthly irrigation demand. For stormwater, continuous model simulations calculated the ability for on-site roof runoff to meet (or partially meet) toilet flushing and irrigation demands.

(Continued)

BOX 3-1 Continued

TABLE 3-1-1 Minimum Monthly Irrigation Demands to Meet Turfgrass Evapotranspiration Deficit

	Irrigation Requirements (Mgal/month/100-acre, medium-density, residential area)					
	Los Angeles, CA	Seattle, WA	Lincoln, NE	Madison, WI	Birmingham, AL	Newark, NJ
Jan	0.00	0.00	0.28	0.00	0.00	0.00
Feb	0.00	0.00	0.99	0.00	0.00	0.00
Mar	0.57	0.00	1.6	0.00	0.00	0.00
Apr	4.2	0.00	0.70	0.00	0.00	0.00
May	5.2	2.3	0.00	1.4	0.02	0.50
Jun	6.3	3.5	1.3	0.00	2.6	1.7
Jul	7.2	5.3	4.3	0.68	0.24	0.24
Aug	7.1	3.5	0.56	0.50	0.79	2.0
Sept	5.4	0.96	3.0	1.4	0.53	0.00
Oct	3.7	0.00	2.2	0.00	0.00	0.00
Nov	0.88	0.00	0.00	0.00	0.00	0.00
Dec	0.00	0.00	0.72	0.00	0.00	0.00
Total annual	40.5	15.6	15.6	4.0	4.2	4.5

NOTES: These calculations are determined by subtracting average monthly precipitation (1995-1999, except for Lincoln, which used 1996-1999) from average monthly evapotranspiration rates. These data may underestimate irrigation requirements when rainfall is clustered in a few days within a month for some years, but these variations are expected to be evened out with the multi-year analyses. SOURCE: See Appendix A.

density, residential area adopted the specified graywater use practices. These scenario analyses were intended to provide an estimate of the maximum possible potential for demand reduction that can be achieved through graywater reuse and does not reflect what can be realistically achieved in the near future. See Box 3-1 for a full description of the scenario assumptions. The results were calculated by a simple spreadsheet analysis, described in Box 3-1.

The results of the scenario analysis of a medium-density residential community based on climate data from 1995 to 1999 (1996-1999 for Lincoln) (see Figure 3-3 and Table 3-1) show that average potable water demand reductions ranging from 13 percent (Los Angeles) to 26 percent (Madison, Wisconsin) are possible with graywater reuse for both irrigation and toilet flushing (Scenario 3). Such savings assume that indoor and outdoor water use habits are unchanged by this new low-cost water source, an assumption that remains untested (see Box 3-2 for discussion of key uncertainties, including behavioral factors). Reductions in potable water demand resulting from the use of graywater for irrigation vary widely with climate. Among the six cities analyzed, the lowest potential irrigation demand savings are noted for Newark, New Jersey, where there is only an irrigation demand for 4 months of the year (see Figure 3-3). Los Angeles has the highest irrigation demand of the study cities, resulting in 11 percent potable water savings, although only 17 percent of the minimum irrigation demand for turfgrass in Los Angeles is met through graywater reuse in the irrigation-only scenario. A scenario using more acreage of native vegetation would enable graywater to meet a greater percentage of irrigation demand. Additionally, areas with year-round irrigation requirements would result in greater savings than those with limited seasonal irrigation requirements. Results from these hypothetical scenarios demonstrate that use of graywater for toilet flushing decreases indoor use by 24 percent across all regions, although this savings as a fraction of overall water use is dependent on the amount of water used for irrigation (see Table 3-1).

Laundry-to-landscape systems result in even lower potential water savings, because graywater is derived only from the washing machine. Table 3-2 shows average potential water savings for laundry-to-landscape systems in the six cities analyzed, although savings would be less with water-conserving washers and greater with older washers that use more water.

Downstream Impacts from Graywater Use. In many areas of the western United States, it is important to consider impacts to downstream water users in conjunction with water savings because of considerations of water rights and environmental uses (see Chapter 8). Graywater use for toilet flushing is a nonconsumptive use of water, because water used for toilet flushing will flow either to a wastewater treatment plant or septic tank, without any significant losses in downstream flows. On the contrary, using graywater for irrigation results in evapotranspiration losses and is considered a consumptive use of water. If existing landscape irrigation with potable water is replaced with graywater, then the property would have the same effect on downstream water availability as it did prior to graywater use assuming that the wa-

BOX 3-2 Assumptions and Uncertainties in the Committee's Scenario Analyses

There are a number of uncertainties associated with the committee's evaluation of potential for graywater and stormwater to reduce potable water demand based on the assumptions, data, and calculations used. The committee did not intend for the analysis to be used as a definitive estimate of on-site water availability. The analysis instead provides a means to compare the potential of graywater and roof runoff under a similar set of assumptions in different locations in the country, with varying hydrology and land-use patterns. Large-scale projects should consider similar analytical approaches tailored to specific site conditions to support decision making. Key uncertainties and limitations of the analysis are discussed below.

- **Expected variability in potential potable water savings.** The committee's analysis produced only a 5-year average of potential water savings, rather than annual values that illustrate year-to-year variability. Evapotranspiration rates were only available for long-term average monthly conditions, and rates for the actual months during the calculation period were not available. Year-to-year variations in evapotranspiration are expected to be moderate (but probably not as great as the year-to-year rainfall variations), and therefore actual potable savings are likely to be greater in some years and less in others.
- **Limited rainfall period and differences from long-term conditions.** Rainfall conditions during the calculation period were somewhat different from the complete data set (but not shown to be significantly different by statistical tests; see Appendix A). The 5-year rainfall conditions for Birmingham, Newark, Madison, and Lincoln (4 years) were quite similar to the long-term records, although Seattle had greater rains over that period compared to long-term conditions. The largest differences were apparent for Los Angeles. During the 1995-1999 analysis period, Los Angeles had greater rains and wider variation compared to long-term conditions (see Figure A-1 in Appendix A), including 2 years of unusually high winter rainfall, which likely reduced the average irrigation demand in some winter months below the likely long-term average and overestimated the average annual stormwater available. However, even with this additional rainfall, Los Angeles could not satisfy much of the household water use for indoor and outdoor demands considering reasonable-sized storage tanks with its long dry periods. Large-scale, city-wide analyses would require design-oriented evaluations with longer rainfall records, but Los Angeles projects may have less stormwater available for recharge compared to the calculations in the report.
- **Domestic water use and the impacts of water conservation.** The base domestic water use rates used in this report when calculating potential water savings are also subject to variations and trends. The per capita water use values used are slightly lower than those recently reported by DeOreo et al. (2016). The per capita rates were calculated using the same household water use data as DeOreo et al. (2016) but assuming three residents per household. The data should be reasonable for the comparative purposes of this analysis, based on the assumptions stated (medium-density residential with 12 residents per acre [40 per ha], conservation irrigation of turfgrass), although the water savings reported will be lower than similar calculations using the most recent data on per capita water use (DeOreo et al., 2016). At an individual household level, water use will vary based on the use of water-savings fixtures and appliances, the number of household members, the percentage of time the house is occupied, the type and extent of vegetation requiring irrigation, irrigation rates and frequency, and other behavioral factors. Most individuals irrigate at rates much higher than the conservation irrigation rates assumed in this analysis, which may underestimate potential potable water savings in locations having moderate to large rainfall amounts. However, water conservation trends may also impact water savings. Based on two studies of roof runoff capture systems in Australia, Beal et al. (2015) observed significantly reduced potable water savings in homes with water restrictions or that recently experienced severe water restrictions. In one study, the potable water savings in areas with no or low-water restrictions were approximately three times larger than water savings observed in areas with moderate or severe water restrictions. A second study in a semi-arid climate reported the lowest savings in communities that had recently experienced severe water restrictions (Beal et al., 2015). Large and significant beneficial use projects should be supported by measured local rates and expected future conditions as part of the design process.
- **Behavioral factors associated with on-site water use.** A significant uncertainty is associated with behavioral factors associated with a "free" on-site water supply that affect water savings. The committee calculated potential water savings, based on several assumptions, including that water use habits do not change with the installation of a graywater or stormwater system. However, this assumption has not been proven, and two laundry-to-landscape pilots have shown that potable water use can actually increase after installation (see Box 2-1). Mukheibir et al. (2013) suggest multiplying theoretical water savings of roof runoff capture systems by a "functionality factor" of 0.5 to 0.7 when estimating actual potable water savings to account for behavioral factors, poor installation, and operational issues. Beal et al. (2015) summarized the results of an Australia study of roof runoff capture systems (1,300-gallon [5,000-liter] tanks plumbed for indoor nonpotable household uses) that resulted in actual water savings of only 0.3 to 0.6 times the theoretical savings.
- **Other limits of the analysis.** The analysis reports an average irrigation demand in Lincoln, Nebraska, in the winter time (November through February account for about 14 percent of the total annual demand), when under usual considerations during typically sub-freezing conditions, surface irrigation is unlikely. These calculations were based on 4 years of rain records, which indicated some winter months having less precipitation than other years. The long-term average winter evapotranspiration values reflect winter irrigation during some mild winters, but this is not expected to be the case for all years and would be rare for most homeowners. Thus, the analysis may over-estimate the irrigation demand and total potable water savings in Lincoln for typical winters.

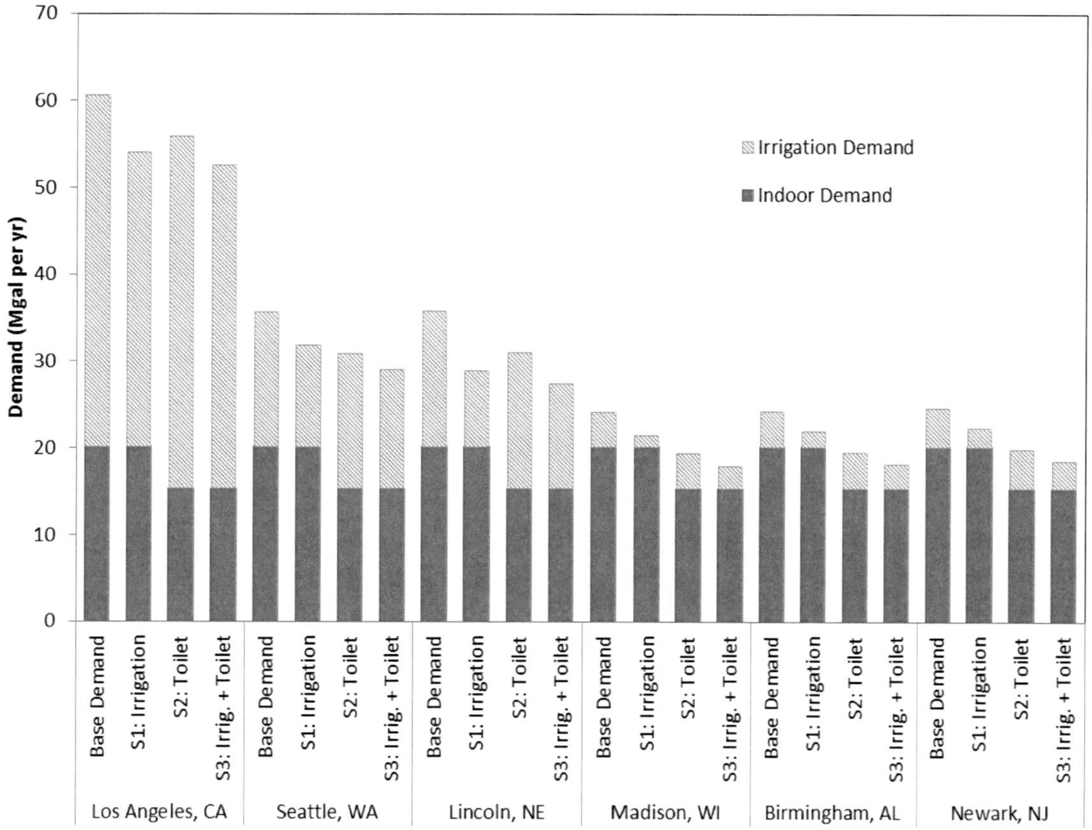

FIGURE 3-3 Estimates of demands for a hypothetical 100-acre, medium-density, residential area of each study city. Base demand reflects typical U.S. indoor water use based on three residents per household (see Box 3-1) and irrigation necessary to barely satisfy the evapotranspiration deficit for turfgrass for that location. Scenario 1 (S1) assumes 100 percent adoption of graywater use for irrigation, and irrigation is supplied only to meet the evapotranspiration deficit. Scenario 2 (S2) assumes 100 percent adoption of graywater reuse for toilet flushing. Scenario 3 (S3) reflects 100 percent adoption of graywater reuse for combined toilet flushing and irrigation. Note that these irrigation volumes are based on the minimum water required to maintain turfgrass (conservation irrigation), and actual irrigation rates in these locations may be significantly greater.

TABLE 3-1 Potential Potable Water Savings in Six Cities from Various Graywater Use Scenarios Based on a 100-Acre, Medium-Density, Residential Area

		Volume Potable Water Savings			Potable Water Savings		
	Base demand (Mgal/yr)	S1: Irrigation use only (Mgal/yr)	S2: Toilet flushing (Mgal/yr)	S3: Irrigation and toilet flushing (Mgal/yr)	S1: Irrigation use only (%)	S2: Toilet flushing (%)	S3: Irrigation and toilet flushing (%)
Los Angeles, CA	60.7	6.7	4.8	8.1	11	7.9	13
Seattle, WA	35.7	3.9	4.8	6.7	11	13	19
Lincoln, NE	35.8	6.8	4.8	8.4	19	14	23
Madison, WI	24.2	2.7	4.8	6.3	11	20	26
Birmingham, AL	24.3	2.3	4.8	6.2	9.5	20	25
Newark, NJ	24.7	2.3	4.8	6.2	9.3	20	25

NOTE: These savings assume that indoor and outdoor water use habits are unchanged by this new low-cost source of water, an assumption that remains untested.

TABLE 3-2 Potential Potable Water Savings Calculated for Six Cities from Laundry-to-Landscape Systems Based on a 100-Acre, Medium-Density, Residential Area

	Laundry to Landscape	
	Volume Potable Water Savings (Mgal/yr)	Potable Water Savings (%)
Los Angeles, CA	2.5	4.1
Seattle, WA	1.4	3.9
Lincoln, NE	2.8	7.7
Madison, WI	1.1	4.6
Birmingham, AL	1.1	4.5
Newark, NJ	1.1	4.4

NOTE: These savings assume that indoor and outdoor water use habits are unchanged by this new low-cost source of water, an assumption that remains untested.

ter supply is from the same watershed and not imported (the details are discussed in Box 3-3). However, if irrigation with graywater exceeds the prior irrigation rates, then impacts would be felt on downstream users. Thus, graywater use for toilet flushing, as a nonconsumptive use of water, does not pose the water rights issues that graywater irrigation does.

Graywater and Water Conservation. The committee's estimates of potential potable water savings associated with the use of graywater (Figure 3-3) are based on the most recently available data on indoor water use in North America. Trends in indoor water use in the United States show substantially decreased water use over the past 10-15 years (Figure 3-4) from 69 gpcd (260 lpcd; Mayer et al., 1999) to 59 gpcd (220 lpcd; DeOreo et al., 2016). Water use data collected from efficient new homes (36 gpcd [140 lpcd]; DeOreo et al., 2011) suggest continued reductions in indoor water demand. Meeting targets for water efficient homes will decrease the amount of graywater available for use, although graywater is projected to continue to meet toilet demand even in a highly water efficient home (Figure 3-4). For comparison, water use in Germany (32 gpcd; leaks not included) is closer to meeting high-efficiency targets than are U.S. water users in 2014. Shower water use remains relatively constant over all samples, including older U.S. data and high-efficiency new homes (Figure 3-4), suggesting a relatively steady source of graywater. In contrast, significant reductions in laundry water use have been documented between 1999 and 2014, and further reductions are possible, reducing the contributions of a key graywater source.

As indoor water conservation fixtures continue to be installed, there is likely to be an impact on graywater quality. Despite the decrease in water used, use of personal care and cleansing products are likely to be used in similar amounts and pathogenic organisms would be loaded similarly. For example, graywater quality in a peri-urban area of Durban, South Africa, where water use was limited to 52 gphd (200 lphd), graywater chemical constituents were greater than those reported for European and U.S. homes by a factor of 2-10 (Salukazana et al., 2005). To date, a consistent trend of increasing concentrations of graywater constituents has not been observed in the United States despite decreased indoor water use over the past 10-15 years. Graywater quality is so variable (Eriksson et al., 2002; see Chapter 4), that such a trend would be difficult to detect. Nonetheless, aggressive indoor water conservation practices could render graywater of a quality in terms of organic matter and salt concentrations not suitable for use for irrigation and could increase the extent of treatment needed for all end uses as a result of increases in concentrations of pollutants due to lower dilution rates. Additional research could improve the understanding of the implications of water conservation trends on the quality and quantity of graywater and impacts on the cost-effectiveness and feasibility of specific uses (see Chapter 10).

Commercial Use

In general, commercial water use is highly variable (Dziegielewski et al., 2000; see Figure 3-5), and most commercial facilities do not generate enough graywater to render its use for either irrigation or toilet flushing worthwhile. Gleick et al. (2003) reported that restroom and laundry water represented only 16 and 2 percent, respectively, of California's commercial and industrial sectors water use. Of the restroom water, only 11 percent was graywater (showers and hand washing basins), with the remainder used by toilet and urinal flushing. In contrast, landscaping water accounted for 35 percent of the total commercial and industrial water use, which is the largest water demand (including industrial process water). Overall, the amount of graywater generated is substantially less than the amount that could be used for irrigation or toilet flushing. Some commercial facilities and offices have on-site showers and/or facilities (e.g., hotels, fitness, and aquatic centers). Use of graywater can be beneficial at such facilities as long as showers or laundry facilities are frequently used. Although the quantity of graywater generated at hospitals is likely enough to meet toilet demands, on-site recycling of graywater at hospitals is not recommended because of the potentially large load of pathogens and high sensitivity of the population in contact with the treated graywater.

A study of water use was conducted at an office building located in Fort Collins, Colorado, occupied by nearly 1,700 employees daily that had an on-site gym with show-

BOX 3-3 Effects of Graywater Use in Existing Developments on Downstream Water Availability

Some state laws protect the availability of water to downstream users who hold appropriated water rights (see Chapter 8). Thus, it is important to understand the potential impact of increased graywater use on the availability of water to downstream users. Graywater reuse at any scale can have impacts on downstream return flows to the hydrologic cycle (Figure 3-3-1; summarized in Table 3-3-1).

When graywater is used for toilet flushing, potable water demand is reduced by the amount of graywater used for flushing. This use results in an equivalent reduction of the wastewater volume generated by the household. There is no change in downstream water flows at a larger scale, because the reduction in wastewater generated is in balance with reduced water demand by the household. The mass of waste constituents discharged remains the same although less water is available for dilution, resulting in increased concentrations in the discharged wastewater along with reduced flow volumes.

If graywater is used for lawn and garden irrigation in an existing development, then the effects on downstream users are dependent upon whether the amount of water applied to landscaping remains constant after a graywater system is installed. Assuming the landscaping and the total amount of irrigation water applied are unchanged, potable water demand and the volume of wastewater generated are reduced by the volume of graywater used for irrigation. Use of graywater for irrigation results in the same water losses from evapotranspiration and from recharge to groundwater, as would have been the case for potable water applied to existing landscaping that the graywater replaced. Thus, there should be no net effect on downstream water availability because the reduction in the volume of wastewater generated is in balance with the reduced potable water use, and the evaporative and recharge losses from the use of graywater for irrigation should be the same as in the base scenario. Because a portion of household waste constituents contained in graywater is diverted from the sewer system, the mass load to the sewer system is slightly reduced, although the concentration of wastewater constituents increases because less water is available for dilution.

If installation of a graywater irrigation system results in a greater demand for irrigation water use on a property (such as from additional plantings that require more irrigation than landscaping being replaced), then this could result in a situation where there is little or no reduction in potable water use. If plantings requiring irrigation are substantially increased, then it is possible that potable water use actually increases compared to water use prior to the graywater installation. Under a scenario of expanded landscaping where a greater volume of water is being used for irrigation, there is enhancement of losses due to evapotranspiration (a consumptive use) as well as additional groundwater recharge (a nonconsumptive use) compared to the base case. However, under efficient irrigation, evapotranspiration (consumptive) losses typically far exceed groundwater recharge (Bijoor et al., 2014). Thus, assuming efficient irrigation methods, increased evaporative effects would reduce water availability for downstream users. This scenario results in the same effects on the wastewater volume, concentrations, and loads as graywater irrigation of existing landscape (see Table 3-3-1).

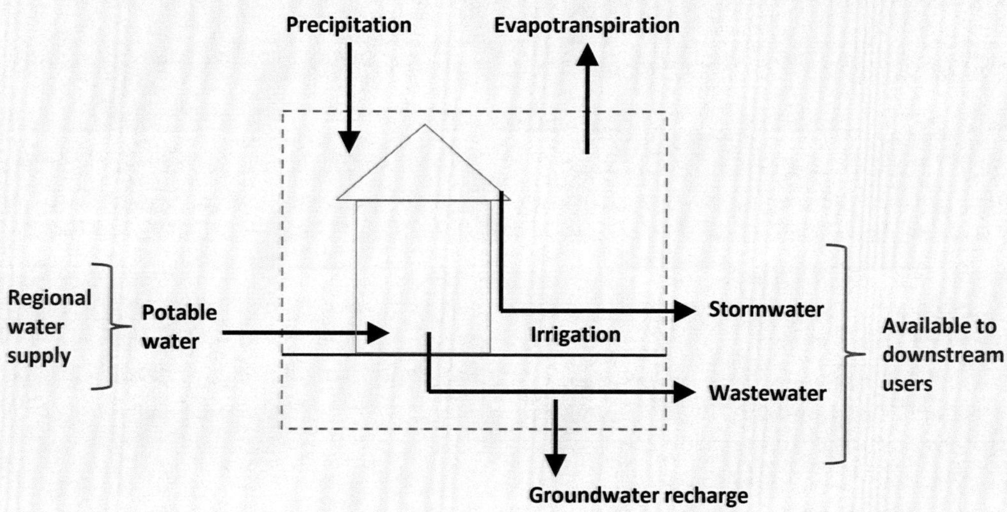

FIGURE 3-3-1 Typical hydrologic cycle at a household scale. When considering impacts to downstream water availability, scenarios can be assessed to evaluate relative changes to inputs and outputs. Water inputs to the property include potable water supply and precipitation. Water outputs include wastewater (graywater and blackwater), stormwater, evapotranspiration, and groundwater recharge from infiltrated water (precipitation and applied irrigation water that is not evapotranspired or incorporated into plant biomass). Consumptive uses—evapotranspiration and that incorporated into plant biomass—reduce the amount of water available to those downstream. At the neighborhood scale, inputs and outputs are composed of the collection of inputs and outputs from individual properties.

(Continued)

BOX 3-3 Continued

TABLE 3-3-1 Summary of Water Budget Effects from Graywater Adoption at Existing Developments

	Water Quantity Effects			Water Quality Effects	
	Potable water demand	Wastewater volume generated	Water available to downstream water users	Wastewater constituent concentration	Wastewater constituent load
Graywater reuse for toilet flushing	Reduced	Reduced	No change	Increased	No change
Graywater reuse for existing irrigated landscape	Reduced	Reduced	No change	Increased	Slightly reduced
Graywater reuse for increased irrigation (e.g., expanded landscaping)	May be slightly reduced, unchanged or increased	Reduced	Reduced	Increased	Slightly reduced

NOTES: These calculations are determined by subtracting average monthly precipitation (1995-1999, except for Lincoln, which used 1996-1999) from average monthly evapotranspiration rates. These data may underestimate irrigation requirements when rainfall is clustered in a few days within a month for some years, but these variations are expected to be evened out with the multi-year analyses. SOURCE: See Appendix A.

ers that were perceived to be used often (Vandegrift, 2014). However, an analysis of building water use showed that the showers actually contributed less than 1 percent of the total building water use, while the toilets accounted for 10 percent. This study only evaluated one such building, so broad conclusions cannot be made. However, the data indicate that showers in office buildings may not contribute enough graywater to render graywater reuse projects feasible. On-site laundry machines in offices, hotels, and other commercial facilities may result in enough graywater generation to render reuse feasible. In this same study, 79 percent of water generated at a fitness facility was estimated to be graywater, which would easily meet the toilet demand with much excess water available for irrigation. In addition, a hotel was found to produce a substantial quantity of graywater, with an estimated 25 percent of water use in laundry, showers, and sinks (Vandegrift, 2014). Site-specific analysis of the availability of commercial graywater for onsite use is therefore necessary to determine the potential for water savings.

STORMWATER

This section explores the amount of stormwater potentially available for various beneficial uses at different scales in a community, from the smallest on-site capture of roof runoff for irrigation of surrounding landscaped areas, to large-scale community collection of stormwater in regional impoundments to augment the water supply. The amount of stormwater generated greatly depends on the amount of rainfall in the area, the land development characteristics, and the effectiveness of the stormwater collection system. The amount of the stormwater that can be effectively used by the different beneficial uses is based on complex interactions of timing of the rainfall and the desired use patterns, the ability to collect and store the runoff, and coordination with other uses and supplies. Thus, estimating the availability of stormwater to address water demands across the United States is much more complex than for graywater. This committee presents (1) an approach to identify the amount of water available from different source areas and land uses and (2) the results of an original analysis to approximate potential water savings from household-scale stormwater capture for various uses in medium-density, residential development in six different locations in the United States.

Factors Affecting the Quantity of Stormwater from Different Areas

Key factors that affect the quantity of stormwater available for beneficial use are rainfall and land development characteristics. The total amount of rainfall and the distribution of rain depths for different periods of the year vary dramatically throughout the country, affecting the quantities of runoff available for different beneficial uses. Land uses also vary by region, affecting the quantity of runoff available for large-scale stormwater collection for beneficial uses.

Rainfall Characteristics

Important factors affecting stormwater runoff quantities across the United States include the amount of the rainfall, depth of individual events, and seasonal patterns of the rains. The six locations examined by the committee represent a range of climatic conditions in the United States and are not intended to represent all of the conditions in each region, or all regions.

- Los Angeles, California, in the Southwest, having a median rainfall of about 12 inches per (30 cm) year over the long-term record (17 inches [43 cm] average during the 5-year calculation period)
- Seattle, Washington, in the Northwest, having a median rainfall of about 37 inches (94 cm) of rainfall per year (42 inches [110 cm] average during the 5-year calculation period)
- Lincoln, Nebraska, in central United States, having a median rainfall of about 26 inches (66 cm) of rainfall per year) (28 inches [71 cm] average during the 4-year calculation period)
- Madison, Wisconsin, in the Great Lakes region, having a median rainfall of about 32 inches (81 cm) of rainfall per year (30 inches [76 cm] average during the 5-year calculation period)
- Birmingham, Alabama, in the Southeast, having a median rainfall of about 54 inches (140 cm) of rainfall per year (50 inches [130 cm] average during the 5-year calculation period)
- Newark, New Jersey, in the East Coast region, having a median rainfall of about 43 inches (110 cm) of rainfall per year. (44 inches [110 cm] average during the 5-year calculation period)

Rainfall can vary greatly from year to year, and the committee's analyses examined 4-5 years of data for each city (1995-1999). The seasonal distributions of rains at these six locations vary greatly, as shown on Figure 3-2. The West Coast locations (Los Angeles and Seattle) experience most (or all) of their rains during the winter months, the central and Great Lakes locations (Lincoln and Madison) experience more rain during the summer months, and the Southeast and the East Coast locations (Birmingham and Newark) experience more evenly distributed rains throughout the year. One of the main challenges with effective beneficial uses of stormwater, therefore, is matching water needs with available stormwater runoff, which may necessitate significant storage.

The depth of individual rainfall events also varies across the country, necessitating varied designs to appropriately capture stormwater. Medium-size rains (from about 0.5 to 2 inches [1.3 to 5 cm]) account for the majority of the annual runoff from the most common land uses in all areas, but higher percentages of the rainfall occur as runoff as the rain depth increases. For example, about 900 gallons (3,400 liters) of roof runoff would be produced from a typical 1,500 ft^2 (140 m^2) roof area during a 1-inch (2.5 cm) rainfall. In comparison, a more common 0.25-inch rain would produce about 200 gallons (760 liters) of runoff for this same roof area. Median rains in these six areas between 1995 and 1999 range from about 0.1 (Seattle) to 0.25 inches (Birmingham),

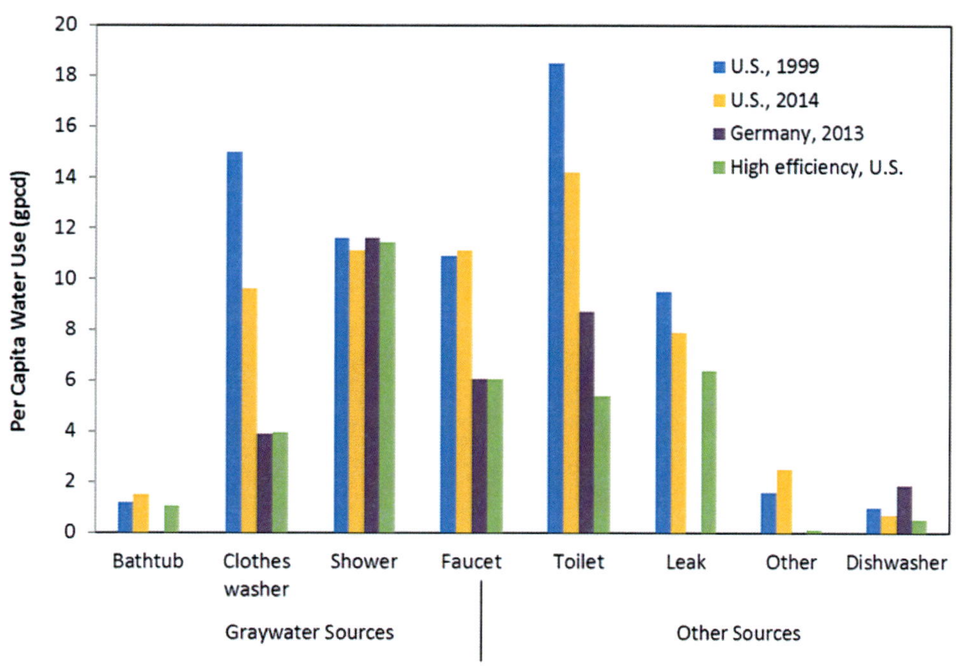

FIGURE 3-4 Average indoor water use in the United States in 1999 (69 gpcd) and 2014 (59 gpcd), Germany in 2013 (32 gpcd), and efficient new homes (36 gpcd). NOTES: Data collected in Germany do not include leaks, bathtubs, or other sources; thus, total water use figures are not comparable. Fixture data for high efficiency homes based on average household data and 3 persons per household. SOURCES: Data from Mayer et al. (1999); DeOreo et al. (2011; 2016); http://de.statista.com/statistik/daten/studie/224682/umfrage/trinkwasserverbrauch-in-deutschen-haushalten.

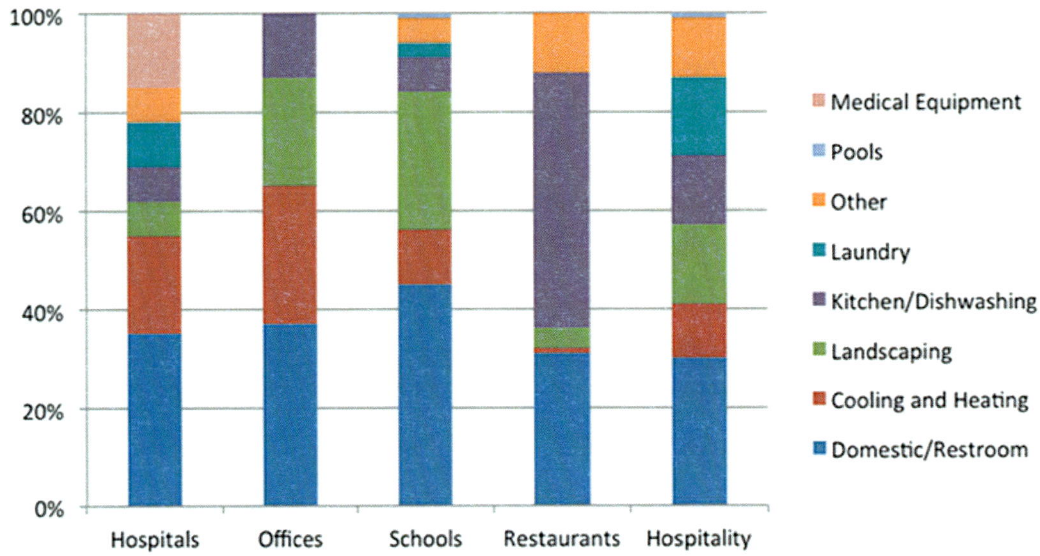

FIGURE 3-5 Variability in commercial water use. Note that in commercial restrooms, toilet flushing represents the majority of water use, limiting graywater production. Among commercial facilities, those with showers and large laundry facilities, such as fitness centers, laundries, and hotels, tend to generate the most graywater. SOURCE: EPA (2015).

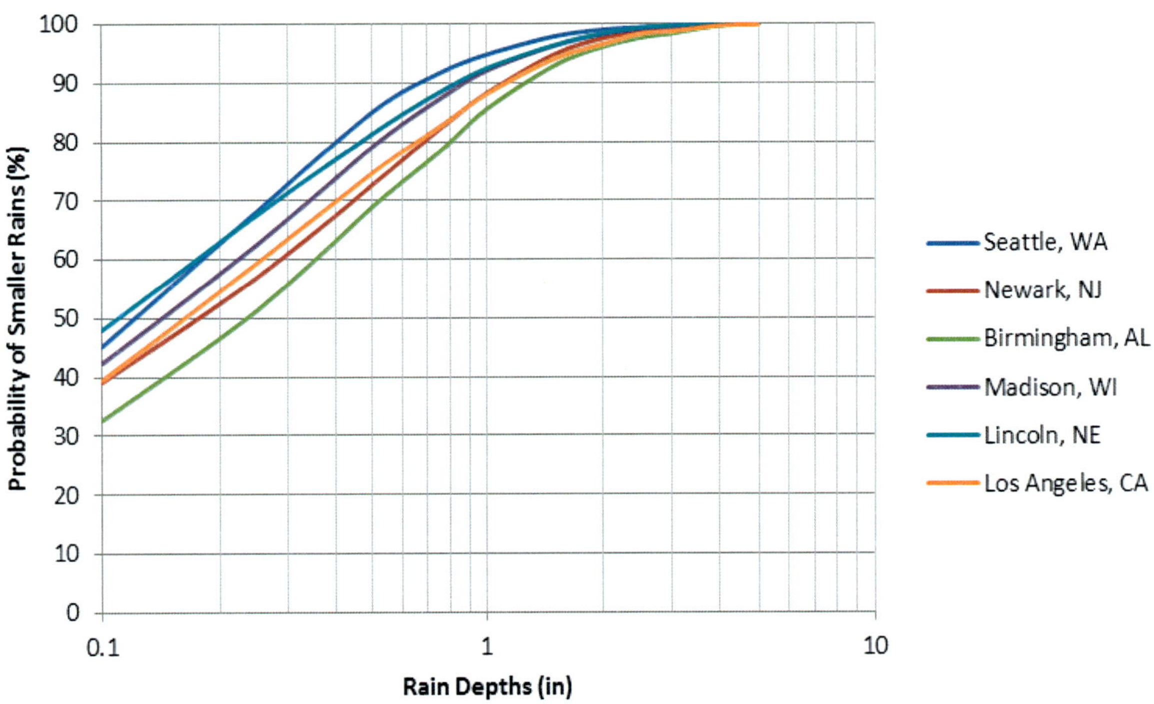

FIGURE 3-6 Distribution of different rain depths for different U.S. locations, shown as the probability of a rainfall of a certain depth or smaller. SOURCE: Based on National Oceanic and Atmospheric Administration (NOAA) hourly rainfall data supplied by EarthInfo (Santa Monica, California) (5-year periods, depending on completeness of rain record: all were from 1995 through 1999, except for the Lincoln calculations, which were from 1996 through 1999 due to many missing 1995 rainfall records).

TABLE 3-3 Summary of Major Land Use Characteristics

Land Use Category	Average Percentage Directly Connected Impervious Area (coeff. of variation)	Average Percentage Partially Connected Impervious Area (coeff. of variation)	Average Percentage Pervious Area (coeff. of variation)
Commercial	80 (0.3)	2 (2.8)	19 (1.0)
Industrial	54 (0.3)	21 (0.4)	24 (0.5)
Institutional	50 (0.4)	9 (0.9)	41 (0.3)
Open space	10 (1.2)	11 (1.3)	79 (0.3)
Residential	24 (0.6)	12 (0.5)	64 (0.2)
Freeway and highway	32 (1.2)	27 (1.2)	41 (0.3)

NOTES: Represents more than 100 research sites in the United States. The coefficient of variation is the ratio of the standard deviation to the average value, an indication of the spread in the data. SOURCE: Pitt (2011a).

and some locations (e.g., Birmingham, Newark, and Los Angeles) have a greater percentage of heavy rainfall (greater than 1 inch) (Figure 3-6).

Land Development and Source Area Characteristics

Directly connected impervious areas (e.g., roofs, streets, and paved parking areas connected directly to the drainage system) are usually responsible for most of the runoff in developed urban areas and are therefore the major source areas of runoff available for capture for beneficial uses. Partially connected impervious areas (e.g., roof drains or paved areas draining to pervious areas before entering the drainage system) contribute smaller amounts of runoff that occur later times during larger rains, while the pervious areas contribute small flows and only after substantial rain has occurred. However, pervious areas can be important sources of runoff in residential areas and other land uses where landscaped and undeveloped land comprise a large portion of the land area (Figure 3-7). Average percentages of directly connected impervious areas, partially connected impervious areas, and pervious areas for six major land use categories from locations throughout the country are shown in Table 3-3.

Mathematical models can be used to calculate the stormwater runoff available for different rainfall events, surface materials, and land characteristics, highlighting both the quantities available for beneficial use and the associated storage challenges (Figure 3-8). Table 3-4 shows the calculated associated annual stormwater runoff yields for the six locations examined. These runoff amounts and the fraction of rainfall that is converted to runoff (or runoff coefficient, see Table 3-5) vary significantly for different land uses and locations. However, the extent to which conventional water demand can be reduced will depend upon the amount of storage provided, the water demand, and the timing of rainfall relative to water demand.

The feasibility of household- or building-scale stormwater capture for irrigation purposes depends on the ratio of the roof area to the area to be irrigated. Areas having relatively small roofs and large landscaped areas may not be able to supply sufficient quantities of water to meet irrigation demands, depending on the rainfall patterns and storage tank sizes. Areas having large roof areas compared to the adjacent landscaped areas (e.g., commercial buildings) could have abundant water for irrigation, in which case excess stormwater could be made available for other beneficial uses.

Committee Scenario Analysis of Stormwater Availability for Household-scale Water Uses

Roof runoff offers the most suitable source for building-scale stormwater collection for beneficial uses because of its generally better water quality (see Chapter 4), high runoff yield per unit area, and elevation above storage tanks and irrigated land (which reduces energy use). To assess the capacity for rooftop runoff to address on-site water demands, the committee analyzed four scenarios in each of six geographical locations in the United States (Los Angeles, California; Seattle, Washington; Lincoln, Nebraska; Madison, Wisconsin; Birmingham, Alabama; and Newark, New Jersey), parallel to the graywater analysis (detailed in Box 3-1):

- **Base scenario:** Potable water is used for all household water needs (indoors and outdoors), using the same base scenario assumptions as described in the graywater analysis discussed earlier in this chapter. The land surface characteristics represented by medium-density, residential development in each of the six locations (e.g., percentage landscaped area [needed for irrigation calculations] and roof area [needed for runoff quantities]) vary based on site characteristics and are outlined in Appendix A.

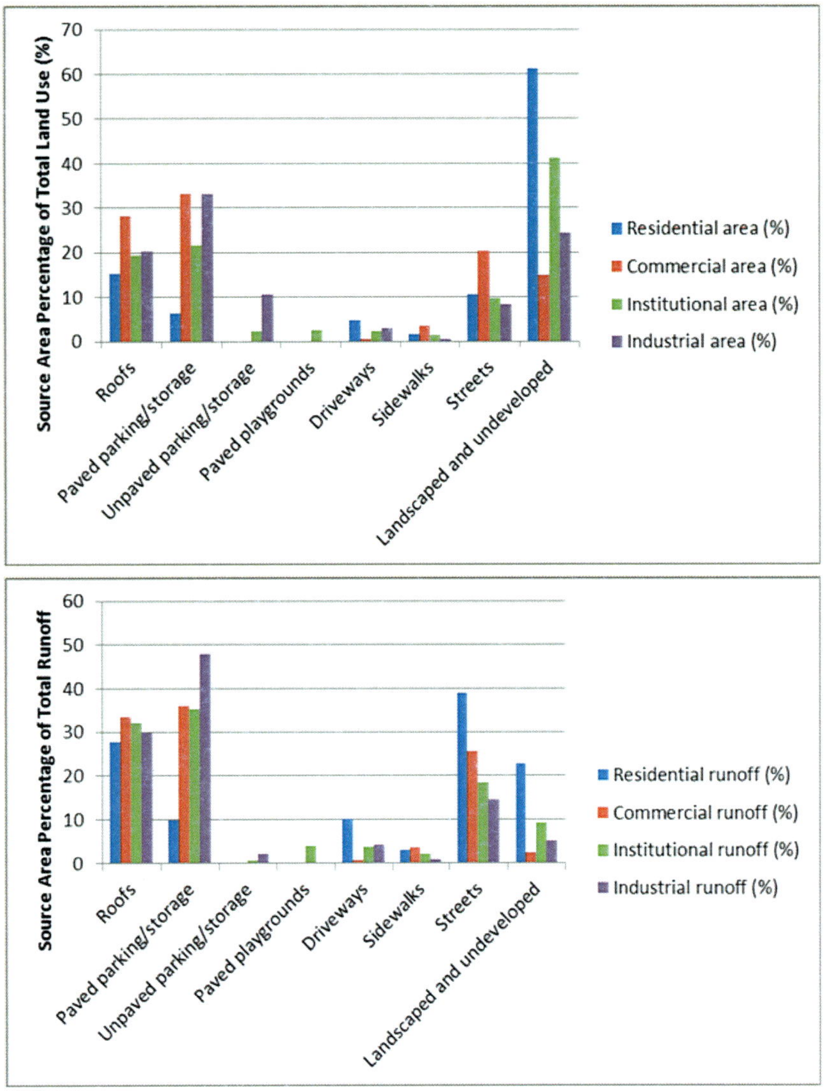

FIGURE 3-7 Source area percentages (top) and percentage contribution to overall runoff (bottom) for different land uses at Madison, Wisconsin. The residential area land use is represented by medium-density development characteristics. In general, low-density residential areas would have few roofs and more landscaped areas, while high-density residential areas would have more roofs and fewer landscaped areas. NOTE: Although the land use data shown in Figure 3-7 (top) are specific to Madison, Wisconsin, the distributions are generally similar to other geographic regions of the country. However, the runoff data in the bottom figure are more variable by geographical area than are the land use characteristics because of differences in the rainfall patterns. For example, the Southeast and East Coast have a greater abundance of larger rains with more runoff from landscaped areas in relationship to impervious areas. SOURCE: Source areas from the standard land use files used by the nonpoint source section of the Wisconsin Department of Natural Resources as maintained by the U.S. Geological Survey. (http://wi.water.usgs.gov/slamm).

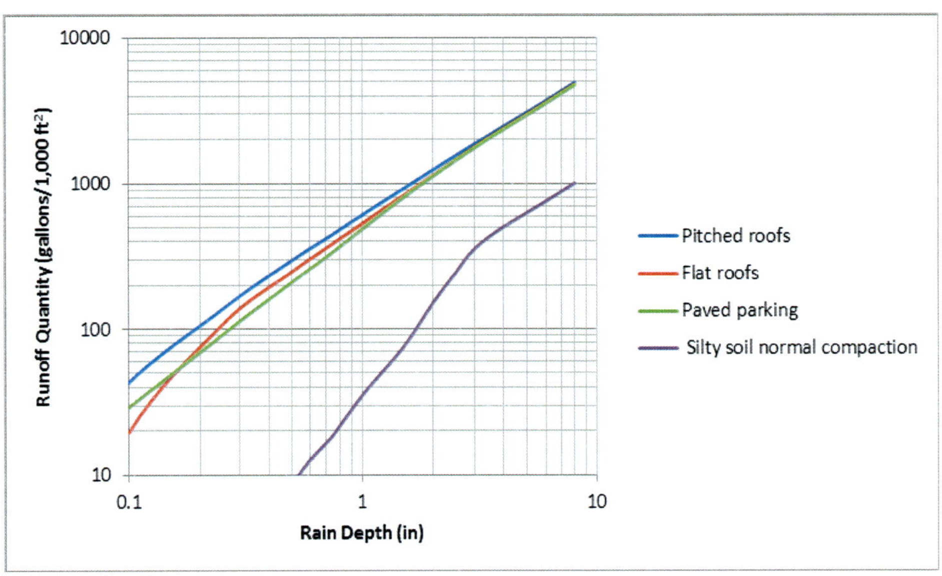

FIGURE 3-8 Runoff quantity for different rain depths at four different source areas. SOURCE: Appendix A describes the modeling calculations used to determine the runoff amounts from the different source areas.

- **Scenario 1 (irrigation):** Rooftop runoff is captured and used to irrigate landscaped areas and offset potable water demand to meet the evapotranspiration deficit for turfgrass.
- **Scenario 2 (toilet flushing):** Rooftop runoff is captured and used for toilet flushing.
- **Scenario 3 (irrigation and toilet flushing):** Both irrigation and toilet flushing demands are addressed using captured rooftop runoff.

The capacity to capture and use stormwater at the building scale is strongly dependent upon storage tank size. The optimum tank size is best determined through continuous simulation of stormwater model analyses for a given location. However, for simplicity, only two stormwater storage volumes were considered for each of the above scenarios, using continuous simulations for 4 or 5 years:

- 70 gallons (260 liters) per household, representing two rain barrels at 35 gallons each, and
- 2,200 gallons (8,300 liters) per household, representing a single larger tank (8 ft [2.4 cm] diameter and 6 ft [1.8 m] tall).

These tank volumes reflect commonly used household-scale, rooftop-runoff capture systems, although larger tanks are certainly possible and may offer additional benefits.

Given the number of factors that affect stormwater runoff and availability for beneficial use, assessments of stormwater availability are most effectively conducted using a continuous stormwater model. The committee performed its analyses using the continuous stormwater model WinSLAMM, using similar procedures as described by Pitt et al. (2011) to assess the potential contributions of stormwater to reduce conventional water supplies in six cities in different regions of the country. WinSLAMM[3] (Pitt, 1997) was selected for these analyses because of the committee's familiarity with the model and the similar analyses conducted previously by Pitt et al. (2011). Specifically, WinSLAMM was able to conduct continuous, long-term analyses considering different water uses, storage tank volumes, and rainfall records with minimal pre- and post-processing. The model calculated the amount of water available in the storage tanks and how much of the irrigation demand could be satisfied during a 5-year period (1995 through 1999 for all areas, except for Lincoln, where 1996 through 1999 rains were used). Details of the committee's analysis are described in Appendix A. Other models, such as the U.S. Environmental Protection Agency's (EPA's) SWMM model[4] or SUSTAIN model[5] may be used to perform these analyses, but they do not incorporate the features needed for these analyses without modifications. However, for large-scale projects, it is always worthwhile to use complementary models having different approaches to obtain better insights and understandings of complex systems.

Beneficial Use Scenarios

As in the graywater analysis, the committee considered two beneficial uses of on-site nonpotable water—irrigation and toilet flushing. In addition to the factors described in Box 3-1, specific stormwater considerations are discussed below.

Irrigation. Three primary irrigation strategies are relevant to the use of stormwater for irrigation: conventional irrigation, conservation irrigation, and land application. Each strategy involves a different approach to the rate at which water is applied to landscaped areas. In conventional irrigation, a consistent, average amount of water is applied to plants on a fixed schedule, independent of water losses due to evapotranspiration or infiltration. Conservation irrigation strategies apply only the minimum amount of water needed to meet the vegetation demand, considering local deficits between evapotranspiration requirements and available rainfall. Finally, land application strategies are intended to maximize the use of stormwater for irrigation by applying water, when available, at the maximum rate that does not produce runoff while preventing damage to plants. Land application, therefore, provides substantial stormwater runoff reduction benefits, while providing potable water conservation benefits that are often similar (depending on climate) to other stormwater capture systems used only for irrigation. Irrigating to barely satisfy the evapotranspiration deficits (conservation irrigation) is typically recommended when minimizing water use and is the focus of the committee's analysis, although many homeowners use much more water than recommended under conservation irrigation.

One of the primary considerations for determining conservation irrigation requirements involves comparing the time series of evapotranspiration and available rainfall for an area (as shown in Figure 3-2). The methodology for the analysis and calculations of irrigation demand are summarized in Box 3-1 and documented in more detail in Appendix A. Table 3-1-1 (Box 3-1) shows irrigation demands for conventional warm season turfgrass for residential areas by month for the six locations examined.

---

[3] See http://winslamm.com. Robert Pitt, committee member, is one of the developers of WINSLAMM.

[4] See http://www2.epa.gov/water-research/storm-water-management-model-swmm.

[5] See http://www2.epa.gov/water-research/system-urban-stormwater-treatment-and-analysis-integration-sustain.

Toilet Flushing. At all locations examined in this scenario analysis for medium-density, residential land use, the available roof runoff on an average annual basis is greater than the toilet flushing water requirements (see Box 3-1 and Table 3-4). However, the ability to use this water is dependent on storage facilities (see Chapter 6), especially in arid areas where seasonal rainfalls result in greatly uneven amounts of runoff throughout the year. The committee analyzed the capacity for two rooftop capture systems to meet toilet flushing demand throughout the year at the six locations to bracket typical conditions. These scenario analyses were designed to examine the potential for stormwater capture at the household scale to address common nonpotable water demands (irrigation and toilet flushing) and to allow direct comparison with the graywater scenarios considered earlier. The committee recognizes that small-scale rooftop capture systems (e.g., rain barrels) are not typically used for toilet flushing, but the scenario could represent other year-round nonpotable uses. The larger storage tank option could enable more effective use of the roof runoff for toilet flushing, as well as other potential nonpotable uses not considered in this scenario.

Scenario Analysis Results

The calculated water savings associated with each of the beneficial use scenarios for the six locations are summarized in Tables 3-5 and 3-6. Table 3-7 shows the percentages of roof runoff used for these beneficial use options, of most interest to stormwater managers interested in runoff volume reductions. The use of irrigation plus toilet flushing in the central United States can use most (74 percent) of the roof runoff for beneficial uses when the larger tank volume is considered. The other locations use from about 31 to 62 percent of the total roof runoff for conservation irrigation plus toilet flushing when the large water storage tank option is used. Some of this roof runoff could therefore be made available for shallow groundwater recharge using rain gardens, but this infiltration was not considered in these analyses. The

TABLE 3-4 Calculated Annual Runoff Quantities (in gallons/year/acre) for Different Land Uses in Six Cities, 1995-1999

	Los Angeles, CA	Seattle, WA	Lincoln, NE	Madison, WI	Birmingham, AL	Newark, NJ
Commercial	320,000	730,000	490,000	560,000	940,000	820,000
Industrial	250,000	630,000	460,000	450,000	610,000	710,000
Medium-density residential	210,000	380,000	260,000	270,000	310,000	490,000

NOTE: Calculated using WinSLAMM (see Appendix A). Calculations are based on 100 percent of a single land use type. Lincoln data represents only 1996-1999. See Appendix A for methods used to derive these data.

TABLE 3-5 Calculated Runoff Coefficients for Different Land Uses in Six Cities, 1995-1999

	Los Angeles, CA	Seattle, WA	Lincoln, NE	Madison, WI	Birmingham, AL	Newark, NJ
Commercial	0.70	0.65	0.65	0.70	0.64	0.69
Industrial	0.55	0.55	0.60	0.56	0.42	0.60
Medium Density Residential	0.46	0.34	0.35	0.34	0.21	0.42

NOTE: The runoff coefficient (Rv) values represents the fraction of the rainfall that is converted into runoff. Calculations are based on 100 percent of each land use type. Source area components for each land use type are determined by local data for each location. Roof runoff represents approximately 25 to 30 percent of the total residential area flows. Lincoln data represents only 1996-1999. See Appendix A for methods used to derive these data.

TABLE 3-6 Potential Potable Water Savings in Six Cities Based on a 100-Acre, Medium-Density, Residential Area Using Two 35-gallon Rain Barrels per Household

	Volume Potable Water Savings				Potable Water savings		
	Base use (Mgal/yr)	S1: Irrigation use only (Mgal/yr)	S2: Toilet flushing (Mgal/yr)	S3: Irrigation and toilet flushing (Mgal/yr)	S1: Irrigation use only (%)	S2: Toilet flushing (%)	S3: Irrigation and toilet flushing (%)
Los Angeles, CA	60.7	0.6	0.8	1.1	1.0	1.2	1.8
Seattle, WA	35.7	1.1	2.3	3.0	3.1	6.4	8.3
Lincoln, NE	35.8	1.8	1.5	2.3	5.0	4.2	6.3
Madison, WI	24.2	0.9	1.7	2.1	3.8	7.0	8.6
Birmingham, AL	24.3	0.6	1.6	1.8	2.6	6.4	7.3
Newark, NJ	24.7	0.9	2.1	2.5	3.5	8.6	10

use of two rain barrels per house only allows about 14 to 19 percent of the roof runoff to be used because of the limited storage provided.

The results are also presented in Figures 3-9 to 3-10 to illustrate the total amount of water demand per household, separated by indoor and outdoor uses. For the base condition and the irrigation or toilet flushing scenarios, the indoor and outdoor water uses can be shown separately. However, for the combined irrigation and toilet flushing use, the flows are withdrawn from the tank as a combined demand in the model, and separate data are not available.

Regional Differences in Potable Water Demand Reduction. Figures 3-9 and 3-10 show the domestic water use for all six cities examined in the committee's scenario analyses. The least potable water savings are shown for Los Angeles, California, located in the arid Southwest. Even with the 2,200-gallon storage tank, water use is reduced by very small amounts (up to 5.4 percent), because of the poor alignment of periods of rainfall and irrigation demand and the normal, long, dry periods during the summer months that outlast the availability of stored roof runoff. Only 42 percent of the roof runoff water is captured with the larger tank in the modeled scenario, showing the impact of intense rainfalls that overfill the tank before it can be used. In areas with similar climate conditions, shallow groundwater recharge using on-site rain gardens and/or larger capture and reuse options (such as regional groundwater recharge projects) may be needed to achieve greater contributions to local water supplies and reductions in stormwater runoff, especially considering the very large variability of rainfall in the area. In the arid Southwest, outdoor irrigation demand is very high for typically used landscaping plants (because of the large evapotranspiration requirements and limited rainfall). Significant water conservation potential is possible through the use of native plants that do not rely on applied irrigation water. Compared to stormwater, graywater is a larger and more consistent source of water in the arid Southwest.

Seattle, Washington, located in the northwest, has a similar seasonal precipitation pattern as Los Angeles with irrigation demands concentrated in the summer when rainfall is lower (Figure 3-2). The longer-duration and less-intense rains and smaller evapotranspiration demands in Seattle allow for more efficient capture of roof runoff for beneficial uses. However, even with large storage tank use and both toilet flushing and conservation irrigation, the potential water demand reduction is only 15 percent (Table 3-6).

At Lincoln, Nebraska, located in the central United States, on average, precipitation occurs simultaneously with periods of irrigation demand (see Figure 3-2), which results in greater opportunities for capturing roof runoff for on-site irrigation. The potential potable water savings for these scenarios ranges from about 11 to 26 percent when using a 2,200-gallon tank, with 21 percent savings for irrigation only (the largest irrigation savings by volume and by percentage of the six sites; see Table 3-8).

The precipitation in Madison, Wisconsin, also matches irrigation demands reasonably well, with irrigation demand in only 4 months of the year, which limits potential water savings from conservation irrigation with captured stormwater (see Figure 3-2 and Table 3-1-1). With 2,200-gallon tanks, water use savings of up to 28 percent are possible. Because of the low irrigation demand in this area, toilet flushing with captured runoff offers greater potential potable water savings (18 percent) compared to irrigation alone (13 percent).

At Birmingham, Alabama, located in the Southeast, and at Newark, New Jersey, located on the East Coast, the irrigation requirements are relatively modest compared to the large amounts of rainfall in these areas (see Figure 3-2). The largest beneficial use potential for roof runoff in these locations is associated with toilet flushing (17 and 18 percent). The maximum potential potable water savings is about 24

TABLE 3-7 Potential Potable Water Savings in Six Cities Based on a 100-Acre, Medium-Density, Residential Area Using One 2,200-gallon Stormwater Tank per Household

	Volume Potable Water Savings				Potable Water Savings		
	Base use (Mgal/yr)	S1: Irrigation use only (Mgal/yr)	S2: Toilet flushing (Mgal/yr)	S3: Irrigation and toilet flushing (Mgal/yr)	S1: Irrigation use only (%)	S2: Toilet flushing (%)	S3: Irrigation and toilet flushing (%)
Los Angeles, CA	60.7	2.4	2.7	3.3	4.0	4.5	5.4
Seattle, WA	35.7	2.8	4.2	5.5	7.8	12	15
Lincoln, NE	35.8	7.6	3.9	9.2	21	11	26
Madison, WI	24.2	3.2	4.3	6.8	13	18	28
Birmingham, AL	24.3	2.4	4.3	5.8	9.7	18	24
Newark, NJ	24.7	3.2	4.2	6.9	13	17	28

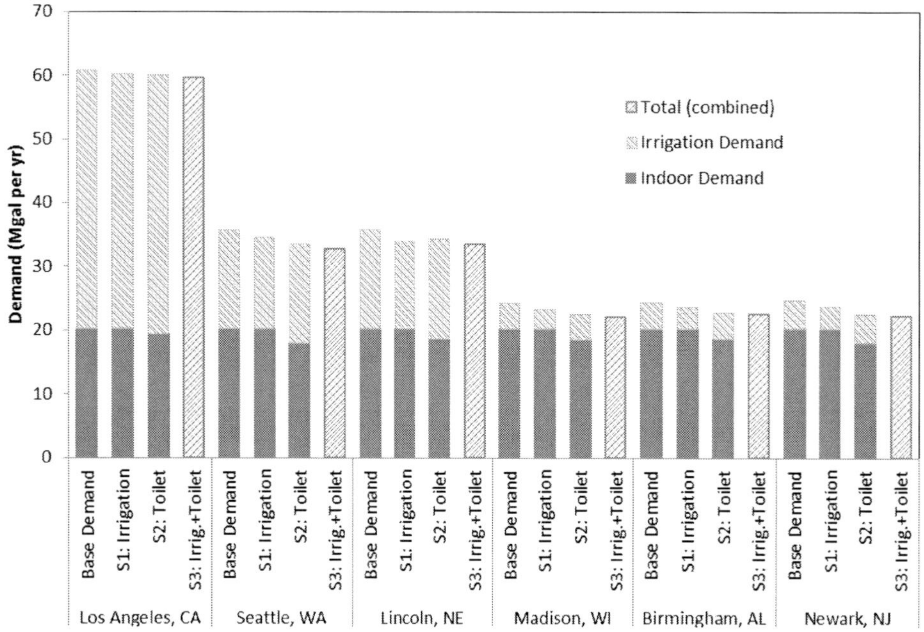

FIGURE 3-9 Effects on water demand from the beneficial use of roof runoff for conservation irrigation and/or toilet flushing using two 35-gallon rain barrels per house, based on a typical 100-acre, medium-density, residential development.

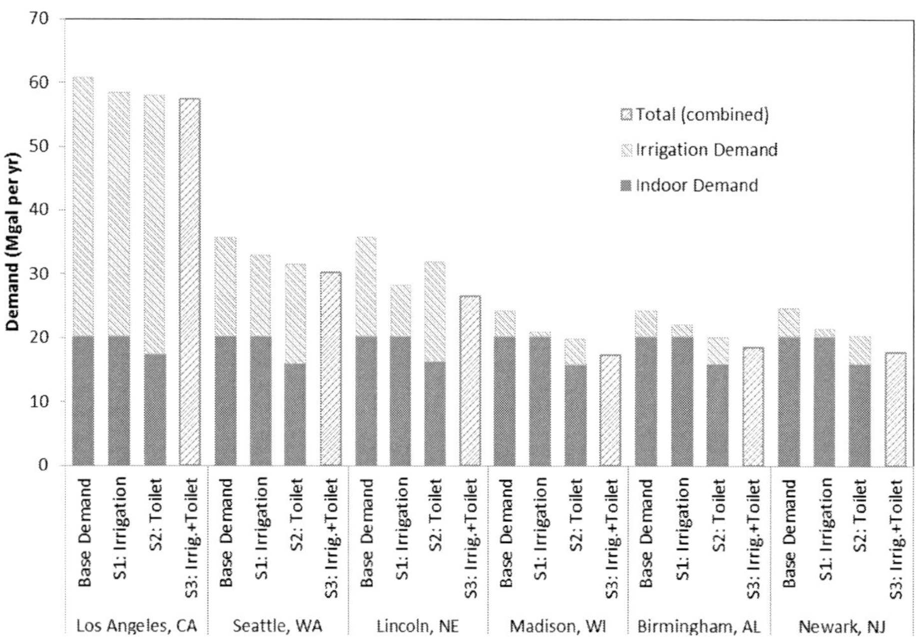

FIGURE 3-10 Effects on water demand from the beneficial use of roof runoff for conservation irrigation and/or toilet flushing using one 2,200-gallon stormwater capture tank per house, based on a typical 100-acre, medium-density, residential development.

TABLE 3-8 Percentage Reduction in Roof Runoff in Six Cities Using Two Rain Barrels or One 2,200-gallon Stormwater Tank per Household

	Reduction in Roof Runoff Using Two 35-gallon Rain Barrels			Reduction in Roof Runoff Using a 2,200-gallon Stormwater Tank		
	S1: Irrigation use only (%)	S2: Toilet flushing (%)	S3: Irrigation and toilet flushing (%)	S1: Irrigation use only (%)	S2: Toilet flushing (%)	S3: Irrigation and toilet flushing (%)
Los Angeles, CA	7.7	10	14	31	35	42
Seattle, WA	6.1	13	17	16	24	31
Lincoln, NE	14	12	18	61	31	74
Madison, WI	8.4	15	19	29	39	62
Birmingham, AL	5.7	14	16	21	38	52
Newark, NJ	4.9	12	14	18	24	39

NOTE: Based on analysis considering 100-acre, medium-density, residential area.

percent in Birmingham and 28 percent in Newark when using the larger storage tanks and with both toilet flushing and irrigation uses (only 10 and 13 percent, respectively, with irrigation alone).

Tank Size. Larger water storage volumes result in more potable water savings, but the differences are related to the magnitude and timing of precipitation relative to the demand. Figure 3-11 shows the effects of increasing water storage tank sizes on the annual domestic water savings based on simulated capture of rooftop runoff in medium-density, residential areas in Los Angeles. In this case, all of the annual roof runoff was used for irrigation when the total storage tank volume for the area was about 11 Mgal (42,000 m^3) of storage per 100 acres of medium-density, residential area (a very large storage volume corresponding to about 10 tanks of 8 ft diameter and 6 ft tall per house). The annual domestic water savings in this maximum situation is about 7.7 Mgal (29,000 m^3) per 100 acres. For this maximum roof runoff storage amount, the corresponding total stormwater runoff reduction for this area is about 37 percent (the percentage of roof runoff to the runoff from the whole area). For this arid area, more reasonably sized water storage tanks result in much less of the roof runoff being available for on-site use.

Based on the scenario analysis for irrigation use only, the 2,200-gallon tanks can result in savings of 2.5 to 4.8 times more potable water than the use of two 35-gallon rain barrels, for the six regions examined, but the large water tanks are 32 times larger than the two rain barrels. Rain barrels saved the largest amount of potable water (5 percent) in Lincoln, Nebraska, which has a low but near-year-round irrigation demand, and the lowest amounts in Los Angeles (where the storage needs are huge) and Birmingham (where the runoff volumes are huge) (Table 3-6). Ultimately, the selection of tank size is dependent upon local climate conditions, stormwater runoff area relative to on-site water demands, site conditions, overall objectives, costs, and benefits (see Chapter 7). Many factors affect the interaction of storage volumes and domestic water savings (most notably demand vs. availability patterns and roof areas vs. landscaped areas), requiring continuous simulations for site-specific analyses, as was conducted for these analyses. The scenarios examined here show typical ranges of conditions and resulting expectations over a broad range of geographical conditions.

Runoff Reduction. Runoff reductions are a common goal of most stormwater management plans and options to use that runoff on site and prevent its discharge during rain events. This can be an important secondary benefit of beneficial use strategies. The results of these scenario analyses show that tank capture and beneficial use of roof runoff lead to maximum reductions (assuming use for irrigation and toilet flush-

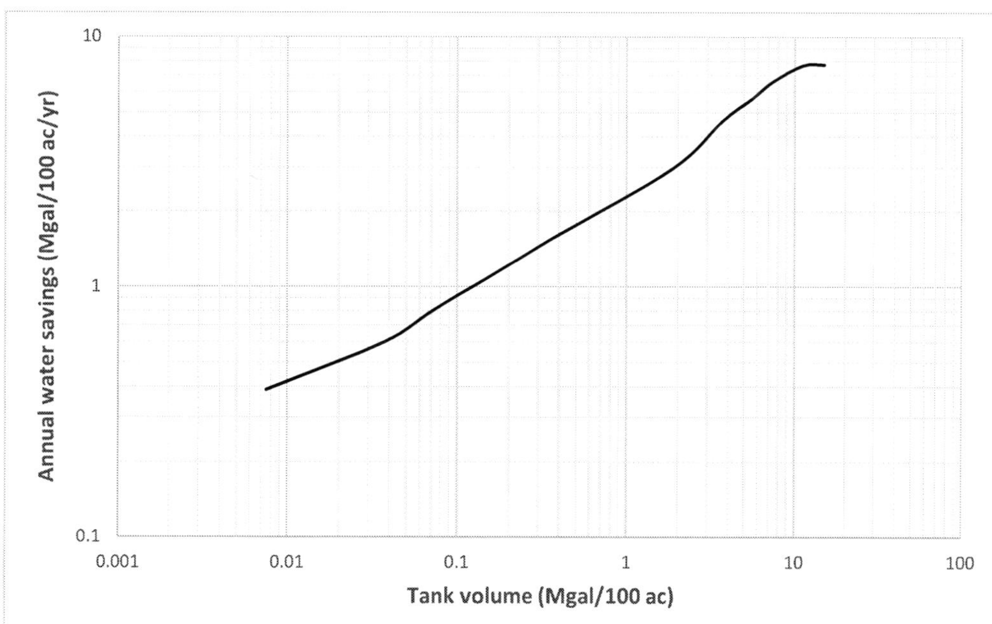

FIGURE 3-11 Example plot showing increasing water savings for a typical, medium-density, residential area in Los Angeles with increasing storage volumes. Two 35-gallon rain barrels per home (70 gal) correspond to 0.036 Mgal of storage per 100 acres, and a 2,200-gallon stormwater tank per home (8 ft diameter and 6 ft tall) corresponds to 1.14 Mgal of storage per 100 acres. To maximize rooftop runoff collection in Los Angeles, tank sizes would need to be nearly 10 times larger than the single 2,200-gallon tank.

ing) between 31 percent (Seattle) and 74 percent (Lincoln; see Table 3-7) of annual roof runoff in residential areas. Rain barrels result in significantly smaller maximum reductions in runoff, ranging from 14 percent (Los Angeles and Newark) to 18 percent (Lincoln); for irrigation only, rain barrels reduce runoff by 5 to 14 percent (Newark and Lincoln, respectively). Factors influencing the magnitude of runoff reduction include the timing of rainfall relative to irrigation and toilet flushing demand as well as the ratio of rooftop area to irrigated land in each area (which varies depending on local development patterns, see Table A-2 in Appendix A). Again, because of the nature of these analyses using multiyear modeling, some years will have greater benefits than these estimates, while other years will have smaller benefits.

Other Land Uses. Water savings from irrigation uses are typically much smaller in industrial and commercial areas compared to the above residential area examples because they have less irrigated landscape areas compared to roof areas. However, the large roof areas offer potential for other nonpotable uses, such as cooling water, toilet flushing, or industrial uses. In most regions of the country, high-density, residential and institutional areas result in the greatest water savings potentials from irrigation for similar storage tank sizes, although the water savings vary (greatest in the central and East Coast areas, and the least in the Southwest).

Water Savings Compared to Shallow Groundwater Infiltration. Rain gardens are popular on-site controls that have relatively low costs compared to other options and can significantly reduce the discharges of roof runoff from homes. The water infiltrating in a rain garden contributes to shallow groundwater recharge, which, depending on the local hydrogeology, may or may not eventually augment regional drinking water supplies. In analyses of rain gardens in Kansas City, Missouri, Pitt et al. (2014) reported that 10 Mgal/yr per 100 acres (38,000 m^3/yr per 40 ha) could be infiltrated with rain gardens in a community having about 600 homes per 100 acres and 39 inches of rain per year. Only a small fraction of the incoming water was lost through evapotranspiration in the rain garden (usually less than 10 percent) because of the large amount of water applied to relatively small areas. This total groundwater infiltration is about the same as the maximum potable water savings from the committee's scenario analysis for Lincoln, Nebraska (9.2 Mgal/yr per 100 acres [35,000 m^3/yr per 40 ha] for toilet flushing and irrigation uses; Table 3-6), although water supply savings would depend on whether groundwater supplies are under stress from excessive withdrawals and whether shallow groundwater infiltration projects ultimately recharge deeper aquifers used for water supply.

Stormwater Availability from Neighborhood-scale Projects

Neighborhood-scale stormwater capture for beneficial use commonly mixes stormwater flows originating from several areas located close together. The most common situation is collecting gutter flows in areas that are several acres to a few hundred acres in size. The captured stormwater is either stored in large subsurface tanks for nearby nonpotable use (e.g., irrigation, toilet flushing, washwater, aesthetic water features [see Chapter 2]) or used for aquifer recharge.

Overall, the potential water savings from neighborhood-scale stormwater capture is related to the available stormwater storage volume and the groundwater infiltration rate, source area, land development types, and correlation of water demand with rainfall. Several examples of neighborhood-scale projects are highlighted in Chapter 2. In the Sun Valley neighborhood of Los Angeles, a new neighborhood stormwater capture and infiltration project is expected to add 300 Mgal/yr (900 AF/yr or 1.1 million m^3/yr) to the groundwater supply based on drainage from a 929-acre area (3.8 km^2; Box 2-6). The new stormwater capture project for irrigation of the National Mall in Washington, DC, is anticipated to save 7.5 Mgal/yr (28,000 m^3/yr; Box 2-5). Because neighborhood-scale stormwater capture projects are centrally managed, they offer the opportunity to reduce conventional water use over a larger area without necessitating investment and maintenance by individual homeowners, simplifying implementation and increasing the reach of stormwater capture programs.

The stormwater flows for neighborhood-scale projects are greater compared to just capturing roof runoff. Table 3-4 summarizes average annual runoff quantities from 1995 to 1999 for the most common three land uses (i.e., commercial, industrial, and medium-density residential) for six locations in the United States. Flow from a given mixed land use area would need to be calculated based on the percentages of various land use types (see Appendix A). For a simplified example, Los Angeles in the Southwest has total runoff ranging from 210,000 to 320,000 gallons/yr/acre (2,000 to 3,000 m^3/yr/ha), depending on land use (Table 3-4). Even if sufficient storage is not available to capture some periods of very high flows, most of the annual runoff could be retained for beneficial uses at the neighborhood scale. These flows provide roughly four times the annual volumes needed to meet toilet flushing demand for a medium-density residential community[6] (48,000 gallons/yr/acre or 450 m^3/yr/ha), leaving substantial water available for other nonpotable uses. Yet, total

[6] Assuming a population density of about 12 persons/acre and average toilet flushing water use of about 11 gallons/person/day.

capture of stormwater in the Los Angeles area could supply roughly only one-half of the outdoor irrigation requirements for medium-density, residential areas having turfgrass (410,000 gallons/yr/acre or 3,800 m^3/yr/ha). For the Lincoln and Newark scenarios, the available flows shown in Table 3-4 provide more than enough water to meet medium-density, residential toilet flushing (48,000 gallons/yr/acre) and irrigation needs (150,000 gallons/yr/acre [1,400 m^3/yr/ha or 450 m^3/yr/ha] in Lincoln, Nebraska; 45,000 gallons/yr/acre [420 m^3/yr/ha] for Newark, New Jersey; see Appendix A). All of the stormwater used for these beneficial uses would directly decrease demand on the normal public water supply system. However, this would require substantial investments in infrastructure (e.g., storage, treatment, and a dual-water distribution system to deliver the water to the buildings, plus substantial building modifications to accommodate a dual-water system). The design of such systems is discussed in Chapter 6, and the costs and benefits are broadly discussed in Chapter 7.

Stormwater Availability from Regional-Scale Projects

Regional-scale systems for the beneficial uses of stormwater collect runoff from many different land uses in relatively large areas. These may incorporate the complete community, ranging from several to many square miles in area. Given the large scale of these projects, the typical applications of the captured stormwater are aquifer recharge through large infiltration basins designed to recharge water supply aquifers or surface impoundments used to augment the conventional water supply. The amount of stormwater potentially available from regional-scale projects has been estimated for the entire Los Angeles basin as part of a stormwater conservation study (RMC, 2014). This study is intended to provide an understanding of the potential benefit of additional stormwater capture systems that could be implemented across the basin by the Los Angeles County Flood Control District and its partners. Considering centralized stormwater capture—such as spreading basins for managed replenishment/recharge of local groundwater basin in an engineered facility—the estimate for the next 20 years is 239,000 AF/yr (295 million m^3/yr) increasing potentially to 494,000 AF/yr (609 million m^3/yr) by 2095.

Calculations of stormwater availability at a regional scale could be similar to those in the prior section on neighborhood-scale. For the Los Angeles area, average annual runoff ranges from 210,000 to 320,000 gallons/yr/acre (2,000 to 3,000 m^3/yr/ha), based on land use and data from 1995-1999 (Table 3-4). Assuming a population density of about 12 persons/acre and a total indoor water use of about 46 gallons/person/day, the total water demand would be about 200,000 gallons/acre. Therefore, if entirely captured and stored for later use through a regional stormwater infiltration system, based on 1995-1999 data, then Los Angeles stormwater could supply all of the total annual domestic indoor water demand, although outdoor water requirements would not be met. Under wetter East Coast and Southeast conditions, stormwater at a regional scale could supply at least twice the water demand for indoor use, providing sufficient additional water for irrigation and other beneficial uses.

The runoff volumes shown in Table 3-4 are modeled runoff averages based on five years of data, and annual runoff would be expected to vary widely. In addition, extensive infrastructure would likely be needed to collect, store, treat, and deliver this water to the points of use (see Chapter 6). Not all of the water that is used for groundwater recharge is withdrawn, and non-recovered groundwater and seepage and evaporation losses would also need to be considered in the water supply evaluations using more complex regional groundwater modeling. Water rights laws may also restrict regional scale water capture unless water utilities can secure the water rights to stormwater that is recharged into a water supply aquifer (see Chapter 8).

Downstream Effects of Stormwater Use in Existing Developments

On-site beneficial uses of stormwater can raise water rights concerns for downstream users. Therefore, it is important that, when water rights present an issue (see Chapter 8), the impact of potential beneficial uses of stormwater on various components of the local water budget (Figure 3-3-1) and its effects on downstream water availability be understood in existing developments. For new developments, there will be increased runoff from the development compared to the undeveloped landscape that further complicate any evaluation of impacts to downstream water availability and would necessitate more detailed water budget analysis to determine the overall effects of on-site stormwater capture. When stormwater is used for toilet flushing, potable water use is reduced by the volume used for flushing, as is the amount of stormwater runoff from the property. Therefore, there is no net effect on regional hydrology in terms of downstream water availability because reduction in stormwater runoff is balanced by reduction in potable water demand (see Table 3-9), assuming that the water supply is from the same watershed. If potable water supply is provided by imported water, then widespread stormwater use for nonconsumptive applications could reduce the need for water imports. In such a case, downstream water availability in the basin with stormwater use could be reduced, although the basin from which the water is imported would see increased water flows.

TABLE 3-9 Summary of Water Budget Effects from the Beneficial Use of Stormwater at Existing Developments

	Water Quantity Effects				Water Quality Effects
	Potable water use	Wastewater volume generated	Off-site stormwater runoff	Water available to downstream water users	Stormwater contaminant loading
Stormwater use for toilet flushing	Reduced	No change	Reduced	No change	Reduced
Stormwater use for irrigation of existing landscape[a]	Reduced	No change	Reduced	No change	Reduced
Stormwater use for irrigation of expanded landscape	May be slightly reduced, unchanged, or increased	No change	Reduced	Reduced	Reduced
Stormwater irrigation of existing landscape at increased rates	Reduced	No change	Reduced	Dependent on local hydrogeologic conditions	Reduced

[a]Assumes irrigation rates throughout the year are the same before and after installation of the stormwater capture system.

If stormwater is captured for landscape irrigation in existing development, then potable water use is reduced by the volume of stormwater used for irrigation, assuming the amount of water applied to landscaping is not increased after installation of the stormwater capture system. Use of stormwater for irrigation results in the same water losses to the hydrologic cycle from consumptive use because of evapotranspiration and recharge to groundwater, as would have been the case for potable water applied to existing landscaping that the stormwater replaced. The amount of stormwater runoff from the property is reduced by the amount used for irrigation, which can result in decreased flows to local receiving water bodies or to a regional stormwater collection system by the amount used for irrigation. However, there is no net effect on regional hydrology in terms of water availability within a watershed because reduction in stormwater runoff is balanced by reduction in potable water demand, again if the domestic water is not imported.

If the use of stormwater for irrigation results in a greater demand for irrigation water use on a property (such as from additional plantings), then this could result in little or no reduction in potable water use (i.e., potable water continues to be used for irrigation but is supplemented by stormwater). With an expanded area of irrigated landscape, there would be increased evapotranspiration losses compared to the base case, resulting in reduced flows to downstream users. If the area of irrigated landscape is unchanged but the irrigation rate increases, then the downstream impacts are less clear. There could be increased evapotranspiration losses but groundwater recharge would likely also increase, and these gains and losses to downstream users would need to be compared to the base conditions of evapotranspiration and groundwater recharge in the environment.

CONCLUSIONS

The following conclusions are based on the committee's scenario analyses of graywater and stormwater uses for conservation irrigation of turfgrass, toilet flushing, or both, as well as other assessments of potable water savings in the literature. The committee's analyses primarily focused on on-site capture and use of graywater or stormwater at the medium-density, residential scale, using 1994-1999 precipitation data. However, the potential for neighborhood and regional stormwater capture was also considered. These analyses were not intended as a definitive assessment of potential potable water savings and should not be considered as such, given the assumptions of the analysis and the inherent uncertainties (see Box 3-2). However, broad lessons can be learned from a comparative analysis of the results.

Water savings associated with beneficial use of stormwater are dependent on tank size and the amount and timing of precipitation relative to water demand. Substantial, potential, household-scale water savings (24 to 28 percent) from the capture and use of roof runoff were calculated for scenario analyses in four of the six cities analyzed using one moderately sized (2,200-gallon) storage tank per house. These cities—Lincoln, Nebraska; Madison, Wisconsin; Birmingham, Alabama; and Newark, New Jersey (all located in the Midwest or East Coast)—have year-round rainfall closely matching irrigation demands. In contrast, the scenario analysis showed lower potential potable water savings for Los Angeles and Seattle (5 and 15 percent, respectively). In much of the arid West, the timing and intensity of rainfall limits the capacity of stormwater collection to reduce potable water use at the household scale. Very small stormwater water storage volumes provide much lower water savings benefits (less than 2 percent in Los Angeles to up to 10 percent in Newark using two 35-gallon rain barrels per house, for example).

Neighborhood- and regional-scale stormwater capture projects can contribute significantly to urban water supplies. This is especially important for arid climates in which stormwater can be stored in aquifers for use during drought or the dry season. The committee's scenario analyses showed that average 1995-1999 stormwater runoff for medium-density, residential developments in Los Angeles would be roughly sufficient to meet indoor, residential, water needs in those areas. However, extensive infrastructure for large-scale collection, treatment, and storage or infiltration would likely be needed.

Graywater reuse offers the potential for substantial potable water savings and could provide a reliable source of water for arid regions. Based on the committee's scenario analyses, graywater reuse provides greater potable water savings than does stormwater capture (based on a 2,200-gallon tank per house) in Los Angeles and Seattle, because graywater provides a steady water source during summer months with little or no rainfall. Additionally, the analyses showed that graywater can more effectively meet toilet flushing demand compared to stormwater in all cities analyzed. Graywater use for toilet flushing has been demonstrated to achieve potable water savings as theoretically expected, but water savings associated with graywater irrigation at the household scale have not been demonstrated with confidence. Little is known about the impact of installing on-site nonpotable water systems on human water use behavior, which points to the need to study behavioral responses to conservation measures.

Beneficial use of graywater is more appropriate for residential and multi-residential applications than for commercial application. Most commercial facilities do not generate enough graywater to justify use for toilet flushing or irrigation. Even offices that have showers on site are not likely to generate enough graywater to meet end use demands (toilet or irrigation). Some commercial applications for which graywater use may be appropriate include fitness facilities, hotels, and laundromats.

If water conservation is the primary driver for stormwater and graywater investments, then strategies that reduce outdoor water use should first be examined. In arid regions, potential potable water savings for residential and multi-residential use of stormwater and/or graywater are significant, but small relative to today's outdoor water demand. Although use of graywater or roof runoff for toilet flushing can reduce indoor demand by up to 24 percent, the potential annual reduction in domestic water demand for the Los Angeles area estimated in the committee's scenario analyses averaged only 13 percent for graywater reuse (and significantly less for stormwater capture, even using large tanks). These limited reductions in total demand result from the very high irrigation demand in arid regions. Irrigation demand accounted for 67 percent of the total demand in Los Angeles in the hypothetical scenarios examined here. Thus, the largest reductions in water demand in arid regions would be provided by approaches to reduce or eliminate irrigation demand, such as lawn-removal rebate programs and the use of xeriscaping and other types of climate-appropriate, low-water-use landscapes. In these circumstances, graywater could be used to supply irrigation water to meet specific small irrigation needs. Otherwise, graywater and stormwater may help facilitate the continued use of landscaping that is not sustainable in the long term and inappropriate for local climate conditions.

On-site use of nonpotable water for toilet flushing is a proven way to achieve potable water savings that does not impact the water availability to downstream users. Reductions in flows to sanitary sewers and stormwater runoff are balanced by reductions in potable water demand, assuming these flows occur in the same watershed. Water savings and effects on regional water availability from irrigation with graywater or stormwater in existing developments depend on whether irrigation rates or the area of irrigated landscape increases compared to the base case. If irrigation rates and landscaped areas are not increased, then potable water savings would occur that do not impact water availability to downstream users. If the irrigated landscape is expanded (or the supplemental irrigation water rate increased), then potable savings may be reduced and additional consumptive losses would occur from evapotranspiration, potentially impacting downstream users.

4

Quality of Graywater and Stormwater

Stormwater and graywater can contain a wide variety of contaminants, including inorganic (e.g., metals, nutrients, and salts) and organic (e.g., industrial chemicals, pesticides, household chemicals) chemicals and microorganisms. The most common applications of graywater and stormwater involve nonpotable uses, such as irrigation, washing, and toilet flushing, but these uses can be associated with human contact and inadvertent ingestion exposures. Groundwater recharge of stormwater can also impact potable uses, and environmental impacts on plants and soils in irrigated areas are also possible (Australian SCEW, 2009; NRMMC et al., 2009a). Therefore, when considering the potential applications of graywater and stormwater to conserve conventional water supplies, it is important to understand water qualities, how they vary under a range of conditions, implications on various beneficial uses, and strategies to reduce the concentrations of harmful contaminants by source control and treatment.

This chapter describes what is known about the quality of graywater and stormwater sources, and issues that water quality may present to certain end uses. Treatment options are available for uses that necessitate improved or more consistent water quality (see Chapter 6). This chapter also discusses source area controls to manage water quality. Chapter 5 presents the human health and environmental risks associated with graywater and stormwater uses, which then can inform decisions regarding additional treatment needed (see Chapter 6) and affect project costs (see Chapter 7).

GRAYWATER QUALITY

Graywater may contain elevated levels of chemicals and disease-causing microorganisms (pathogens), but the quality of graywater can vary greatly from location to location based on the contributing sources (e.g., laundry, showers, baths), the amounts and types of chemicals used or disposed there (e.g., detergents, bleach, solvents, cleansers, personal care products), and the health of the residents in the source area. At smaller scales (i.e., households), these factors can result in widely variable contaminant concentrations in graywater, although larger-scale projects (i.e., large multi-residential developments) would likely have more consistent water quality, because variations in contaminant loads are averaged over many more contributing households. Table 4-1 presents general ranges of common physical, chemical, and microbial water quality constituents for graywater, considering a range of possible sources. Graywater quality varies substantially, and the ranges of water quality measurements provided in Table 4-1 are intended to provide a general idea of concentrations of each constituent that can be expected in graywater. The data are from Eriksson et al. (2002), which is the most comprehensive summary of graywater quality, and other sources where the committee deemed the number of persons contributing to the system and number of samples collected appropriate. Table 4-1 also includes water quality data from kitchen sources, which show high levels of solids, organic matter, and indicator organisms, demonstrating why kitchen water sources are typically excluded from graywater collection systems.

Many different types of microorganisms capable of causing human illness may be present in human fecal material. Depending on the specific sources contributing to the graywater, these microorganisms may be present. For example, laundry wash and rinse water, as well as shower and bathtub water may contain fecal material and therefore create the potential for pathogenic (disease-causing) microorganisms to be present in graywater. These microorganisms can be generally grouped into the following types: viruses, bacteria, protozoa, and helminths. The most common health concern caused by waterborne pathogens associated with direct exposure to graywater is gastroenteritis, although many other illnesses, including hepatitis, encephalitis, and myocarditis, may result from exposure to enteric pathogens. Fecal indicator bacteria (such as total coliform bacteria and fecal coliform bacteria) are often used as surrogates for the presence of pathogenic microorganisms, and their concentrations are therefore often used to set treatment requirements for different end uses. However, the detection of indicator bacteria does not necessarily mean that pathogenic organisms are also present. Because pathogens are only excreted by infected individuals (Ashbolt et al., 2001), the greater the number of people contributing to graywater, the greater the likelihood of the presence of a range of pathogens. However, even in waste streams to which a small number of people contribute, when an infected individual is excreting pathogens, the concentration can be very high because of the relative lack of dilution. Perhaps more importantly, the absence of indicator microorganisms does not necessarily mean an absence of pathogenic microorganisms, because many pathogens are more persistent than the indicator microorganisms. Microorganisms capable of causing skin infections, such as *Staphylococcus aureus* and *Pseudomonas aeruginosa*, may also be present in graywater (Casanova et al., 2001). Table 4-2 summarizes reported concentrations of microbial indicators and pathogens in graywater.

The amount of organic matter is measured as total organic carbon (TOC), and biochemical oxygen demand (BOD_5) is a measure of its degradability. These parameters can indicate the risk of oxygen depletion due to microbial degradation of organic matter during transport and storage of graywater, potentially resulting in hypoxia and sulfide production. High levels of biodegradable organic matter in graywater limit the potential for graywater to be stored or used for toilet flushing or in ornamental fountains without treatment, because of the likelihood of microbial growth.

Other pollutants present in graywater include xenobiotic organic chemicals, consisting of personal care products and household chemicals, although only a limited number of studies have comprehensively investigated the occurrence of these chemicals in graywater sources (Donner et al., 2010; Eriksson et al., 2002). Although trace organic chemicals, such as pharmaceuticals, household chemicals, and endocrine-disrupting chemicals, have been reported to occur in domestic raw wastewater at the household or neighborhood scale (Conn et al., 2010; Teerlink et al., 2012), the composition of graywater differs from wastewater sewage and usually exhibits lower concentrations of pharmaceutical residues and high-

TABLE 4-1 Chemical and Microbial Quality of Untreated Graywater from Individual and Combined Sources

Parameter	Bathroom	Laundry	Kitchen Sink and Dishwasher	Graywater Combined (excludes kitchen water)
Physical				
Temperature (°C)	29	28-32	27-38	
Turbidity	28-240	14-210		15-140
Total suspended solids (TSS), mg/L	54-200	120-280	240-2,400	
Total dissolved solids (TDS), mg/L	140-1,300			310-930
Electrical conductivity (μS/cm)	82-250	190-1,400		
Chemical				
pH	6.4 – 8.1	8.1-10	6.3-7.4	6.7-7.6
Alkalinity	24-67	83-200	20-340	150-200
BOD_5 (mg/L)	26-300	48-380	1,000-1,500	125-250
COD (mg/L)	100-630	13-720	3.8-1,400	250-430
Total organic carbon (mg/L)	30-100	100-280	600-880	
Sodium absorption ratio				2.3 - 6
Boron (mg/L)				0.1-1.6
Chloride (mg/L)	9.0-19	9.0-90		22-34
TN (mg/L)	5-17	6-21	0.3-74	0.6-5.2
TP (mg/L)	0.1-4	0.1->100	68-74	
PO_4 (mg/L)	0.94-49	4-170	13-32	4-35
NH_4 (mg/L)	<0.1-15	0.04-11	0.005-6	0.15-3.2
NO_3 (mg/L)	0.28-6.3	0.4-2	0.3-5.8	0-4.9
Anionic surfactants (mg/L)	21	92	6	
Microbial				
Total coliform/100 mL	$10^{2.7}$-$10^{7.4}$	$10^{1.9}$-$10^{5.2}$	10^{7}-10^{9}	$10^{7.2}$-$10^{8.8}$
Pseudomonas aeruginosa/100 ml				1.99 x 10^{4}
E. coli/100 mL	$10^{1.6}$-$10^{3.4}$	$10^{1.5}$-$10^{3.9}$	$10^{5.4}$-10^{9}	
Cryptosporidium spp.	no detection	no detection		

NOTE: Graywater as defined in this report does not include kitchen water.
SOURCES: Birks and Hills (2007); Casanova et al. (2001); Christova-Boal et al. (1996); Donner et al. (2010); Eriksson et al. (2002); Gross et al. (2007); Mehlhart (2005); Nolde (1999); Ottoson and Stenstrom (2003); Rose et al. (1991); Sharvelle et al. (2013); Sheikh (2010); Weingaertner (2013).

er concentrations of personal care products and antimicrobial chemicals found in hand soap (Etchepare and van der Hoek, 2015; Table 4-3). These chemicals pose possible concerns for irrigation uses (see Chapter 5).

Major ions such as sodium, chloride, and boron can also adversely affect vegetation if present at elevated concentrations. Sodium has been reported to be elevated in soil irrigated with graywater compared to potable water sources (Negahban-Azar et al., 2012). Boron in graywater can be derived from some laundry detergents and cleaning agents.

Source Control of Graywater Quality

Some practices can minimize graywater quality issues. To reduce adverse effects on irrigated plants and soils, liquid rather than powdered detergents should be used to prevent high sodium loads, and boron-containing detergents and cleaning agents should be avoided. In addition, use of products containing antimicrobial compounds is not recommended when graywater is to be applied for irrigation. Materials containing large amounts of organic matter that would exert a high oxygen demand or interfere with the disinfection process and toxic ingredients (e.g., paints, solvents) should not be poured down the drain into a graywater collection system.

The potential presence of human pathogens is a concern for irrigation, as well as for toilet flushing (see Chapter 5). The risks associated with these pathogens can be reduced by implementing such measures as not washing feces-soiled clothing or diapers in laundry machines that drain to the graywater system or diverting laundry water that is used for this purpose to the sewer.

STORMWATER QUALITY

The quality of stormwater is highly variable over time and space. Stormwater can be derived from a wide variety of

TABLE 4-2 Pathogenic and Indicator Microorganisms in Untreated Graywater

Microorganism	Range Reported	Positive Samples (%)	Mean	Standard Deviation
Pathogens				
E. coli O157:H7 (per L)	ND	0		
Salmonella (MPN/L)	detected	13		
Legionella pneumophila	ND	0		
Legionella non-*pneumophila*	ND	0		
Campylobacter (per L)	ND	0		
Giardia (cysts/L)	0.5-1.5	63		
Cryptosporidium (oocysts/L)	ND	0		
Enterovirus (per 10 L)	ND	0		
Indicator organisms				
Total coliforms/100 ml			2.2×10^7	9.0×10^7
E. coli/100 ml			3.9×10^5	2.4×10^6
Fecal enterococci/100 ml			2.5×10^3	4.8×10^3

NOTE: Eight pathogen samples were taken over 3 months from a graywater collection tank that received water from baths, showers, and sinks from 18 units of an apartment building that primarily housed married students.
SOURCE: Birks and Hills (2007).

TABLE 4-3 Maximum Concentrations of Trace Organic Chemicals Reported in European Graywater and Municipal Wastewater Effluents

Chemical	Class	Graywater (µg/L)	WWTP Effluent (µg/L)
Salicylic acid	Pharmaceutical	1.5	777
Caffeine	Stimulant	0.5	43.5
Benzophenone	Personal care product	4.9	0.23
Galaxolide	Personal care product	19.1	2.77
Tonalide	Personal care product	5.8	0.32
Triclosan	Antimicrobial	35.7	6.88
4-Nonylphenol	Surfactant	38	7.8
4-Octylphenol	Surfactant	0.16	1.3
Bisphenol A	Plasticizer	1.2	4.09
Diethyl phthalate	Plasticizer	38	2.58

SOURCE: Etchepare and van der Hoek (2015).

source areas and land uses, ranging from rooftops and open spaces to industrial areas and high-traffic roadways. The concentrations of contaminants in stormwater will also vary depending on the building materials in the catchment area, the size of the drainage area, the intensity of the storm event, and environmental and seasonal factors. Small rain events (i.e., less than 0.5 inches [1.2 cm]) generally include most of precipitation events by number, but they produce a small percentage of annual runoff volumes. The largest rains (i.e., greater than 2 inches [5 cm]) also supply a relatively small percentage of total annual flows and pollutant discharges, although heavy rains can mobilize high concentrations of solids and sediment-associated pollutants. Most of the total annual stormwater flows and pollutant discharges (frequently more than 75 percent by mass) occur from intermediate rainfall events (i.e., 0.5 to 2 inches).

Compared to graywater, an even wider array of contaminants can be found in stormwater because of the diversity of source areas. Primary contaminants of concern for beneficial uses include metals, organic chemicals (including herbicides, industrial chemicals, and petroleum-derived chemicals), pathogens, salts, nutrients, and suspended solids. Table 4-4 provides an overall summary of these contaminants, measured at stormwater outfalls at the neighborhood or regional scale, during more than 9,400 storm events. These data reflect a variety of land uses, with about 46 percent from residential areas, 19 percent from commercial areas, 17 percent from industrial areas, 8 percent from major transportation areas, 6 percent from open space areas, and 2 percent from institutional areas. For comparison, these data are presented next to U.S. Environmental Protection Agency (EPA) drinking water maximum contaminants levels and water reuse guidance for irrigation. Individual contaminant classes are discussed further in the sections that follow.

Nutrients and Organic Matter

When discharged into surface waters nutrients and organic matter can cause algal blooms and low oxygen conditions, thereby harming aquatic life. Sources of nitrogen and phosphorus to stormwater include atmospheric nitrogen deposition, fertilizer runoff, animal feces, and combined sewer overflows. In stormwater beneficial use scenarios, excess nutrients can foster algal growths in stormwater storage facilities or surface water features, such as ponds or fountains. Most storage tanks are designed to be opaque to restrict sun penetration and associated algal growths. Although nitrate poses human health concerns at high concentrations in drinking water, most stormwater concentrations are well below the 10 mg/L maximum contaminant level (see Table 4-4).

Excessive biodegradable organic matter can contribute to severe odor problems in storage systems and cause nuisances when the water is used, but organic matter in stormwater samples tends to be low. Less than 10 percent of samples in the National Stormwater Quality Database exceeded the recommended water reuse criteria for irrigation for BOD_5, reflecting elevated levels of biodegradable organic matter.

Suspended Sediment

Suspended sediment (total suspended solids [TSS]) conveys particle-associated contaminants (e.g., phosphorus, metals, some organic contaminants and pathogens) (Characklis et al., 2005; Jartun et al., 2008; Murakami et al., 2005). In stormwater beneficial use scenarios, suspended solids can clog irrigation systems and can result in reduced water clarity, causing aesthetic concerns when the water is used in toilet flushing or ornamental water features. Greater than one-than half of the National Stormwater Quality Database observations exceed the recommended guidance for TSS, and the mean detected level is approximately 5 times greater than the recommended 30 mg/L level for irrigation use (Table 4-4).

Salt

Salts represent a concern for irrigation uses and groundwater infiltration. Plants have different salt tolerances, and high chloride concentrations can severely damage some plants. Excessive sodium concentrations (especially in relation to calcium and magnesium) can cause an elevated sodium adsorption ratio (SAR). High SARs dramatically inhibit water infiltration in soils when the sodium interacts with even small amounts of clay.

In northern areas, de-icing chemicals are major sources of salt to stormwater and could pose a risk to groundwater quality in stormwater infiltration projects. In a U.S. Geological Survey (USGS) occurrence study of sodium and chloride in groundwater in 19 northern U.S. states, Mullaney et al. (2009) detected chloride contamination (above the EPA secondary criteria of 250 mg/L) in 1.7 percent of drinking water wells and exceedance of the sodium advisory level (20 mg/L) in nearly 47 percent of public-supply wells and 34 percent of domestic wells. Mullaney et al. (2009) determined that de-icing salts were the predominant source. Enhanced recharge of stormwater without attention to salt concentrations could exacerbate this problem.

Pathogens

Pathogenic microorganisms in stormwater are typically derived from animal wastes (e.g., *Salmonella* and *Campylo-*

TABLE 4-4 Stormwater Quality Data at Neighborhood/Regional Outfalls from the National Stormwater Quality Database, Version 4 (March 17, 2105 version)

			All Locations Combined (9,052 total events)					
Constituent	EPA Drinking Water MCL[a]	Irrigation Use Guidance[b]	Average of detected values[c]	5th percentile of all values[c]	50th percentile of all values (median)[c]	95th percentile of all values[c]	# of observations	% detected
pH	6.5 to 8.5[c]	6 to 9	7.3	6.1	7.3	8.6	3,179	100
Total dissolved solids (TDS) (mg/L)	500[c]		140	25	80	370	4,120	99
Chloride (mg/L)	250[c]		26	1	6.2	92	869	84
Total suspended solids (TSS) (mg/L)		30	140	7	63	510	7,637	99
Turbidity (NTU)			39	4	19	120	936	100
BOD_5 (mg/L)		30	14	2	8.3	42	5,152	95
COD (mg/L)			79	6.3	51	240	5,214	96
TOC (mg/L)			16	3	8.6	52	678	100
Ammonia (mg/L as N)			0.77	<0.1	0.28	2.1	2,946	72
Nitrate (mg/L as N)	10		0.97	<0.1	0.58	2.7	1,028	92
Nitrite (mg/L as N)	1		0.17	<0.1	<0.1	0.38	714	64
Total phosphorus (mg/L as P)			0.4	0.05	0.24	1.1	7,943	97
Microorganisms								
Fecal coliforms (MPN/100 mL)		<200[d]	60,000	65	4,600	200,000	2,168	92[e]
Fecal streptococci (MPN/100 mL)			73,000	500	19,000	300,000	1,317	94[e]
Total coliforms (MPN/100 mL)	5.0% positive		260,000	300	24,000	1,600,000	282	77[e]
E. coli (MPN/100 mL)			5,900	23	1,200	28,000	139	98[e]
Metals								
Arsenic, total (µg/L)	10	100	5.9	<5	1	8	2,367	34
Barium, total (µg/L)	2,000		55	2	21	110	582	66
Cadmium, total (µg/L)	5	10	3.5	<1	0.35	5	4,002	40
Chromium, total (µg/L)	100	100	12	0.5	4	25	2,266	57
Copper, total (µg/L)	1,300[f]	200	33	0.5	13	94	5,836	89
Iron, total (µg/L)	300[c]	5,000	2,700	17	470	6,500	608	86
Lead, total, since 1984 (µg/L)	15[f]	5,000	34	<5	8	100	4,960	74
Nickel, total (µg/L)		200	14	<5	4	30	2,090	51
Zinc, total (µg/L)	5,000[c]	2,000	200	5.9	91	560	6,563	96
Organic Contaminants								
Oil and grease, total (mg/L)			21	0.2	3	37	2,256	68
Total petroleum hydrocarbons(mg/L)			3.9	0.3	1.8	9.6	295	65
2-Chloroethylvinlether (µg/L)			3.4	0.2	2.4	5	624	58
Dichlorobromoethane (µg/L)			0.85	<1	0.55	1.6	116	36
1,2-Dichloroethane (µg/L)	5		1.5	0.05	0.15	1.8	247	21
Methlenechloride (µg/L)			12	<1	<1	14	457	20

[a]EPA maximum contaminant levels (MCLs) for drinking water (see http://water.epa.gov/drink/standardsriskmanagement.cfm).
[b]EPA 2012 Water Reuse Guidelines (EPA, 2012a).
[c]Indicates secondary MCL.
[d]Fecal coliform/100 mL (not MPN).
[e]Most bacteria values that are not quantified exceeded the upper limit of the analytical method (over-range).
[f]Action level.
NOTE: Most of these data were obtained from the municipal stormwater permit program (MS4), with additional data from the National Urban Runoff Program (EPA, 1983) and various research projects. Although the database contains sampling sites in all nine rain zones in the United States, most of the sampling data has been collected from the upper Midwest and Northeast, mid-Atlantic, Southeast, Southwest, and Northwest areas. Because these data include multiple data sets, the concentrations reported may reflect different sampling points and times for different constituents. The median and percentile values are calculated considering both the detected values and nondetected or over-range values, with no data substitutions, while the average values are for only the detected values. See Maestre (2005) for detailed discussions of effects of the multiple data sets on the overall statistics. Highlighted values exceed either the drinking water MCL or the irrigation guidance values.
SOURCES: Maestre et al. (2015); National Stormwater Quality Database, version 4 (March 17, 2015, updates; see http://www.bmpdatabase.org/nsqd.html).

bacter from birds), although leaking sewer systems or poorly functioning septic tanks can also introduce human waste into stormwater. As with graywater, given the number of possible organisms and the difficulty quantifying their occurrence, indicator bacteria (e.g., total coliform bacteria, fecal coliform bacteria, fecal streptococcus, *E. coli*) are often monitored instead of pathogens. However, as discussed previously, indicator bacteria may be a poor analog for human pathogens, particularly if the organisms are not derived from wastewater, as is often the case for stormwater (Clary et al., 2014). Data presented in Table 4-4 illustrate that indicator bacteria were detected in high numbers from stormwater outfalls at the neighborhood scale or larger. Mean fecal coliform bacteria concentrations exceeded the guidance value for irrigation water by more than 300-fold, while the 95th percentile fecal coliform value exceeded this guidance value by 1,000-fold.

Limited data are available on the occurrence of actual human pathogens in stormwater (Bambic et al., 2011; Page and Levett, 2010; Page et al., 2013; Vanderzalm et al., 2014), and the data can be highly variable. In a review of the literature, O'Shea and Field (1992a) cited studies that reported that some of the disease-causing microorganisms isolated from urban stormwater and streams include enteroviruses (e.g., poliovirus, coxsackieviruses, and echovirus) and bacteria (e.g., *Pseudomonas aeruginosa*, *Staphylococcus aureus*, *Campylobacter*, and *Salmonella*) (Olivieri et al., 1977; Pitt and McLean, 1986; Qureshi and Dutka, 1979). In three studies, some with limited sampling, *Salmonella* was the most commonly detected pathogenic bacterium in urban stormwater, with 3-80 percent positive detections reported, while the other bacterial pathogens were detected in less than 10 percent of the samples (Kinde et al., 1997; O'Shea and Field, 1992b; Schroeder et al., 2002). Two studies reported 0-10 percent of stormwater samples with detectable *Cryptosporidium*, while none had detectable concentrations of *Giardia* (Schroeder et al., 2002; Wohlsen et al., 2006). Page et al. (2013) reported protozoa detections in 50 percent of all stormwater samples for a catchment area in Australia. Bambic et al. (2011) reported that viruses are rarely detected in municipal stormwater. Adenovirus, rotavirus, enterovirus, human polyomavirus, and hepatitis A were monitored and detected in 0-5 percent of stormwater samples (Brownell et al., 2007; Grohmann et al., 1993; O'Shea and Field, 1992b; Rajal et al., 2007; Schroeder et al., 2002), although three studies reported positive adenovirus occurrences in 10-59 percent of municipal stormwater samples (CREST, 2009; Jiang et al., 2007; Page et al., 2013).

Ahmed and Toze (2015) reviewed the literature on the microbiological quality of roof runoff. Most of the studies focused on indicator organisms, and many of the results are reported as presence or absence of the organisms, rather than pathogen concentrations (Ahmed et al., 2014; NRMMC et al., 2009a). The most commonly tested bacterial pathogen appears to be *Campylobacter*; with reported occurrence ranging from 0 to 125 samples (Simmons et al., 2001) and 45 percent of 27 samples tested (Ahmed et al., 2008). *Salmonella* have also been detected, generally at lower frequencies than *Campylobacter* (Ahmed and Toze, 2015). Four of six samples tested for *Shigella* and *Vibrio* were positive in a study conducted by Uba and Aghogho (2000). Roof runoff has also been tested for the protozoan pathogens, *Giardia* and *Cryptosporidium*, with the highest frequency of detection being reported by Crabtree et al. (1996): 23 and 45 percent occurrence, respectively, in 45 samples. No reports of virus detection in roof-captured rainwater were found. The results of studies that reported concentrations are summarized in Table 4-5. Most of the pathogen studies used molecular methods to detect the microorganism, so it is not possible to directly infer health risk from the presence of the pathogens.

Many of the studies of the presence of pathogenic microorganisms in stormwater have focused on the same organisms. Because it is not possible to monitor water for the hundreds of potential pathogens that it may contain, researchers typically choose a few that they believe are representative of the other pathogens. There are different reasons for the choices of the different organisms. For example, enteroviruses are often chosen to represent enteric viruses, because the methodology for monitoring these viruses is standardized and well established. Rotaviruses may be chosen to represent a "worst-case" scenario for viruses, because they have an extremely low infectious dose (i.e., very few organisms are required to cause infection, so the presence of just a few rotaviruses may signify a potential public health risk). *Cryptosporidium* is often a target of interest because of the very large (and widely publicized) outbreaks of drinking-water-borne disease that it has caused in the United States, Europe, and Australia. *Salmonella* is often chosen because it is one of the most commonly detected bacteria in wastewater.

The limited data available exhibit wide ranges of occurrences and concentrations. This may be due to the size of the sample analyzed (e.g., only 5 or 10 percent of a sample may be analyzed because of methodological constraints), the temporal variability of infection in the population, and/or limitations in analytical methods. Statistical comparisons and descriptive characteristics of these data would therefore require a much larger number of samples.

Metals

Metals in stormwater are most commonly detected when source areas include industrial storage areas, highways, streets, and parking areas. Additionally, roofing drainage systems, conveyance systems, and water storage tanks are often made of metallic materials or components, including aluminum, lead, zinc, and copper, which can affect water quality in stormwater capture systems (see Box 4-1).

Chronic aquatic life criteria have been established by the EPA to protect aquatic life in receiving waters. Copper, lead, cadmium, and zinc are a concern for projects with surface reservoirs or wetland features, because their mean concentrations exceed criteria for aquatic organisms (see Table 4-6; Davis et al., 2001). Some plants may be sensitive to some metals, but irrigation guidelines mostly focus on metal uptake in plants that may be consumed, such as in household gardens. Although most metal concentrations detected in stormwater are below published irrigation use guidelines, elevated levels of iron could also pose concerns for irrigation use (see Table 4-4).

Because of frequent exceedances of the EPA maximum contaminant level (MCL) (Table 4-4), lead, iron, and cadmium pose human health concerns if stormwater is consumed in large quantities (Sabin et al., 2005). Median concentrations exceeded the recommended drinking water standards for iron. Since the removal of lead from gasoline, about 35 percent of recent lead observations exceeded the drinking water MCL (National Stormwater Quality Database, version 4). The 95th percentile values in stormwater outfalls are close to the drinking water standards for arsenic and cadmium.

Organic Chemicals

A wide array of organic contaminants can also be detected in stormwater, including pesticides, industrial chemicals and solvents, and petroleum-derived chemicals. Compared to total suspended solids, nutrients, and metals, however, relatively little information is available on concentrations of organic chemicals in urban stormwater (Grebel et al., 2013). Polynuclear aromatic hydrocarbons (PAHs) are frequently detected because of releases from automobile exhaust and paving materials (see also Table 4-7). A number of chemicals used in industrial manufacturing and consumer products have been detected in stormwater, such as tire additive chemicals (e.g., benzotriazoles), plastic additives (e.g., bisphenol A, phtalates), and flame-retardants (i.e., perfluorochemicals and organophosphates) (Stachel et al., 2010). A recent study in Arizona focusing on perfluorinated chemicals concluded that secondary wastewater effluents and stormwater runoff from downtown areas exhibited similar perfluorooctane sulfonate (PFOS) concentrations ranging from 10 to 1,000 ng/L (Quanrud et al., 2010).

Also present in stormwater are herbicides (e.g., diuron, glyphosphate, 2,4-Dichlorophenoxyacetic acid [2,4-D]) and pesticides (e.g., pyrethroids, fipronil) used in residential and commercial properties and along transportation corridors (Blanchoud et al., 2004; Gan et al., 2012; Gilliom et al., 2007; Weston et al., 2009). A recent study of stormwater herbicides and insecticides conducted by the Montana Department of Agriculture reported 29 different pesticides in stormwater samples (Table 4-8). The most common groups of pesticides detected were the phenoxy herbicides (e.g., 2,4-D, 2-methyl-4-chlorophenoxyacetic acid [MCPA], and methylchlorophenoxypropionic acid [MCPP]) and herbicides used as soil sterilants (e.g., diuron, glyphosate, prometon, tebuthiuron, and triclopyr). Only two pesticides (i.e., 2,4-D and malathion) were detected at levels that exceeded aquatic life benchmarks, and none exceeded drinking water standards.

TABLE 4-5 Reported Numbers of Indicator and Pathogenic Microorganisms in Rooftop Runoff

Microorganism	Rooftop Runoff Storage Tank	Rooftop Runoff Toilet Bowl[a]
E. coli (cfu/100 ml)	ND-990	ND-54,000
Enterococci (cfu/100 ml)	ND-110	ND-110
Pseudomonas aeruginosa (cfu/100 ml)	<1-20	<1-870
Aeromonas sp. (cfu/ml)	<10-30	<10-4,400
Legionella pneumophila	ND	ND
Legionella non-*pneumophila*	ND-detected	ND-detected
Campylobacter	ND-detected	ND-detected
Campylobacter (by qPCR; cells/L)	ND-110	ND-110
Mycobacterium avium	ND-detected	ND
Salmonella (by qPCR; cells/L)	ND-7,300	ND
Giardia (cysts/L)	ND	ND
Giardia lamblia (by qPCR; cells/L)	ND-580	ND-40
Cryptosporidium (oocysts/L)	ND-50	ND-10

[a]Samples were also taken from a toilet for which untreated roof-runoff was used for toilet flushing. No pathogens were detected in toilet bowls containing domestic tap water.

SOURCES: Ahmed et al. (2012); Albrechtsen (2002); Despins et al. (2009).

BOX 4-1 Effects of Stormwater Capture Materials on Water Quality

Roofing materials, pipes, and storage tanks can, in some cases, significantly degrade the quality of stormwater, leading to high concentrations of zinc, copper, and lead. For all scales of stormwater capture, suitable materials need to be selected to minimize these problems.

Zinc

Galvanized materials, including zinc-based roofing materials and drainage systems, are the largest sources of zinc in stormwater runoff. Runoff from roofs with galvanized steel components, such as roofing sheets, flashing, or gutters, can exceed 10,000 µg/L (Table 4-1-1). Factory-coated galvanized materials have much less zinc releases than do exposed galvanized material (Clark et al., 2008a). Ogburn (2013) reported that elevated zinc concentrations were detected after short (0.5 to 27 hours) and long exposures (1 to 3 months) to galvanized steel. When numeric discharge limits are proposed for zinc by regional water quality control boards (usually about 100 to 200 µg/L), roof runoff samples frequently exceed these limits.

Copper

In general, the highest copper runoff concentrations are typically observed in runoff from exposed copper materials. However, as shown in Table 4-1-1, very high copper concentrations have been observed from other materials (e.g., copper-based paints). Some studies have found relatively constant copper concentrations in runoff over time , while others observed higher concentrations in runoff from new copper materials compared to older copper materials (Clark et al., 2008a). Runoff from clay tile and cedar shake roofs can have very high copper concentrations, possibly from copper algaecides added to these roofing materials (Clark et al., 2008a; Gromaire-Mertz et al., 1999; Zobrist et al. 2000).

Lead

Galvanized steel and PVC can be sources of lead in roof runoff (Table 4-1-1). Increased material age and exposure time and lower pH levels have been associated with higher lead releases (Ogburn, 2013). Additionally, lead can be released into stormwater from materials painted with older lead-based paints (Davis and Burns, 1999).

TABLE 4-1-1 Concentrations of Zinc, Copper, and Lead in Roof Runoff Based on Roof Material Type

Roof Materials	Runoff Concentration (µg/L)
Zinc	
New uncoated galvanized steel	500-10,000
Old uncoated galvanized steel	1,000-38,000
Coated galvanized steel	200-1,000
Uncoated galvanized aluminum	200-15,000
Coated galvanized aluminum	100-200
Other (aluminum, stainless steel, titanium, polyester, gravel):	<200
Copper	
Uncoated cooper	2-175
Uncoated galvanized steel	<3
Clay tiles	3-4,000
New asphalt shingles	10-200
New cedar shakes	1,500-27,000
Aged/patinated copper	900-9,700
Lead	
Uncoated galvanized steel	1-2,000
Coated and uncoated galvanized aluminum	<0.1-6
Painted materials	<2-600

SOURCES: Zinc data: Clark et al. (2008a,b); Faller and Reiss (2005); Förster (1999); Gromaire-Mertz et al. (1999); Heijerick et al. (2002); Mendez et al. (2011); Schriewer et al. (2008); Tobiason (2004); Tobiason and Logan (2000); Zobrist et al. (2000). Copper data: Clark et al. (2008a); Gromaire-Mertz et al. (1999); Karlen et al. (2002); Wallinder et al. (2009); Zobrist et al. (2000). Lead data: Clark et al. (2007); Davis and Burns (1999); Förster (1999); Gromaire-Mertz et al. (1999); Good (1993); Gumbs and Dierberg (1985); Mendez et al. (2011); Schriewer et al. (2008).

TABLE 4-6 Stormwater Quality Data at Neighborhood/Regional Outfalls Compared to Chronic Aquatic Life Criteria

Constituent	EPA Chronic/Aquatic Life Criteria[a]	Median[b]	5th percentile[b]	95th percentile[b]	Approximate Percentage Exceeding EPA Chronic Aquatic Live Criteria
Chloride (mg/L)	230	6.2	1.0	92	1
Arsenic, dissolved (μg/L)	150	0.62	<5	2.7	0
Cadmium, dissolved (μg/L)	0.25	<1	<1	1.0	35
Copper, dissolved (μg/L)	BLM[c]	7	0.8	40	n/a
Iron, dissolved (μg/L)	1,000	60	<100	930	4
Lead, dissolved (μg/L)	2.5	1.0	<5	15	20
Mercury, dissolved (μg/L)	0.77	<1	<1	0.06	<1
Nickel, dissolved (μg/L)	52	2.0	<5	11	<5
Zinc, dissolved (μg/L)	120	55	6	450	30

[a]See http://water.epa.gov/scitech/swguidance/standards/criteria/aqlife/index.cfm.
[b]The median and percentile values are calculated using detected and nondetected values with no substitutions for nondetected or over-range values.
[c]Chronic aquatic life criteria for copper are calculated using the biotic ligand model (BLM), which considers water quality parameters, including hardness, that affect copper bioavailability. See http://water.epa.gov/scitech/swguidance/standards/criteria/aqlife/copper.
SOURCE: National Stormwater Quality Database, version 4, updated March 17, 2015 (http://www.bmpdatabase.org/nsqd.html).

Comprehensive data on organic contaminants in stormwater have been collected by the state of Washington (see Table 4-7). No information relating to land uses or other site features are available, although it is expected that much of these data are associated with mixed land use areas. In many cases, the organic contaminants reported in these stormwater samples (including many organochlorine pesticides, PAHs, and polychlorinated biphenyls [PCBs]) exceeded the state's proposed human health water quality criteria (set according to drinking water uses). In some cases, even median and minimum reported concentrations exceeded the proposed water quality criteria. At these concentrations, organic chemicals can pose hazards to aquatic life if stormwater is stored in surface reservoirs or wetlands. Additionally, toxic organic chemicals can cause cancer risks at significant exposures, particularly when drinking water supplies are augmented with stormwater through groundwater recharge (see Chapter 5). Some persistent organic chemicals, such as PFOS, do not have a strong sorption potential (Higgins and Luthy, 2006) and may percolate through soil with concentrations essentially unchanged (Quanrud et al., 2010). Very similar results were reported in an occurrence study of organic chemicals in stormwater across different stormwater catchment sites in Australia (Vanderzalm et al., 2014). Among the chemicals targeted, herbicides and notably simazine were the most detected organic chemicals, but at no site did the 95th percentile of herbicide analytes exceed the Australian drinking water guideline.

The occurrence of pollutants in stormwater can change over time based on changes in the use of certain chemicals and materials in the drainage area. This is especially true for anthropogenic compounds such as pesticides. Figure 4-1 illustrates the decreasing concentrations of Diazinon in stormwater in Fresno, California, following a ban on the pesticide (NRC, 2009a).

Source Area and Land Use Effects on Stormwater Quality and Source Control

Source areas, such as roofs, parking lots, streets, and landscaped areas, have a significant effect on stormwater quality, but there are limited data describing the quality of stormwater originating from specific source areas. There can also be large variations in contaminant concentrations within a source area type associated with building or construction materials and the activities conducted at a particular site. Most of the data on the impacts of specific source areas on stormwater quality has been collected for studies on the effects of roofing materials on runoff quality (see Box 4-1). However, stormwater runoff samples compiled from research sites in Wisconsin and Michigan showed significant differences in TSS, phosphorus, copper, and zinc by source area type (Figure 4-2) (J. Horwatich, USGS, personal communication, 2015). These data included residential, commercial, and light industrial land uses. Figure 4-2a shows significantly lower levels of TSS and phosphorus in roof runoff compared to lawns, undeveloped areas, and paved surfaces such as parking lots, driveways, and streets. In contrast, lawns show the lowest levels of zinc (Figure 4-2d). Zinc concentrations in roof runoff ranged widely because galvanized metals are common materials in roof flashings. Roofs and lawns show the lowest average concentrations of

TABLE 4-7 Observed Stormwater Contaminants in the State of Washington

Pollutant	# of Observations	% Detected	Reported Results (μg/L)			Proposed State Drinking Water Limits (μg/L)
			Minimum	Median	Maximum	
Metals						
Antimony	50	32	0.52	1	50	14
Arsenic	275	83	0.17	1	30	10
Copper	1,495	92	0.001	6.4	12,300	1,300
Nickel	141	71	0.24	2.2	30	156
Selenium	74	15	0.5	0.70	120	141
Zinc	1,653	98	0.017	49	21,000	2,347
Organochlorine Pesticides						
4,4'-DDD	38	39	0.000064	0.0025	0.88	0.00036
4,4'-DDE	47	57	0.00012	0.013	0.88	0.00025
4,4'-DDT	46	57	0.00049	0.0096	1.8	0.00025
alpha-HCH	38	32	0.000093	0.0025	0.5	0.0039
beta-BHC	38	16	0.00012	0.0025	0.44	0.014
Dieldrin	38	42	0.000064	0.0038	0.88	0.000061
Endosulfan sulfate	48	58	0.0003	0.012	6.07	0.93
Endrin	38	13	0.00011	0.0025	0.88	0.034
gamma-HCH (Lindane)	48	25	0.00049	0.0025	0.44	0.019
Heptachlor epoxide	38	47	0.00012	0.0025	0.44	0.000045
Isophorone	31	13	0.03	0.06	10	8.4
PAHs						
Benzo(a)Anthracene	658	24	0.002	0.08	11	0.0028
Benzo(a)Pyrene	862	22	0.004	0.1	15	0.0028
Benzo(b)Fluoranthene	503	27	0.0052	0.1	13	0.0028
Benzo(k)Fluoranthene	499	22	0.0075	0.1	13	0.0028
Chrysene	786	38	0.003	0.1	16	0.0028
Dibenzo (a,h) anthracene	786	10	0.003	0.1	10	0.0028
Fluoranthene	781	50	0.005	0.1	33	16
Indeno (1,2,3-cd) pyrene	786	22	0.003	0.1	10	0.0028
Pyrene	781	55	0.0054	0.1	26	331
Volatile Organic Compounds—BTEX						
Benzene	209	10	0.13	1	190	1.2
Ethylbenzene	209	11	0.1	1	65	934
Toluene	210	15	0	1	460	4,132
Trichloroethylene	87	11	0.02	0.17	2	2.7
Other Organics						
PCBs	15,277	28	0.000002	0.00004	0.28	0.00017
Pentachlorophenol	769	23	0.02	0.5	60	0.28
2,4-Dimethylphenol	31	10	0.06	1	10	87
Phenol	83	14	0.01	0.1	10	10,690
Phthalates						
Bis(2-Ethylhexyl) phthalate	669	51	0.024	1.2	41	1.8
Butylbenzyl phthalate	623	14	0.018	0.59	10	215
Diethyl phthalate	619	23	0.024	0.67	10	4,332
Dimethyl phthalate	623	11	0.021	0.5	13	96,386
Di-n-Butyl phthalate	623	15	0.023	0.5	10	455

NOTE: Only compounds having greater than 10 percent detection frequencies of concentrations within the reporting range limits are shown. The proposed drinking water limits shown are the state's human health water quality criteria as contained in the proposed amendments to the Water Quality Standards for Toxicants as published in September 2014.

TABLE 4-8 Herbicides and Insecticides Observed in Helena and Billings, Montana, Stormwater

	Total # of Samples	% Detected	Median (μg/L)	Max. (μg/L)	Human Health Drinking Water Standard (μg/L)	Lowest Aquatic Life Benchmarks[a] (μg/L)
Herbicides						
2,4-D	23	100	1.23	27	70	13.1
Bromacil	23	35	0.02	0.26	90	6.8
Chlorsulfuron	23	30	0.007	0.036	1,750	0.055
Chlopyralid	23	22	0.029	0.8	3,500	56,500
Dichloprop	23	39	0.0029	0.099	n/a	n/a
Diuron	23	91	0.042	0.92	10	2.4
Glyphosate	19	58	0.0029	0.01	700	1,800
Imazapic	23	57	0.0015	0.0081	4,000	n/a
Imazapyr	23	96	0.021	0.53	21,000	24
MCPA	23	100	0.093	2.2	4	20
MCPP	23	100	0.14	4.6	7	14
Picloram	23	22	0.13	0.39	500	550
Prometon	23	100	0.025	0.61	100	98
Simazine	23	22	0.003	0.023	4	36
Sulfomuterun	23	17	0.056	0.2	2,000	0.48
Tebuthiuron	23	78	0.0015	0.0024	500	50
Triclopyr	23	96	0.014	3	350	100
Insecticides						
Imidacloprid	23	30	0.015	0.05	400	1.05
Malathion	23	17	0.036	1.1	100	0.035

[a]Aquatic life benchmarks listed here reflect the lowest benchmark values considering acute and chronic effects on fish, acute and chronic effects on invertebrates, acute effects on nonvascular plants, and acute effects on vascular plants.
NOTE: Sixteen samples were collected from four locations during four storms in Helena, and seven samples were collected from four locations during two storms in Billings. These samples were collected from streams and storm drains that received stormwater and represent mixed land use areas. The samples were analyzed for 148 pesticides.
SOURCE: Montana Department of Agriculture (2011).

copper, although some high concentrations were observed in roof runoff (Figure 4-2c). Heating, ventilation, and air conditioning (HVAC) condensers are commonly constructed of copper and as such can be a source of copper in runoff when located on rooftop areas.

Bacteria levels in source areas vary widely (see Clary et al., 2014; Pitt et al., 2005a,b,c) but are notably increased by the presence of animals (see Figure 4-3). As an example, E. coli and enterococci levels in roof runoff vary dramatically depending on the extent of squirrel and bird activity in trees above the roofs and possibly the season. If roofs are not shaded by trees (which provide habitat for squirrels and birds), then bacteria levels are much lower (Shergill and Pitt, 2004). The presence of contaminated materials or inappropriate connections with sewage can also increase bacteria concentrations in stormwater. Overall, roof runoff is generally the preferred source area for beneficial stormwater uses based on water quality, but treatment, especially for bacteria, may be necessary to meet beneficial use guidelines, and roofing materials should be considered (see Box 4-1).

Land uses (e.g., residential, industrial, institutional) may contain multiple source area types (e.g., roofs, paved surfaces, landscaped areas), and although water quality is widely variable within a given land use, some general water quality trends can be observed among various land use types. National Stormwater Quality Database outfall data (see Table 4-4) were analyzed to identify statistically significant groupings of the data by land use categories (Pitt and Maestre, 2014). An example is shown in Figure 4-4 for copper, which showed significantly elevated concentrations in runoff from freeways and industrial land uses, compared to residential, institutional, commercial, and open land uses, although copper concentrations are extremely variable in all three land use groupings. In an analysis of Wisconsin stormwater quality by source area, only a few samples had copper concentrations in excess of the 200 μg/L irrigation guidance, and these were associated with runoff from streets and highways (Figure 4-2). If specific contaminants pose a risk for the desired beneficial use of stormwater, then land use effects on those contaminants should be understood and less-impacted areas can be selected to reduce the contaminant load at the source and reduce the level of treatment required.

For neighborhood- and regional-scale beneficial use projects in existing developed areas, source control strate-

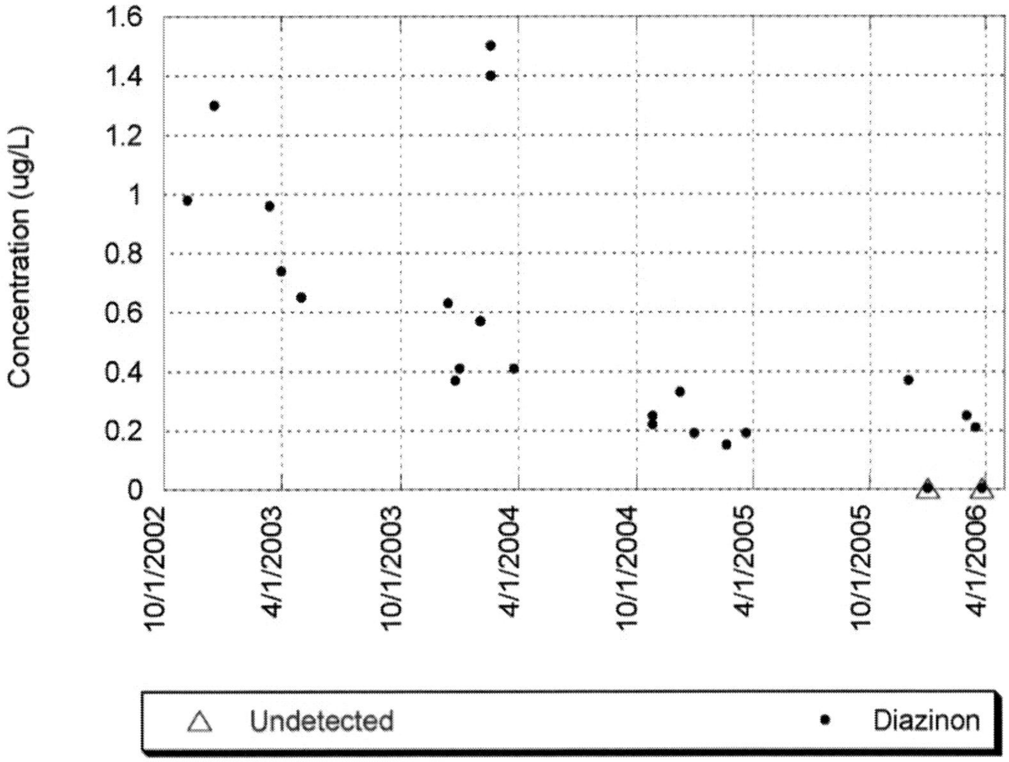

FIGURE 4-1 Trend of the organophosphate pesticide Diazinon in stormwater discharges in Fresno County, California, following its nonagricultural ban starting in 2002. SOURCE: Reprinted, with permission, from Brosseau (2007). Copyright 2006 by Fresno Metropolitan Flood Control District.

gies can be used to focus stormwater collection on the least contaminated source areas and land uses, considering exposed materials and on-site activities. Industrial or freeway land uses should probably be avoided, unless extensive water quality treatment is incorporated. Areas mostly comprised of residential, open space, and institutional areas are generally most suitable, although areas with low animal activity would be preferred for beneficial uses when bacteria are a concern. Even in areas having the best quality stormwater, there are still constituents that may cause concern with some beneficial uses requiring treatment before use. Chapter 5 discusses methods to evaluate exposures and risk to guide treatment design.

Materials Management for Source Control

Where stormwater is captured for subsequent beneficial use, an important element of source control is the management of roofing, drainage, and tank materials, particularly in new construction. There is an increasing trend in the use of metal roofs (mainly galvanized) in residential areas for increased service life, aesthetics, and fire protection. However, metal roofs can release significant amounts of zinc, copper, and lead over both short and long time frames and under a wide range of pH and salinity conditions (see Box 4-1). There may also be use of copper flashing and gutters in high-end residential and commercial areas, which is not advised because of the associated copper releases to stormwater, particularly in coastal areas. Factory-applied coatings on the galvanized metals result in greatly reduced metal releases, while homeowner applied coatings (and painting) are not as durable and these surfaces have large metal releases within a few years of application (Clark et al., 2008a). The use of lead flashing even occurs in new construction in some areas [1]and would be a significant source of lead in roof runoff for those buildings.

Galvanized materials are also not advised for roof runoff capture and storage because of the substantial zinc releases. Storage tanks and other components made from concrete, high-density polyethylene, and vinyl materials can instead be used in stormwater capture systems without elevated metal releases (Ogburn, 2013).

[1] See http://marsmetal.com/sheet-lead/roofing-and-flashing for example.

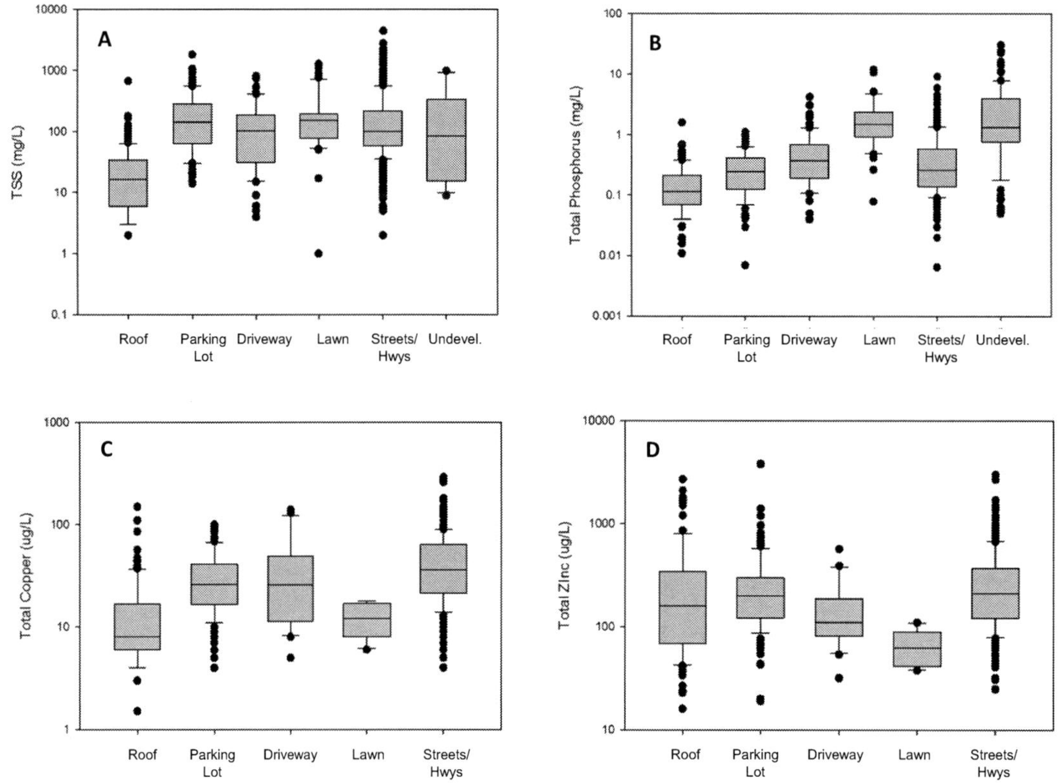

FIGURE 4-2 Box and whisker plots of: (a) total suspended solids (TSS), (b) total phosphorus, (c) total copper, and (d) total zinc, by source area type. The boxes designate the 25th, 50th, and 75th percentiles, while the end of the whiskers indicate the 5th and 95th percentiles. The dots are observations that are less than and greater than the 5th and 95th percentile values. NOTES: The stormwater data collected by the Wisconsin Department of Natural Resources and the USGS between 1991 and 1997 for multiple research studies were compiled and analyzed as a single data set. The data include sampling locations at roofs, streets, driveways, parking lots, lawns, and undeveloped areas in residential, commercial, and light industrial land uses (Bannerman et al., 1983; Corsi et al., 1999; Holmstrom et al., 1995, 1996; Roa-Espinosa and Bannerman, 1995; Steuer et al., 1997; Waschbusch et al., 1998). The data include the following numbers of event samples: 158 from roofs, 141 from parking lots, 70 from driveways, 41 from lawns, 418 from streets and highways, and 12 from undeveloped areas (J. Horwatich, USGS, personal communication, 2015).

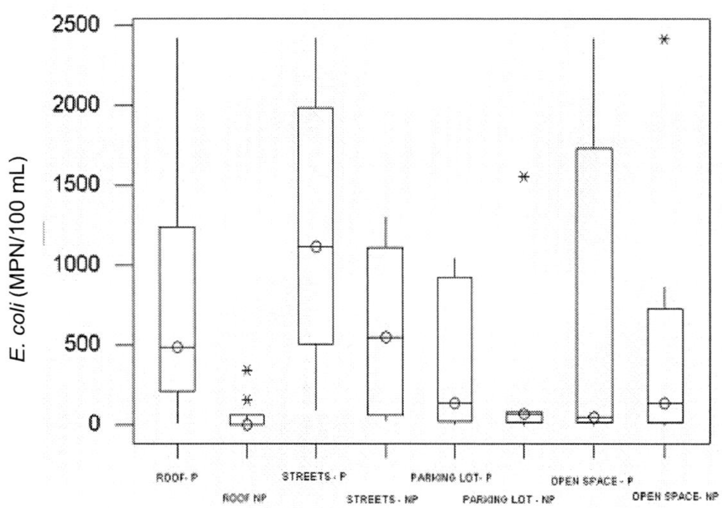

FIGURE 4-3 Comparisons of E. coli by source area type affected or unaffected by domestic pets and urban wildlife. The presence of these animals were observed as dogs being "walked" by owners in park areas, and trees having large squirrel and bird populations over roofs or in parking areas and streets. Parallel samples were obtained during the same rains for comparison from multiple locations. SOURCE: Sumandeep and Pitt (2004).

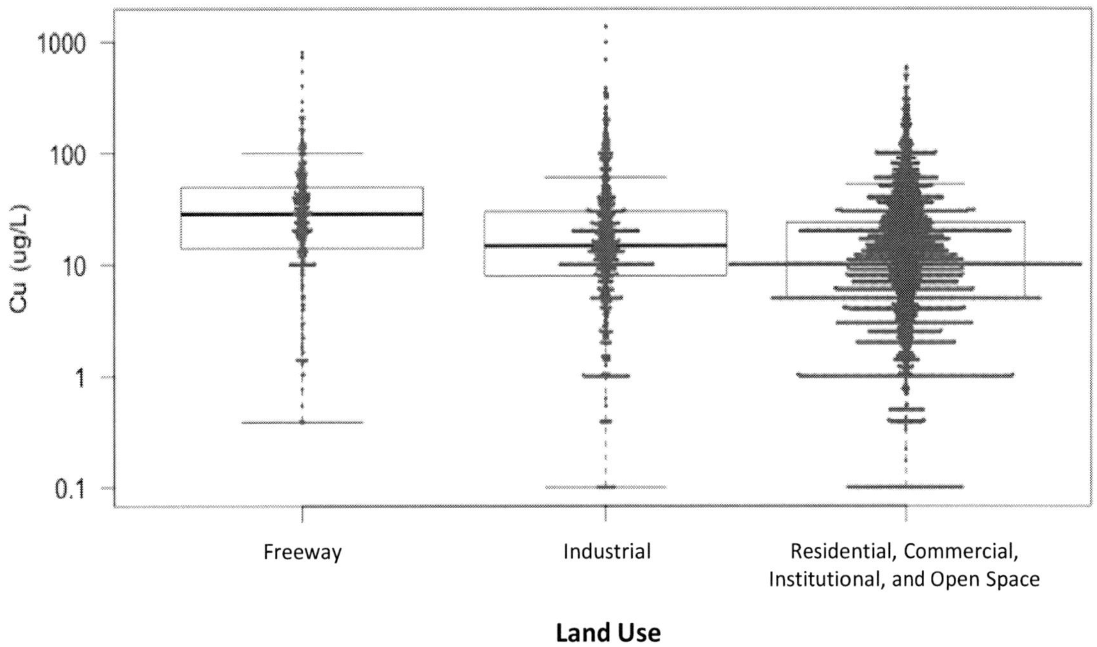

FIGURE 4-4 Significant differences among land use types for copper in stormwater. Data are from the National Stormwater Quality Database (version 4) with significant land use groupings determined by non-parametric Kruskal-Wallis one-way analysis of variance on rank tests along with multiple pairwise comparisons with Dunn's Method. SOURCE: Pitt and Maestre (2014).

A well-known example of materials management for source control is limiting copper in brake pads. In urban watersheds, up to one-half of the copper originates from brake pads.[2] California and Washington recently passed legislation limiting the amounts of copper and other heavy metals in brake pads (Ch. 173-901 WAC; California Health and Safety Code sections 25250.50–25250.65). That legislation inspired similar bills in other states, effectively setting a national standard (Motavalli, 2012).

CONCLUSIONS

The quality of graywater and stormwater determines their potential uses without treatment, but many additional applications are possible with treatment. Chapter 5 discusses a risk assessment approach to determine the appropriate levels of treatment, but this approach requires a clear understanding of source water quality.

Pathogens and organic matter in graywater impact opportunities for beneficial uses without treatment. Considering the source water, human pathogens are likely to occur in graywater, although the specific types and concentrations vary substantially. The occurrence and environmental fate of pathogens in graywater are not yet well understood. Organic matter is present in high enough concentration in graywater to enhance microbial growth, thus limiting the potential uses of graywater without disinfection. Graywater also contains a wide array of personal care products. Sodium, chloride, boron, and other chemicals can impact the quality of graywater for irrigation uses. Best management practices exist for source control of microbial and chemical constituents, and such practices can be implemented at the household scale to reduce concentrations of these constituents in graywater.

Stormwater quality is highly variable over space and time and might contain elevated levels of microorganisms, metals, organic chemicals, and sediments, potentially necessitating treatment to facilitate various beneficial uses. Stormwater quality is a direct function of land use, source area, catchment size, and climatic and seasonal factors. Existing data suggest that most stormwater contains elevated levels of organic matter, suspended sediment, and indicator bacteria. Metals are also commonly found in urban stormwater runoff and may pose concerns for some beneficial uses, including irrigation and surface reservoirs or wetland features. Nutrients may also impact the function of some stormwater applications, such as ornamental water features, without treatment. Despite the enormous spatial and temporal variability of stormwater quality, the data show that there are a number of water quality parameters with at

[2] See http://www.ecy.wa.gov/programs/hwtr/betterbrakes.html.

least 5 percent of samples consistently above the guideline values for irrigation, drinking, or protection of aquatic life. This suggests that in spite of aggregated data showing high variability, the treatment systems required for achieving end uses may be relatively consistent over a wide variety of catchments.

Little is known regarding the occurrence of human pathogens and organic chemicals in stormwater, and additional research is needed to characterize their occurrence and fate. Studies of the presence of microorganisms in stormwater have consistently reported high concentrations of fecal indicator microorganisms across different source areas. In the few studies that analyzed for pathogenic microorganisms in stormwater, they were generally detected, at least in some samples. However, more work is needed to characterize their occurrence and fate, particularly for roof runoff systems where the beneficial use of untreated stormwater is common and raises concerns for uses with the potential for human exposure. More research is also needed to characterize the occurrence of organic chemicals in stormwater and their fate during various uses.

Land uses, contributing areas, and collection materials can be selected that minimize contaminants of concern to optimize stormwater quality and minimize treatment requirements for the designated beneficial use. Even though all land uses have the potential for problematic water quality conditions in runoff, residential areas generally have lower concentrations of these contaminants than do commercial and industrial areas. Local data should be used to help select the best source area. On average, residential roofs have the highest quality runoff of the various source areas, but there are many exceptions. Copper and galvanized metals in roofing, piping, and stormwater capture tanks can create hazardous levels of lead, zinc, and copper in roof runoff. The presence of pets and urban wildlife can cause high levels of indicator bacteria, which may indicate the presence of disease-causing microorganisms. Regional stormwater capture and recharge systems drain large source areas and many land uses, and opportunities for catchment area separation may be limited. With increasing catchment area, stormwater quality will be more difficult to manage and treatment may be required prior to beneficial use. However, targeted source control of contaminants through materials management may be possible at larger scales. For example, stormwater containing high concentrations of road salt should be diverted from stormwater capture and infiltration systems. Limiting copper and other heavy metals in brake pads is another example.

5

Characterizing and Mitigating Human Health and Environmental Risks

The reuse of graywater and stormwater has the potential to significantly impact the use of the potable water supply. However, the potential presence of a variety of contaminants in these water sources raises concerns about their uses. These concerns may be exacerbated by the lack of federal guidelines for the use of these types of waters and inconsistencies among limited existing state and local regulations or guidelines for the beneficial use of graywater and stormwater (see Chapter 8).

The presence of a contaminant does not automatically equate to significant risk. The situation in which the water is used has a significant impact on the potential risk. When there is little potential for human exposure to the water (such as subsurface irrigation with graywater), the risk will be lower than when the same quality of water is used in higher exposure environments (such as spray irrigation with graywater, where the potential for ingestion exists).

In a literature review, the committee could not find any documented reports of adverse health effects from the use of stormwater or graywater. Sharvelle et al. (2013) surveyed health departments from 15 states that allow graywater reuse, who reported no sicknesses resulting from graywater reuse, including Arizona, which has promoted graywater systems for more than 10 years. Although these results help bound the extent of risk—that is, the risk is unlikely to be large—waterborne infectious diseases tend to be underreported (Yang et al., 2012). Given the many potential exposure routes, linking illness with water supply contamination is challenging, particularly for distributed on-site sources such as graywater or stormwater. Therefore, in this chapter, the committee relies on risk assessment strategies to assess the risks of stormwater and graywater use.

When assessing the potential risk posed by graywater and/or stormwater reuse, a number of factors should be considered. These include the fact that potential risks may vary considerably, based on the specific nature of the water, how the water will be used, the potential exposure to the water, and the characteristics of the environment in which the water will be used. It is also critical to understand that it is not possible to reduce to zero the risks associated with the use of these waters, just as the risks associated with the use and consumption of conventional drinking water are not zero. However, an understanding of the risks associated with graywater and stormwater use can inform responsible decision making regarding the integrated use of all available water resources and facilitate communication with stakeholders.

This chapter provides an overview of the most common methods that can be used to assess human health risk associated with graywater or stormwater use. The chapter also summarizes what is known about human health and environmental risks.

OVERVIEW OF THE QUANTITATIVE RISK ASSESSMENT PROCESS FOR HUMAN HEALTH

Different methods can be used to calculate the risk associated with a particular activity or contaminant. In terms of potential impacts on human health, the National Research Council (NRC, 1983) risk assessment method is the most commonly used. The following sections describe the major elements of that process—hazard assessment, exposure assessment, dose-response assessment, and risk characterization—although readers seeking a complete description should consult NRC (1983, 2009c). Once the risk assessment is completed, the risk management phase considers the overall benefits and costs of various risk management approaches as well as social justice and legal issues (Figure 5-1). In its 2009 review of the U.S. Environmental Protection Agency's (EPA) risk assessment process, the NRC reaffirmed the process but stated that more attention should be focused on the design of the risk assessment, especially in the beginning stages of the process. Specifically, there is a

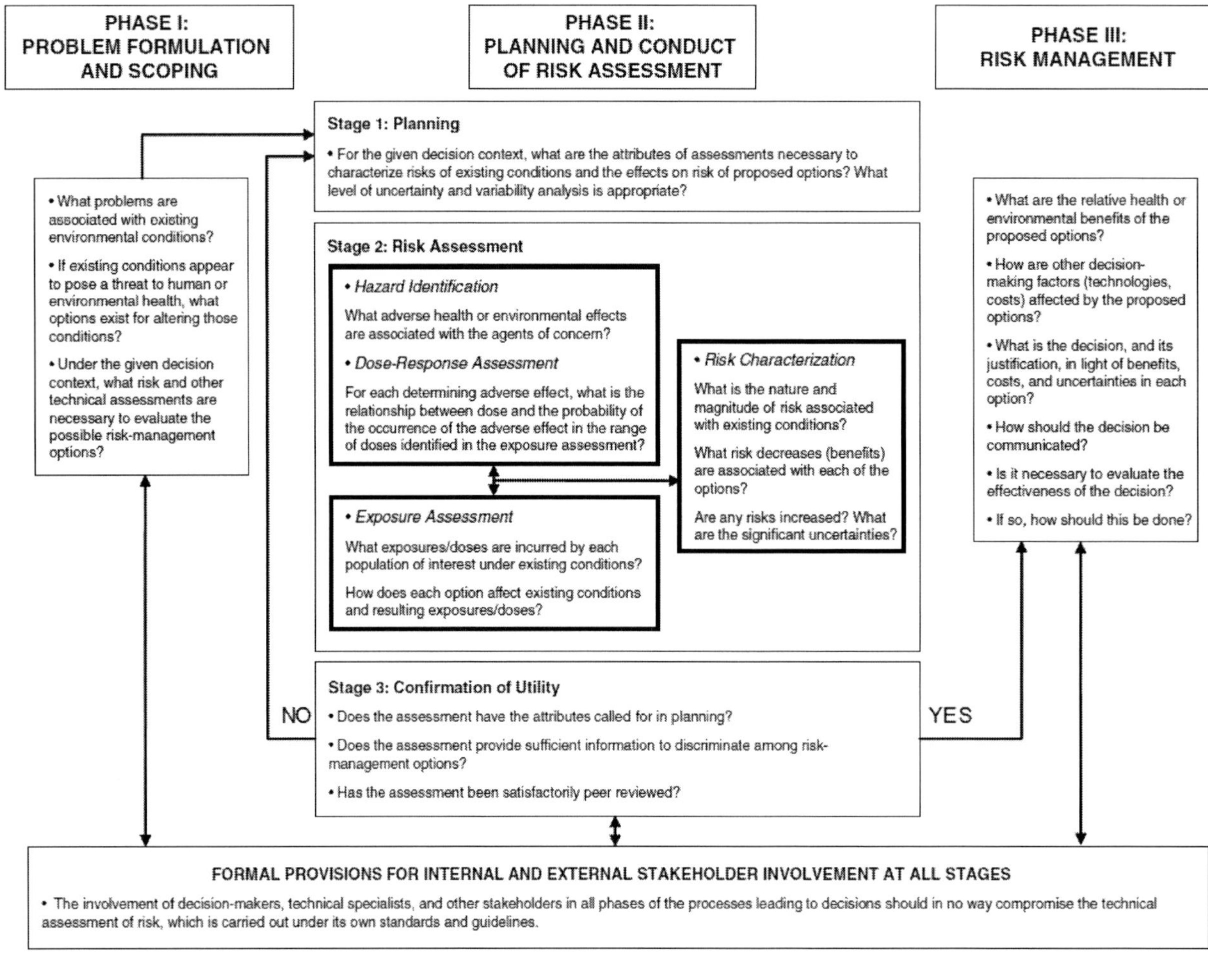

FIGURE 5-1 A framework for risk-based decision making. SOURCE: NRC (2009c).

need to bring all stakeholders into the process to determine the major factors to be considered, to define the decision-making context, and to agree upon the timeline and depth needed to ensure that the right questions are being asked in the context of the assessment. This will increase the likelihood that the outcome of the risk assessment will be more useful and better accepted by decision makers and the regulated community (NRC, 2009c).

Hazard Assessment

The first step in the risk assessment process is hazard assessment, in which the contaminants of concern are identified. Such contaminants will vary, depending on the specific situation, including the source water quality and its end use. In the case of graywater, because it is derived from domestic wastes, the contaminants of most concern include human pathogenic microorganisms and inorganic and organic chemicals found in wash waters. In the case of stormwater, the contaminants of concern include metals, pesticides, other organic contaminants, and pathogens, derived from runoff from streets, roofs, lawns, and industrial and commercial areas. Chapter 4 discusses contaminants of concern in stormwater and graywater.

Under sufficient exposures, chemical and microbial contaminants can cause a range of adverse health effects, either acute or chronic. Acute illnesses occur suddenly and severely after only one or a few exposures to a contaminant and are common after exposure to human pathogenic microorganisms. Acute illnesses associated with exposures to waterborne pathogens include gastroenteritis (stomach flu), skin and respiratory infections, and conjunctivitis (pinkeye). Chronic health effects are long-lasting and typically occur after repeated, long-term exposures, most commonly by ingestion of contaminated water. Examples of possible chronic health effects include various cancers and adverse reproductive outcomes.

Exposure Assessment

The next step in the process involves determination of the nature of an individual's exposure to the hazard(s). Exposure assessment requires knowledge of the amount of contaminant to which the individual is exposed (e.g., the contaminant concentration multiplied by the volume of water ingested), as well as the route(s) and frequency of exposure. This can be one of the most challenging steps in the process, because exposures can vary from day to day, location to location, and person to person.

Contaminant Concentration

As discussed in Chapter 4, different source areas can generate different qualities of stormwater, and source control practices substantially impact the quality of graywater or stormwater. For graywater, factors affecting contaminant concentrations include whether best management practices are followed, such as bypassing laundry water when washing diapers or other soiled clothing, not disposing of hazardous chemicals down the sink, and avoiding storage of untreated water. For stormwater, rooftop rainwater capture tends to result in the highest water quality, but some roof and storage tank materials can leach high levels of metals such as copper, zinc, and lead. Runoff from parks and lawns may contain pesticides and fecal microorganisms. Larger source areas, including more roadways and parking areas, can result in higher concentrations of organic contaminants (see Chapter 4).

There may be hundreds of different types of chemicals or microorganisms in graywater or stormwater. Therefore, it is common to analyze the water for indicator or surrogate contaminants, rather than for each of the contaminants themselves. For example, rather than monitoring for human pathogenic microorganisms, which can be extremely costly and time-consuming, indicator microorganisms, such as fecal coliform bacteria, are commonly used. Although this simplifies the monitoring process, the use of surrogates to determine the risk of a water source can be problematic for a number of reasons.

In the case of indicator bacteria, such as *E. coli*, these microorganisms are present in the intestines of numerous animals as well as humans. Therefore, they are always present in domestic wastewater and are often detected in stormwater (see Chapter 4, Table 4-3). However, pathogenic microorganisms are only present in the intestines when an individual is infected. Additionally, *E. coli* and enterococci have been detected in a variety of environmental reservoirs (Clary et al., 2014; Byappanahalli et al., 2006; Yamahara et al., 2007). Therefore, the detection of indicator bacteria in water does not necessarily mean that the water contains disease-causing microorganisms. Thus, the reliance on indicators alone limits the capacity to accurately assess human health hazards in a water source.

Additionally, wide variation in concentrations of specific pathogens in graywater exists between individual households, and therefore the use of indicator microorganisms (especially from individual households) to predict pathogen concentrations and associated risks can be problematic (O'Toole et al., 2014). The wide variation in pathogens exists for several reasons, including the numbers, ages, and health of the inhabitants; the fact that fecal shedding intensity and duration varies considerably from person to person; and the source of graywater used (e.g., shower, laundry), which can have a significant impact on the concentrations and types of microorganisms present in the graywater.

There are also situations in which pathogens are present in the absence of the indicator microorganisms, particularly for more environmentally stable pathogens such as protozoan parasites. Many waterborne disease outbreaks have occurred in which indicator bacteria were absent but pathogenic microorganisms were present in sufficient numbers to cause illness (e.g., MacKenzie et al., 1994). Overall, the reliance on indicator microorganisms, although practical from an economic and analytical perspective, may not provide accurate information regarding the water's safety for its intended end use. Instead, indicator organisms serve as an imperfect but low-cost screening tool that can indicate possible concerns but cannot be used to prove safety.

Water quality treatment (see Chapter 6) may be applied to graywater and stormwater to reduce the concentrations of contaminants present for end uses with a higher degree of human exposure. Water quality may also degrade over time with extended storage, either from contact with tank materials (e.g., zinc [Hart and White, 2006]) or from the growth of microorganisms (Pitt and Talebi, 2012a).

Route(s) and Volumes of Exposure

Individuals can be exposed to contaminants through many different routes, including ingestion, inhalation, and dermal exposure. Some contaminants pose a potential hazard through many exposure routes, while others are harmful only if exposure is through a specific route. Some microorganisms, such as *Legionella spp.*, are transmitted through inhalation. Others can cause harm through dermal exposure (e.g., *Pseudomonas*, *Schistosoma*) or exposure of the mucous membranes (e.g., adenovirus), resulting in eye and ear infections. For many microorganisms that are transmitted through water (e.g., *Salmonella*, *Cryptosporidium*, noroviruses), ingestion is the primary exposure route of concern. In addition to exposure from the direct ingestion of water, exposure can also result from consumption of food crops that have been spray irrigated with graywater or stormwater that contains pathogens or chemicals. In determining the dose of the hazard to which an individual is exposed, it is critical that all potential exposure routes be considered so that an accurate risk assessment can be made.

The volume of water to which the individual is exposed is often determined by the specific exposure scenario. There are no standard exposure volumes for the different types of possible exposure. The District of Columbia's Department of the Environment (DDOE, 2013) developed exposure volumes based on stormwater use and exposure conditions that vary from 0.01 mL from ingestion of aerosol spray from toilet flushing to 200 mL from ingestion of water in a swimming pool (Table 5-1). Other organizations may assume different exposure volumes.

In addition to the volume of water to which an individual is exposed, certain activities may affect the amount of contaminants present at the point of exposure. For example, foods that are irrigated with stormwater and then consumed uncooked will expose the consumer to a higher number of pathogens compared to the same food if it is consumed after cooking. Likewise, foods that are spray irrigated with graywater or stormwater have a higher probability of containing pathogenic microorganisms than food crops that are watered via subsurface irrigation (with the exception of root crops).

Frequency of Exposure

In the exposure assessment, the frequency with which the individual will be exposed to the hazard is another critical piece of information for consideration. In some situations, there may be only a single exposure; for example, an individual who inadvertently swallows some water while playing in a fountain on a hot day. In other cases, exposure might be on a daily basis, such as in the case of an individual who is exposed to aerosol spray from graywater used for toilet flushing. Frequency of exposure will also vary considerably depending on the water use and exposure conditions. For example, DDOE (2013) assumed that an individual would be exposed to aerosols via toilet flushing more than 1,000 times/yr but would accidentally ingest 100 mL of stormwater as a result of using stormwater for home lawn or garden spray irrigation only a single time per year (Table 5-1). Where data are lacking on the frequency of exposure, conservative assumptions are often made.

Scale Issues

Other considerations related to the exposure of the individual to the hazard include those of scale—both temporal and spatial. The time that elapses from the release of the con-

TABLE 5-1 Exposure Assumptions Based on Stormwater Use and Exposure Conditions Developed by the District of Columbia's Department of the Environment

		Exposure Assumptions	
Stormwater Use	Route of Exposure, Conditions	Volume ingested (mL) in a single exposure	Events per year
Home lawn or garden spray irrigation	Ingestion of aerosol spray from typical watering	0.1	90
	Ingestion after contact with plants/grass	1	90
	Infrequent inadvertent ingestion of stormwater	100	1
Open space or municipal park drip or spray irrigation	Ingestion via casual contact with irrigated grass (picnic, walking pet)	0.1	32
	Ingestion via low-intensity sports on irrigated field (golf, Frisbee)	1	32
	Ingestion via high-intensity sports on irrigated field (baseball, soccer)	2.5	16
	Ingestion on playground by child (frequent hand-to-mouth activity)	4	130
	Indirect ingestion of spray from public fountain with spray element	0.1	130
	Infrequent ingestion of public fountain water from standing pool on hot days	4	130
Home garden drip or spray irrigation	Ingestion of irrigated vegetables and fruit	7	50
Commercial farm produce drip or spray irrigation	Ingestion of irrigated vegetables and fruit	10	140
Home car wash spray application	Ingestion of water and spray	5	24
Commercial car wash spray	Ingestion of water and spray by car wash operator	3	250
Toilet	Ingestion of aerosol spray	0.01	1100
Washing machine use	Ingestion of sprays	0.01	365
Fire fighting	Ingestion of water and spray	20	50

NOTE: In a correctly designed subsurface irrigation system (with no surface ponding and no application to food crops) no water would be ingested.
SOURCE: DDOE (2013).

taminant to the exposure may have a significant impact on the concentration or nature of the contaminant. For example, human pathogenic microorganisms have a finite lifetime in the environment. This may range from less than a day to several months or years, depending on the specific microorganism and the environmental conditions. Chemical contaminants may degrade over time, and depending on the specific chemical, the transformation products could be more or less harmful than the parent compound. Modified risk assessment models have been developed to include dynamic modeling (see, e.g., Eisenberg et al., 2004), as well as to allow for the effects of environmental factors on the concentrations of the contaminants (see, e.g., Whelan et al., 2014)

Spatial scale is also important. If the graywater or stormwater is collected over a large area, then a large number of individuals/households/businesses will contribute to the water's composition. This may result in the water containing microorganisms or chemicals that it would not have contained had the contributing area been smaller. The project's scale will also affect the number of people who may be exposed to potential hazards in the water. At the household scale, graywater use does not significantly increase risk of illness from pathogens because there are many other pathways for spreading communicable illnesses among members of the same household (Maimon et al., 2010). However, untreated graywater used in larger scale (e.g., multi-residential) projects could substantially increase risk, because graywater use creates exposure pathways between infected and uninfected individuals that otherwise did not exist. Therefore, larger-scale projects will be more likely to involve some type of treatment than will projects that occur on an individual homeowner's property (see Chapter 6). A project that occurs over a large spatial scale may also, by necessity, involve a longer time scale, so attenuation of contaminants may occur as a result.

Dose-Response Assessment

After the amount of the contaminant to which the individual will be exposed is known, it is necessary to understand the effect that that amount of the contaminant will have on the exposed individual—in other words, what response will a specific dose produce? Typically, information on the dose-response relationship of a particular contaminant is obtained from published literature values, rather than from conducting a dose-response study for each contaminant in each specific situation. Box 5-1 describes the approaches used to determine dose-response relationships for individual microbial or chemical contaminants.

For both chemicals and microorganisms, however, exposure is commonly to mixtures of the contaminants, rather than to single contaminants. Unfortunately, little is known about the effects on the dose-response relationship of contaminants when an individual is exposed to more than one contaminant at a time. Exposure to a mixture of contaminants can cause effects that are equal to, less than, or greater than that of the individual components, and understanding these effects is the focus of ongoing research (see, e.g., Backhaus, 2014; Jarvis et al., 2014).

Risk Characterization

The next step in the process is to calculate the risk, considering the contaminant of interest, the level and frequency of exposure, and the dose-response relationship. The risk may be expressed in several different ways. For microorganisms, the acute risk of infection, illness, or mortality can be presented. In the case of many chemicals, the risk of cancer over the course of a lifetime is the endpoint of interest.

The different outcomes associated with the contaminants of interest can make it difficult to compare risks and to make decisions regarding possible risk-risk tradeoffs. For example, how does one compare the risk of getting cancer to the risk of getting a norovirus infection? Cancer is typically acquired after long-term exposure to a relatively low dose of a chemical, while hepatitis can be acquired from a single exposure to contaminated food or water. Infection caused by norovirus is typically self-limiting, while cancer can be a short-term or long-term illness.

One way to more easily compare the health effects caused by exposure to different types of contaminants with different health effects is to quantify the burden of disease morbidity and mortality through the use of disability-adjusted life years (DALYs). Per the World Health Organization, "One DALY can be thought of as one lost year of 'healthy' life. The sum of these DALYs across the population, or the burden of disease, can be thought of as a measurement of the gap between current health status and an ideal health situation where the entire population lives to an advanced age, free of disease and disability."[1] DALYs are calculated by adding the number of years of life lost due to disability (for people who are living with the adverse health effect) to the number of years of life lost due to premature deaths across the exposed population. Results for specific scenarios can then be evaluated in the context of acceptable risk targets. For example, Australia developed guidelines for potable reuse (NRMMC et al., 2006) that set a tolerable microbial risk of 10-6 DALYs per person per year (approximately 1 diarrheal illness per 1,000 people per year).

---

[1] http://www.who.int/healthinfo/global_burden_disease/metrics_daly/en.

BOX 5-1 Determining Dose-Response Relationships for Microbial and Chemical Contaminants

Microbes

Microbial dose-response relationships have typically been found to conform to one of two models—the exponential or the beta-Poisson model (Haas et al., 2014). Determination of microbial dose-response relationships can be challenging. Numerous dose-response studies have been conducted using humans (Dupont et al., 1995; Ward et al., 1986), but the number of individuals who can participate in such studies is necessarily very small (fewer than 100), for reasons of logistics and cost, and these "feeding studies" are typically conducted with healthy individuals who are not elderly or very young. This raises concerns about the representativeness of the results when trying to assess the health impacts for more sensitive subpopulations, such as individuals who are immune-incompetent, elderly, or very young. Additionally, several different outcomes, commonly referred to as endpoints, may result from exposure: infection, clinical disease, or death. An individual who is infected but shows no clinical signs or symptoms of disease is still capable of transmitting the organism to another individual, a phenomenon known as secondary spread. In some disease outbreaks, more individuals are infected as a result of secondary spread than from exposure to the contaminated source water.

To account for some of these phenomena, enhancements to the quantitative microbial risk assessment models have been made for particular applications. For example, Parkin et al. (2003) developed a more detailed model to allow for the incorporation of differences in exposure responses of sensitive subpopulations, rather than assuming that all individuals have the same response to the same dose of a contaminant. Soller (2009) described a method to incorporate person-to-person transmission of human pathogenic microorganisms, allowing a calculation of the amount of disease that results from secondary transmission, rather than just the amount that results from primary exposure to the contaminated water.

Chemicals

Chemical dose-response assessments and the subsequent risk assessment for chemical contaminants are typically derived by one of two approaches: threshold-based methods for noncancerous agents and linear methods for cancerous chemicals. For noncarcinogens, the dose-response is described above a threshold level, which is typically expressed as a reference dose or acceptable daily intake. Below this threshold daily dose (or no observed adverse effect level), humans are unlikely to incur additional health risks over their lifetimes. Risk can be screened by comparing contaminant exposures against the reference dose. Cancer risk from mutagenic chemicals does not follow a threshold approach, because a single molecule is capable of causing a mutation that causes cancer. Cancer risk is, therefore, assumed to be linearly proportional to chemical exposure, and is typically described by a linear model (described by a cancer slope factor) that can be extrapolated to low concentrations.

Determination of the dose-response relationship for chemicals with chronic effects can be challenging, because it is generally not possible to conduct a study for the length of time that would be required to observe the health effects. Many times, researchers are limited to short-term studies using relatively large doses of the chemical, where the effects are manifest in a short period of time and at a measurably higher frequency than background. Generally, these studies require the use of animals, which raises questions about the suitability of the specific animal model used as a surrogate for the response of humans. In addition, extrapolation of effects from the high dose/high response seen in the studies to the response that would be expected at the lower concentrations to which people would typically be exposed can be problematic and subject to interpretation.

As a point of reference, the "acceptable" dosage may be compared to drinking water maximum contaminant levels (MCLs), which, for carcinogenic chemicals, are typically set at a level associated with 1 additional cancer death per 100,000 to 1 million persons over the course of a lifetime. This can provide useful information that can be used as guidance when assessing risks from graywater and stormwater uses for which no regulations or guidelines have been established.

Risk Management

The process of calculating the risk from exposure to a contaminant is a scientific process. Once the risk is calculated, the determination of whether that risk is acceptable involves not only science but also technological feasibility, economics, politics, and societal factors. Interested readers should consult NRC (2009c) for a detailed discussion of risk management.

Several cities have taken a tiered approach for managing the risks of harvested stormwater use, with increasing levels of regulation or treatment required with increasing exposures (see Los Angeles County Department of Public Health tiered framework in Table 8-2). The District of Columbia recently developed a quantitative process for tiered risk assessment and management for nonpotable uses of harvested stormwater (DDOE, 2013). The process (shown in Figure 5-2) uses tiered risk-based screening levels for individual chemical and microbial contaminants and four exposure classes (low to severe; see Tables 5-2 and 5-3). The process allows planners to compare typical stormwater concentrations against tiered risk-based levels, with higher contaminant concentrations allowed for activities with low exposure. For cases when stormwater concentrations exceed these screening levels (i.e., the risk is considered unacceptable), the risk management process (Figure 5-2) requires treatment or additional justification for why treatment is not needed. Note that the examples in Table 5-2 are based only on direct human exposure—not ecological risk or indirect human exposure. Consideration of other pathways, including the use of stormwater to recharge an aquifer used for water supply, would include different exposures and could result in different chemical concentration limits.

Australia also adopted risk-based guidelines for managing the beneficial use of stormwater and graywater (NRMMC et al., 2006, 2009). The guidelines include treatment recommendations for different applications based on 95th percentile concentrations from existing pathogen data or conservative estimates where data were lacking. In addition, the risk management framework includes a commitment to responsible use and recommendations for preventive measures, management of system failures, employee and community awareness, evaluation, review, and continual improvement.

QUANTITATIVE RISK ASSESSMENT FOR GRAYWATER AND STORMWATER REUSE

Several quantitative human health risk assessments have been published on graywater and stormwater. The most relevant to the uses considered in this report are presented here.

Graywater

Microorganisms

In a review of onsite reuse of graywater for irrigation, Maimon et al. (2010) attempted to summarize the potential risks from exposures to human pathogenic microorganisms, using rotavirus as the example. Their scenario included exposures from the accidental ingestion of untreated graywater (100 mL; once/yr), the routine ingestion of the graywater from touching irrigated plants (1 mL; 90 times/yr), and in-

TABLE 5-2 Examples from the DDOE Chemical Risk-based Levels for Stormwater Use Based on Human Exposure Category and Comparison with DDOE's Drinking Water Standards

		Direct Human Exposure Category		
Contaminant (μg/L)	Drinking Water Standard	High[a]	Medium[b]	Low[c]
Benzene	0.41	210	550	6,000
Cadmium	18	9,100	25,000	270,000
Helptachlor	0.015	7.5	20	220
Polybrominated biphenyls	0.0022	1.1	3	32
Polychlorinated biphenyls	0.50	250	670	7,300
Trichloroethylene	2.0	1,000	2,700	29,000
Zinc	11	5,500	15,000	160,000

[a]High exposure includes applications such as commercial farm produce drip or spray irrigation, firefighting, and commercial car washes.
[b]Medium exposure includes public fountains, spray irrigation of playgrounds, home garden spray irrigation, home drip irrigation of fruits and vegetables, and home car washing.
[c]Low exposure includes toilet flushing, washing machine use, and open space spray irrigation of parks (non-playgrounds).
NOTES: For each of the exposure classes (grouped by the amount of water ingested), the risk-based levels represent the contaminant concentrations corresponding to a cancer risk of 10^{-6} or a non-cancer hazard index of 1.0. DDOE states that although EPA suggests a discretionary cancer risk level between 10^{-4} and 10^{-6}, it selected this cancer risk level to account for the presence of multiple contaminants.
SOURCE: DDOE (2013).

TABLE 5-3 Examples from the DDOE Microbial Risk-based Levels for Stormwater Use Based on Human Exposure Category

		Direct Human Exposure Category		
Contaminant (μg/L)	Swimming	High	Medium	Low
E. coli (CFU/100 mL)	126[a]	1,714	4,615	50,000
Cryptosporidium[b] (oocyst/L)	NA	0.016	0.033	0.320

[a]RSLs correspond to a risk level of 8 in 1,000 of developing a gastrointestinal disease.
[b]RSLs correspond to a 10-6 risk level of developing a gastrointestinal disease.
NOTE: See Table 3-2 for examples of exposure categories.
SOURCE: DDOE (2013).

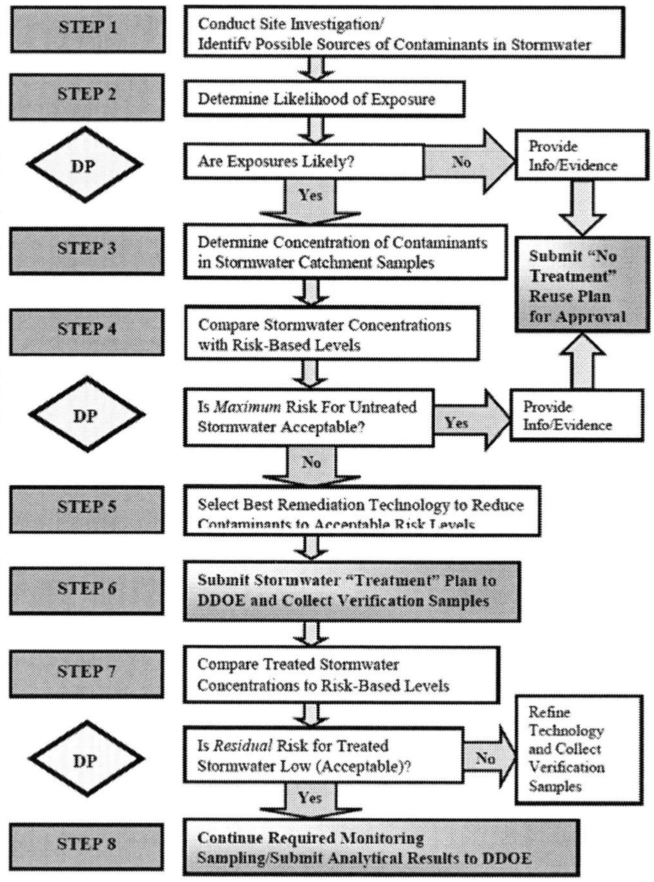

FIGURE 5-2 The tiered risk assessment and management approach for stormwater use implemented by the District of Columbia. SOURCE: DDOE (2013).

gestion of the graywater spray (0.1 mL; 90 times/yr). Using the assumption that the acceptable annual risk of infection was 1.4 x 10-3 infections per person per year, they calculated the acceptable safe dose of rotavirus as 0.0024 viruses per exposure if the exposure occurs from the accidental ingestion of 100 mL of untreated graywater once per year or 0.00014 viruses per exposure if the exposure is to 1 mL and occurs 90 times per year (e.g., from the routine ingestion of the graywater from touching irrigated plants). Using this acceptable dose and the estimated concentrations of rotavirus in graywater (based on three different assumed relationships between rotavirus and measured *E. coli* concentrations), the maximum volume of graywater that can be "safely" ingested in a single exposure occurring once per year was calculated to range from 0.003 mL (assuming 0.8 rotaviruses/mL) to 0.24 mL (assuming a rotavirus concentration of 0.01 rotaviruses/mL). Maimon et al. (2010) concluded that most of the exposure scenarios examined would result in exposure that would exceed the acceptable safe dose and that using graywater for spray irrigation or food crop irrigation would necessitate disinfection to protect against rotavirus.

The committee could not find any risk assessments of exposure from surface drip irrigation (no landscape cover) with graywater. Such exposures would presumably be higher than those of subsurface irrigation, but exposure estimates and the increase in risk at the household scale, where many vehicles of disease transmission already exist, are needed to inform safe design practices.

Using a different approach, Ottoson and Stenstrom (2003) calculated the risks associated with exposure to human pathogens in treated (but not disinfected) graywater.[2] Because fecal indicator bacteria can overestimate the amount of fecal material in the graywater, the authors used measured concentrations of coprostanol (a fecal sterol) at the site as an estimator of the amount of fecal material in the graywater, and estimated pathogen concentrations based on the prevalence of pathogens in the general population. Risks were modeled in three scenarios—direct contact, spray irrigation of sports fields, and daily consumption of groundwater recharged

[2] The treatment system included settling tanks, activated sludge, a biofilter, and surface storage in ponds.

with treated graywater. In the case of groundwater recharge, it was assumed that the microorganisms traveled through a 3-m-thick vadose zone and that reductions in the numbers of microorganisms occurred during transport as well as in the groundwater during retention. The study considered several pathogens, including *Campylobacter jejuni*, *Cryptosporidium parvum*, *Giardia lamblia*, rotavirus, and *Salmonella*, and modeled rates of natural attenuation in the subsurface. These organisms were chosen as conservative representatives for the behavior of fecal bacteria, viruses, and protozoan parasites. Lack of disinfection resulted in sizable risk across a range of exposures, although the risks ranged considerably by pathogen and exposure scenario. The highest risks across all scenarios were associated with rotavirus, ranging from 0.25 probability of infection from a single event with direct contact to 0.63 annual probabilities for both spray exposure and groundwater consumption after 1 month retention (see Table 5-4). Risks from groundwater consumption decreased notably over time and were negligible (less than 10-11) for all organisms after 6 months retention. In addition, the effects of removal during transport through the soil are clearly seen in the much lower risks for the larger protozoan parasites (*Cryptosporidium* and *Giardia*) compared to those for the smaller, more easily transported bacteria and viruses.

Calculating risk based on indicator concentrations presents many challenges. Ideally, risk assessment would be based on direct pathogens data. This effort, however, has been limited by the relative lack of quantitative data on the concentrations of pathogenic microorganisms in graywater and stormwater. One report on the concentrations of pathogens in untreated graywater was located (see Table 4-2; Birks and Hills, 2007), and these data were used to perform example calculations of infection risk infection for two different scenarios in which individuals could be exposed to graywater (see Box 5-2), although these limited data should not be assumed to describe the risk of graywater use generally.

Chemicals

The committee could not find any published chemical risks assessments for graywater. Debroux et al. (2012) recently reviewed the potential human health effects associated with exposure to nonregulated trace organic compounds (including pharmaceuticals and personal care products) in recycled municipal wastewater from potable and nonpotable reuse. They concluded that none of the risk assessments conducted over the past 10 years found any adverse human health effects or significant risks. The risk of ingesting trace organic compounds after uptake into food crops is not well understood and has been identified as a research need for nonpotable reclaimed water (NRC, 2012a). Concentrations of trace organic chemicals in graywater may differ from those in recycled municipal wastewater. Compared to nonpotable recycled water, graywater would contain higher concentrations of personal care products (which are primarily derived from sink and shower water, and therefore would be diluted by other water sources in wastewater) and lower concentrations of pharmaceuticals, which are primarily excreted in urine and occasionally flushed directly down the toilet. In addition, the treatment levels for potable and nonpotable reuse tend to be much greater than those for typical graywater, if treatment is even applied, although the exposure levels for typical potable and nonpotable reuse projects are also much greater. Nevertheless, this analysis provides potentially useful reference information. The risks of long-term, low-level exposures to mixtures of trace organic compounds remain unclear for conventional drinking water sources, and typical nonpotable graywater exposures under best management practices would be lower.

Very limited information is available in the refereed literature on the risks of various uses of stormwater. A study of the risks associated with chemicals and pathogens in stormwater that was treated in a reed bed and then recharged into

TABLE 5-4 Comparative Risks of Infection across Pathogens from Treated, Non-disinfected Graywater

	Risk from Single Exposure		Annual Risk		
		Spray irrigation (1 mL exposure, 26 times/yr)	Groundwater recharge (2 liter/day consumption)		
Pathogen	Direct contact (1 mL/event)		1 mo. retention	3 mo. retention	6 mo. retention
Campylobacter jejuni	0.00158	0.00316	0.00316	3.2×10^{-9}	$<10^{-11}$
Cryptosporidium parvum	0.0000251	0.0002	7.9×10^{-8}	$<10^{-11}$	$<10^{-11}$
Giardia lamblia	0.00000316	0.0000316	1.26×10^{-8}	$<10^{-11}$	$<10^{-11}$
Rotavirus	0.25	0.63	0.63	0.000001	$<10^{-11}$
Salmonella	0.0000501	0.00001	0.01	10^{-11}	$<10^{-11}$

SOURCE: Based on data in Ottoson and Stenstrom (2003).

BOX 5-2 An Example Risk Calculation for Toilet Flushing with Untreated Graywater

From a graywater storage tank, Birks and Hills (2007) collected samples of untreated graywater derived from baths, showers, and hand basins from a subset of apartments in a residence hall and analyzed the samples for microorganisms. Giardia was found to be present at concentrations ranging from 0.5 to 1.5 cysts per liter (see Table 4-3). This information was used to make two sets of calculations regarding potential risk from exposure to untreated graywater. To determine the exposure volume and frequency, the exposure assumptions in DDOE (2013) (see Table 4-1) were used.

In the first example, a calculation of the risk from exposure to Giardia from using the untreated graywater for flushing toilets was made. DDOE (2013) assumes that an individual is exposed to 0.01 ml water from aerosols during a toilet flushing. The risk of Giardia infection was calculated using the following equation:

$$P = 1 - e^{-rN}$$

Where: P = probability of infection
N = number of microorganisms ingested in a single exposure
r = fraction of microorganisms that survive to initiate infection; constant for a given microorganism.
In the case of Giardia, r = 0.0198 (Haas et al., 2014).

Using the highest measured concentration of Giardia cysts, 1.5 cysts/liter:

$$N = 1.5 \text{ cysts/L} \times 0.01 \text{ mL} \times 1\text{L}/1{,}000 \text{ mL}$$
$$= 1.5 \times 10^{-5} \text{ cysts}$$

$$P = 1 - e^{-(0.0198)(1.5E\text{-}5)}$$
$$= 2.97 \times 10^{-7}$$

This is interpreted to mean that an individual has a risk of 1 in 3,367,003 of infection by Giardia after a single exposure to graywater from aerosols created by flushing a toilet.

The calculation can also be performed to determine the annual risk of infection. Using DDOE's assumption that the number of exposures over the course of a year is 1,100:

$$P_{\text{annual risk}} = 1 - (1 - P)^{1100}$$
$$= 1 - (1 - 2.97 \times 10^{-7})^{1100}$$
$$= 3.27 \times 10^{-4}$$

This is interpreted to mean that an individual has a risk of 1 in 3,061 of infection by Giardia over the course of a year from using untreated graywater to flush toilets. Note that this is the risk of infection; the risk of illness would be lower.

To put this risk into context, the amount of treatment that would have to be done on the graywater to achieve what the EPA considers an acceptable risk from microorganisms in drinking water was calculated. The EPA assumes that an acceptable annual risk of infection from microorganisms in drinking water is 1 infection per 10,000 individuals per year—an annual risk of 10-4. The calculation was done using the following process:

$$P_{\text{annual risk}} \text{ risk} = 10^{-4} = 1 - (1 - P)^{1100}$$
$$P = 9.091 \times 10^{-8}$$
$$P = 9.9091 \times 10^{-8} = 1 - e^{-rN}$$
$$N = 4.6 \times 10^{-6} \text{ cysts}$$

This is the number of cysts in the 0.01mL volume assumed to be the exposure volume from flushing toilets. This is equivalent to 0.46 cysts/liter. To achieve the same risk level as is acceptable from drinking water, the concentration of Giardia cysts in the graywater would have to be reduced from 1.5 cysts/liter to 0.46/liter, a reduction of approximately 70 percent.

These calculations are meant only to illustrate what risk might be present in one situation and cannot be extrapolated to other environments and conditions.

BOX 5-3 Quantitative Risk Assessment of a Stormwater Aquifer Storage Transfer Recovery System

Page et al. (2013) conducted a quantitative risk assessment to determine the potential risks to human health and the environment from an aquifer storage transfer recovery (ASTR) project in Salisbury, South Australia. The ASTR system collects stormwater from a catchment area and then transmits the water to an instream basin, a holding/storage basin, and a cleansing reed bed. The final phase involves injecting the water into an aquifer, from which it is ultimately withdrawn via a recovery well after approximately 200 days retention time. The researchers collected and analyzed water samples for numerous water quality constituents including nutrients, pathogens, indicator microorganisms, physical characteristics, major ions, metals, and organic micropollutants. The study focuses on the risk associated with three uses for the harvested stormwater: open space irrigation, nonpotable residential applications (e.g., toilet flushing, laundry), and drinking water. They found that the concentrations of most chemicals met most of the Australian Drinking Water Guidelines.

Page et al. (2013) chose *Cryptosporidium parvum*, *Campylobacter*, and adenovirus as the study pathogens because of their high infectivity, high DALYs per case of infection, and possible prevalence in stormwater. Rather than comparing the final concentrations to drinking water guidelines (as with the chemical contaminants), the researchers used DALYs. This allows for easier comparison of the impacts across microorganisms that can have very different health outcomes—from minor diarrhea to much more severe health effects. Using a guideline value of 10^{-6} DALYs per person per year for drinking water, the researchers determined the reductions in pathogen numbers that would be required to ensure protection of human health. The reduction levels that would be required to meet the tolerable risk level of 10^{-6} DALYs per person per year were as follows:

- For open space irrigation: 1.6 $\log_{10}$ reduction of viruses, 0.6 $\log_{10}$ reduction of protozoa, and 1.2 $\log_{10}$ reduction of bacteria
- For nonpotable residential uses (e.g., toilet flushing, laundry): 2.7 $\log_{10}$ reduction of viruses, 1.8 $\log_{10}$ reduction of protozoa, and 2.3 $\log_{10}$ reduction of bacteria
- For drinking water: 5.8 $\log_{10}$ reduction of viruses, 4.8 $\log_{10}$ reduction of protozoa, and 5.3 $\log_{10}$ reduction of bacteria.

an aquifer prior to recovery and use was performed by a group of scientists at CSIRO (Page et al., 2013). Risks from three pesticides (i.e., diuron, simazine, and chlorpyrifos) and three human pathogens (i.e., rotavirus, Campylobacter, and Cryptosporidium) were calculated. Box 5-3 provides the details of this quantitative risk assessment.

Albrechtsen (2002) published data on Cryptosporidium in roof runoff used for toilet flushing. These data are used in example calculations of risk for toilet flushing with captured rainwater (see Box 5-4). The results suggest a 1 in 2,170 annual chance of Cryptosporidium infection using the data reported and exposure assumptions from DDOE (2013).

These very limited data suggest that the risks of stormwater use should not be taken lightly in applications where human exposures are likely. They represent only two case studies, but they show how a risk framework can be useful for informing project treatment needs.

RISKS FROM SYSTEM FAILURES, CROSS-CONNECTIONS, AND GROUNDWATER CONTAMINATION

The risk assessment calculations discussed in the previous section represent risks incurred under routine exposures, but the risks of treatment failure, cross-connection between potable and nonpotable water lines, or inadvertent groundwater contamination also need to be considered as part of an overall understanding of risk.

Treatment Failures

To date, there have been no reported adverse health effects resulting from failures of graywater reuse systems (Sharvelle et al., 2013). However, failures in highly engineered systems will eventually occur, and the potential impacts of such events must be understood. In systems with substantial human exposures, real-time monitoring or redundant treatment systems can be developed to reduce these risks (see Chapter 6).

Cross-Connections

No graywater or stormwater cross-connections have been reported, but accidental or intentional cross-connections between nonpotable and potable water supplies are a public health concern, particularly for non-disinfected water supplies. There have been several reports in which cross-connections have occurred between potable water systems and nonpotable reclaimed water, some of which resulted in illnesses. For example, in 2007 it was reported that 12 individuals who worked at Melbourne Water became ill after consuming water in an administration building from a tap

BOX 5-4 Risk Calculation for Use of Stormwater for Toilet Flushing

Albrechtsen (2002) conducted a study of seven Danish stormwater collection systems, most of which harvested roof runoff, although some collected pavement runoff. The systems' collection areas ranged from 200 m^2 to more than 5,000 m^2; the stormwater was stored in collection tanks holding from 4 to 115 m^3 of water and was used to supply apartment buildings with 6 to140 units. Samples were collected from the storage tanks and from toilet bowls that contained untreated stormwater. Of the ten samples analyzed for *Cryptosporidium*, one was found to be positive, containing 10 oocysts per liter. (None of the toilet bowls containing potable water was found to contain detectable levels of *Cryptosporidium*, leading the investigators to conclude that the rainwater was the source of the pathogen.) To assess the exposure volume and frequency, the committee used exposure assumptions developed by DDOE (2013). In this example, a calculation to determine the risk from exposure to *Cryptosporidium* from toilet flushing was made. The DDOE assumes that an individual is exposed to 0.01 mL toilet water from ingestion of the aerosol spray generated during flushing. The calculation used the following equation:

$$P = 1 - e^{-rN}$$

Where: P = probability of infection

N = number of microorganisms ingested in a single exposure

r = fraction of microorganisms that survive to initiate infection; constant for a given microorganism.

In the case of *Cryptosporidium*, r = 0.004191 (Haas et al., 2014)

$$N = 10 \text{ oocysts/L} \times 0.01\text{mL} \times 1\text{L}/1000 \text{ mL}$$
$$= 10^{-4} \text{ oocysts}$$

$$P = 1 - e^{-(0.004191)(10E-4)}$$
$$= 4.191 \times 10^{-7}$$

This is interpreted to mean that an individual has a risk of 1 in 2,386,000 of infection by *Cryptosporidium* after a single exposure to aerosols from flushing a toilet containing this graywater.

The calculation can also be performed to determine the annual risk of infection. Using DDOE's assumption that the number of exposures over the course of a year is 1,100:

$$P_{\text{annual risk}} = 1 - (1 - P)^{1100}$$
$$= 1 - (1 - 4.191 \times 10^{-7})^{1100}$$
$$= 4.609 \times 10^{-4}$$

This is interpreted to mean that an individual has a risk of 1 in 2,170 of infection by *Cryptosporidium* after exposure to aerosols from flushing a toilet containing this graywater over the course of a year. This is 4.6 times more than the risk that the EPA considers to be acceptable from microorganisms in drinking water (1 infection per 10,000 people per year) (EPA, 1992).

that had been mistakenly connected to a reclaimed water pipe (Herald Sun, 2007). Another incident in Queensland, Australia, involved 630 homes in a housing development with dual plumbing for toilet flushing and outdoor use (WQRA, 2010). Within 2 days of the nonpotable reclaimed water being delivered to the homes, complaints of foul taste and odor were received from residents and a cross-connection was discovered. Hambly et al. (2012) described several cases in Australia in which cross-connections between potable water and reclaimed water occurred. Guidance and testing programs have been developed to minimize accidental cross-connections in water systems (AWWA, 2009; EPA, 2003) that could be applied to dual-plumbed graywater or stormwater use systems, as appropriate. The California Plumbing Code recommends annual cross-connection inspections and testing for permitted graywater and stormwater use systems.

Groundwater Contamination

Another potential failure that could affect human health is the unplanned recharge of stormwater or graywater into an aquifer used for a drinking water supply. Or planned aquifer recharge may not adequately remove contaminants before they reach a potable aquifer. Box 5-3 describes one published risk assessment for groundwater impacted by stormwater-derived contaminants. Because of the magnitude of the potential health risks associated with long-term ingestion of contaminated water, the committee describes what is known about the potential for groundwater contamination from graywater and stormwater in this section. Chapter 4 discusses contaminants of greatest concern to groundwater infiltration projects.

Graywater Infiltration

There is some concern about graywater constituents leaching to groundwater when graywater is applied for irrigation, particularly in cases where graywater is applied over large areas at rates greater than required based on evapotranspiration. When graywater is used for irrigation, constituents of primary concern for groundwater quality include nitrogen, salts (including sodium, chloride, and boron), pathogens, and organic contaminants from cleaning or personal care products. However, the actual human health risk would depend on many factors, including contaminant concentrations in graywater (see Chapter 4), irrigation rates, potential for contaminant sorption or biodegradation, soil and aquifer characteristics that affect contaminant transport, depth to the water table, and distance to the point of groundwater withdrawal. In a soil-column study, Negahban-Azar et al. (2013) showed that salts have potential to leach through graywater-irrigated soil. This would pose the most concern in soils with high sand content and/or where high infiltration rates are observed. Stevens et al. (2011) noted a risk for salt transport to groundwater, which would pose the most risk in arid climates where evapotranspiration is high. Although leaching of phosphorus is generally not a concern because of limited mobility in soil, leaching of inorganic nitrogen is possible. Surfactants leached from columns ranged from 0 to 20 percent of what was added, resulting in low concentrations in leachate, although surfactant concentration in leachate increased over the 17-month duration of experiments. To reduce these risks, at least two states (i.e., New Mexico, Arizona) recommend that household graywater not be applied in areas where the seasonally high groundwater table is less than 5 ft from the application point (Sharvelle et al., 2013). Source control (best management practices) can also be used to reduce the concentrations of contaminants in graywater used for irrigation (see Chapter 4).

Stormwater Infiltration

As described in Chapters 1 and 2 of this report, there is increased interest in urban stormwater capture and enhanced infiltration through engineered structures as a means to manage urban stormwater and peak flows, reduce non-point pollution, and replenish groundwater supplies. Because urban runoff also contains pollutants, there exists the potential to contaminate groundwater during infiltration, especially in the future if large volumes of urban stormwater are captured for groundwater recharge, thereby increasing human health risks for current or future groundwater users. As outlined in Table 4-4, pollutants in urban stormwater include salts, suspended solids, nutrients (e.g., nitrogen and phosphorus), heavy metals (e.g., copper, lead, chromium, nickel, and zinc), organic compounds from automotive use and biocide applications, and pathogens. The likelihood for these pollutants to migrate through the soil and contaminate groundwater during stormwater infiltration depends on a number of factors including the infiltration rate, permeability and character of the soil or infiltration media, biological activity in the subsurface, depth to the water table, and the properties of the pollutants. With the growing practice of "enhanced infiltration" for groundwater replenishment, there is concern that these practices may put the groundwater at risk from chemical and microbial contaminants (Nieber et al., 2014).

Risk Factors for Chemical Contamination of Groundwater. Chemical pollutants in urban stormwater that are most likely to contaminate groundwater are those that are relatively non-volatile, hydrophilic (dissolve or mix easily in water), ionic, and non-sorbing. Soluble, non-sorbing salts, such as road deicing compounds, will flow with the infiltrating runoff and not be removed during infiltration (Bannerman et al., 2014; Mullaney et al., 2009). The existing literature on the fate of organic compounds in stormwater infiltration systems, however, is much less than for heavy metals (Weiss et al., 2008). Mikkelsen et al. (1996) studied metal movement in percolating stormwater at two infiltration systems in Switzerland, and metal concentrations in the water were found to decrease rapidly to background conditions within 1.5 m of depth. In Perth, Australia, Appleyard (1993) reported sediment concentrations of 3,500 ppm of lead in stormwater infiltration basins because of strong sorption to iron oxides. Nightingale (1987) studied water quality beneath five stormwater recharge basins in Fresno, California. The basins drained single-family residential neighborhoods and captured winter stormwater; sampling at depths up to 26 m showed no contamination except for trace levels of diazinon.

The potential for subsurface transport of metals and sorbing organic chemicals into groundwater depends on the character of the media and whether fine solids are retained by filtration. Stormwater contaminants are unlikely to be removed if stormwater infiltrates directly into coarse media or karst formations with extremely high percolation rates with little or no opportunity for attenuation or filtration (e.g., Stephensen et al., 1999). A study of 15 dry wells—precast concrete structures with open bottoms resting on and surrounded by crushed stone for subsurface disposal of stormwater—found no subsurface changes in water quality for filtered forms of copper, lead, and zinc or for *E. coli* and enterococci, even after percolating through gravel and at least 4 feet of urban subsurface soils (Pitt and Talebi, 2012a). Similarly, groundwater is more at risk from stormwater contaminants in areas where the soil is sandy and the groundwater is shallow (Fischer et al., 2003). In a study of groundwater beneath 16 stormwater detention basins in New Jersey that had sandy and unconsolidated soils, the sampling showed elevated levels of petroleum hydrocarbons, as well as the herbicides metolachlor and prometon. Fischer et al. (2003) concluded that high recharge in urban stormwater basins may impact groundwater even when the constituent concentrations are low.

Risk Factors for Pathogen Contamination of Groundwater. Few studies have examined the efficacy of infiltration practices for removing pathogenic organisms (Weiss et al., 2008). Because pathogens are typically associated with particles, physical straining through the soil or engineered media may remove pathogens just as sand filters are used in water treatment, although the effectiveness will depend on the organism and the porous media properties. Straining is most effective for protozoan pathogens (greater than 3 microns), such as *Giardia lamblia* and *Cryptosporidium parvum*, and larger bacteria (approximately 1-2 microns) than viruses, which are too small (0.02 to 0.08 microns) to be effectively removed by filtration through porous media. However, virus removal during passage through porous media may also occur via attachment to soil particles or aquifer material, depending on the specific characteristics of the virus and the environment (see Schijven and Hassanizadeh, 2000). Clark and Pitt (2007) documented pathogen contamination of groundwater due to infiltration practices that included stormwater sand filters. High bacterial and virus concentrations were found in groundwater on Long Island where the groundwater table was close to the land surface. In contrast, the Orange County Water District performed extensive analysis of viruses and protozoa in groundwater in the Santa Ana River basin, where reclaimed wastewater and stormwater recharges groundwater, and concluded that the surface recharge is not a significant source of pathogens to groundwater (NWRI, 2004; OCWD, 2004).

Inactivation also plays a key role in determining whether pathogens present in the infiltrating water will survive to contaminate the underlying groundwater. A large body of literature describes the numerous factors that affect the length of time that microorganisms can survive in the subsurface (see, e.g., John and Rose, 2005). Typically, microorganisms survive longer at cooler temperatures and near-neutral pH conditions (John and Rose, 2005; Yates et al., 1988). Some bacteria are reported to survive longer in acidic soils and soils with a large amount of organic matter, as long as several months (Pitt et al., 1999). Viruses typically survive longer than bacteria (Yates et al., 1988; Sidhu et al., 2010). As a general rule, protozoan parasites and helminth ova survive longer in the environment than the other types of enteric pathogens because of their environmentally resistant non-metabolically active forms (e.g., *Cryptosporidium parvum* oocysts).

ENVIRONMENTAL RISKS

In addition to health effects on humans, the potential impacts of contaminants in graywater and stormwater on the environment should be considered when significant environmental exposures are likely. Assuming best practice are followed and graywater is not ponded or discharged directly to surface water during irrigation, aquatic organisms should not experience significant contaminant exposures. For stormwater capture projects, ecological exposure scenarios may not be common, but for projects with surface impoundments or wetland treatment cells (see Box 2-6), aquatic life may become an intentional or unintentional component of the project, where they could be impacted by trace metals or organic contaminants (Grebel et al., 2013). Risks could also include algal blooms and low dissolved oxygen associated with elevated nutrients, leading to fish die off in stormwater ponds. The committee found only a few analyses of ecological risks involving graywater (Gross et al., 2005; Maimon et al., 2010), although more work has been done on ecological risks associated with stormwater ponds (reviewed in Tixier et al., 2011).

Stormwater retention ponds and wetlands typically provide entirely new aquatic habitat, enhancing biodiversity in the urban environment (Brand and Snodgrass, 2010; Le Viol et al., 2009). However, in these settings stormwater-derived contaminants in water and sediment can exceed probable-effect levels for aquatic life (VanLoon et al., 2000; Wik et al., 2008) and cause adverse ecological impacts, such as lethal and sublethal effects on embryonic and larval amphibians (Bishop et al., 2000b; Snodgrass et al., 2008). Thus, stormwater ponds tend to be populated by more pollution-tolerant organisms (Wik et al., 2008) and have low species richness (Bishop et al., 2000a). A central issue when considering ecological risk, therefore, is to determine what level of impairment is acceptable in consideration of the new environmental benefits provided. Without stormwater capture features, urban streams are impacted by the same contaminants as well as extremes in flow associated with runoff from largely impervious surfaces. The EPA (1998) developed a framework for ecological risk assessment that can be used to evaluate the probability of adverse ecological effects, which includes comparison of field data to reference sites. Selection of an appropriate reference site in the urban environment is critical because the degree of impairment in the reference site determines the magnitude of calculated risk and the habitat objectives of the stormwater project (Tixier et al., 2011). This challenge has led some researchers to call for new strategies to understand the ecological functioning of stormwater-based habitats to better develop ecological objectives and management measures (Lafont et al., 2007; Tixier et al., 2011).

In addition to assessing potential toxicological effects on aquatic life, graywater or stormwater irrigation projects should be aware of potential water quality impacts to plants and to soil properties. Constituents of greatest concern for stormwater include salinity, sodium, chloride, and metals (EPA, 2012a). For graywater, contaminants of concern include nitrogen, phosphorus, salinity, boron, sodium, chloride, and surfactants (Sharvelle et al., 2013). As discussed in Chapter 4, salinity is a key concern, because of potential negative effects on plant health and because high quantities of sodium relative to calcium and magnesium (measured as the sodium adsorption ratio [SAR]) can impact soil structure, making the soil less permeable and more erodible, particularly soils with a high clay content. In a study of seven U.S. households using graywater for irrigation, soil SAR levels were elevated in graywater-irrigated soil compared to freshwater-irrigated soil, and the soil SAR at a site with more than 30 years of graywater irrigation was 2-22 times greater than at the control site. However, at all sites, the soil SAR was less than 5, low enough to prevent any harmful effect for plants' water uptake (Sharvelle et al., 2012). Salinity is likely to be a greater concern in arid climates with high evapotranspiration rates and fewer rainfall events to flush the soils of salt build-up (Stevens et al., 2011).

Excess concentrations of boron, metals, and surfactants can be toxic to plants, and surfactants in graywater can also cause the soil to become more hydrophobic, impacting plant health (Garland et al., 2000; Gross et al., 2005). A large percentage of the applied surfactants have been shown to biodegrade in the soil, and although surfactant concentrations are elevated in graywater-irrigated soils compared to control sites, surfactants do not accumulate in soils over time (Sharvelle et al., 2012). Boron has been found to accumulate in soils irrigated with graywater for more than 5 years (Negahban-Azar et al., 2012), which may be a concern. In addition, some plants may be more sensitive to contaminant effects than others. For example, Sharvelle et al. (2013) observed reduced growth or adverse plant health effects from graywater irrigation on only 3 species (i.e., avocado, lemon tree, and Scotch pine) out of 22 studied.

Although indicator organisms are present in graywater, graywater-irrigated soil has not been found to consistently contain elevated concentrations of indicator organisms compared to potable water-irrigated soil (City of Los Angeles, 1992; Negahban-Azar et al, 2012; Sharvelle et al., 2012). In these studies, animals were known to contribute to indicator organisms in both graywater- and potable water-irrigated areas, and it was not possible to differentiate the contribution of indicator organisms from graywater from other natural contributions of indicator organisms. Antimicrobial chemicals have been found to accumulate in graywater-irrigated soil (Negahban-Azaer et al., 2012). The effects of antimicrobial chemicals on soils are not well understood, and there is increasing concern that they might contribute to the abundance and persistence of antibiotic resistance in soil microorganisms (Auerbach et al., 2007).

TABLE 5-5 Published Water Quality Guidelines for Irrigation

Hazard	Australian and New Zealand Trigger Values	2012 EPA Water Reuse Guidelines
Boron (mg/L)	0.500	0.75
Cadmium (mg/L)	0.01	0.01
Copper (mg/L)	0.2	0.2
Iron (mg/L)	0.2	5.0
Lead (mg/L)	2.0	5.0
Zinc (mg/L)	2.0	2.0
Salinity (μS/cm)	<950 to 12,200[a]	<700 to 3,000
Chloride	<175 to >700 mg/L	
Sodium	<115 to >460 mg/L	

[a]Depending on crop sensitivity.

NOTE: Trigger values are established to minimize soil build-up of contaminants and prevent direct toxic effects to crops. Australian trigger values for metals reflect the long-term trigger values, assuming tolerance for 100 years.

SOURCES: Australian and New Zealand Environment and Conservation Council and the Agriculture and Resource Management Council of Australia and New Zealand (2000); EPA (2012a).

Guidelines for irrigation quality to minimize adverse effects have been established by the EPA and in Australia and New Zealand (see Table 5-5). Most stormwater outfall samples fall within these guidelines (see Table 4-4), although household-scale projects capturing runoff from roofs with certain materials could exceed guidelines for metals (see Box 4-1). Source control strategies can be used to control boron and sodium in graywater and salts in stormwater (see Chapter 3).

CONCLUSIONS

Although no documented reports of adverse human health effects from the beneficial use of stormwater or graywater have been identified, additional examination of risk is necessary to support safe and appropriate design and implementation of stormwater and graywater use systems. This effort will be especially important as the use of graywater and stormwater becomes more widespread, particularly in water-scarce regions.

Risk assessment provides a means to determine "fit-for-purpose" water quality criteria or treatment needs based on human exposures. Risk from graywater or stormwater is a factor of chemical or microbial concentrations and exposure (typically, the amount or water ingested). Thus, unlike drinking water criteria, which are established based on 2 liters of water consumed per day, criteria for applications with minimal human exposures might allow for much higher concentrations of contaminants in graywater or stormwater and still result in acceptably low health risks. Risk assessment tools provide a ready means for developing such criteria for many chemicals and microbes for which drinking water criteria exist. As nonpotable on-site use of graywater and stormwater becomes more common, additional public health risk communication efforts would be beneficial to help the public understand risk-based treatment objectives and appropriate safeguards.

Considering the low exposures in most nonpotable graywater and stormwater applications, pathogens represent the most significant acute risks. Available risk assessments and the committee's risk calculations using limited, observed pathogen data and various possible exposure scenarios suggest that disinfection is necessary for many uses of graywater, including spray irrigation, food crop irrigation, and toilet flushing, to protect human health. Subsurface landscape irrigation with graywater does not pose significant risk, if best practices are followed, because human exposure is minimized. These findings are consistent with most regulatory guidance (see Chapter 8), although the risk of surface drip irrigation (without landscape cover) at the household scale remains unresolved. Limited data on pathogens in roof runoff suggest that treatment may also be needed, even for low levels of human exposure, such as toilet flushing, although more research on pathogens in roof runoff is needed. Chemicals become of concern in groundwater infiltration projects, where drinking water supplies could be impacted.

Extremely limited data are available on the pathogen content in graywater and roof runoff, which precludes a full assessment of microbial risks. Most water quality monitoring assesses microbial indicator data, and microbial risk assessments are conducted using assumed relationships between the concentrations of indicator microorganisms and pathogenic microorganisms. Consistent relationships between surrogates and contaminants have not been established for graywater or stormwater. Such relationships would be extremely variable at smaller scales but, even at large scales, could differ substantially from traditional indicator-pathogen relationships derived for municipal wastewater. This is a particular concern for roof runoff, which may include microbial indicator organisms from the waste of animals that do not transmit human pathogens. Therefore, the actual concentration of human pathogens in the water—and the associated risk of exposure to that water—may be much higher or lower than that calculated using the concentration of indicator microorganisms.

Enhanced infiltration of stormwater for groundwater recharge poses risks of groundwater contamination and necessitates careful design to minimize those risks. The risk of groundwater contamination from stormwater recharge is related to the contaminants present, any pretreatment processes installed, the capacity for the subsurface soil and engineered media used in the infiltration basin to remove them, and the proximity to groundwater used as a drinking water supply. Dry wells, which directly inject water into the subsurface and surface infiltration through sandy soils do not effectively attenuate chemical contaminants, and treatment prior to injection might be needed to prevent groundwater contamination.

As the uses of graywater and stormwater become more common, care to prevent cross-connections needs to be taken. Cross-connections between potable and nonpotable water systems can expose residents to elevated risks. No reports of adverse health effects from cross-connections between graywater or stormwater systems and potable systems have been documented, but there have been reports of cross-connections between reclaimed water and potable systems that have resulted in illnesses. Regulatory guidance and inspection criteria can help reduce cross-connection risks.

Environmental impacts from the outdoor use of graywater and stormwater generally appear low, but risks depend upon several factors, including water quality, application rates, and plant or animal species exposed.

Effects of irrigation on plant and soil health can occur from salts, boron, and metals, but source control practices and appropriate irrigation rates can reduce these impacts. If not controlled at the source, then long-term build-up of boron or salt can pose risk to plant and soil health, depending on soil and climatic conditions. Constructed stormwater ponds and wetlands typically contain elevated contaminant levels sufficient to impair reproduction among some aquatic species, often leading to a habitat dominated by pollution-tolerant organisms. Such ecological affects may be acceptable, considering the overall environmental benefits provided by such features, including reduced pollution to other surface waters, but the ecological objectives of such projects are often unclear, hindering efforts to limit ecological risks through improved management and design.

6

State of Design Practice for Stormwater and Graywater

Stormwater and graywater are similar in that they are on-site sources of water that, in many cases, can be used beneficially. However, source conveyance, capture, and distribution for end uses are very different. Although treatment for both water sources is simple at the smallest scales and increases in complexity with scale and degree of human contact, graywater and stormwater pose different treatment concerns. This chapter describes general design principles, the state of practice for graywater and stormwater system designs at household, neighborhood, and regional scales, and treatment that may be used to meet specific end use quality objectives.

OVERALL DESIGN PROCESS

The overall planning and design processes for graywater and stormwater systems (including integrated graywater and stormwater systems) for beneficial use have many similarities. Both start with the identification of project objectives. Chapter 1 outlines a range of objectives that may be associated with on-site graywater or stormwater projects, such as providing a cost-effective alternative water supply, diversifying the water portfolio, reducing runoff volume and peak flows, reducing downstream pollutant loads, and conserving water and energy resources. In addition to the overarching project objectives, additional design objectives may drive major choices in system configuration (some common design objectives are included in Table 6-1). It is important to consider all project objectives when deciding whether and how to design a beneficial use project.

Figure 6-1 outlines the general steps for the progression of project design, beginning with overall objectives for the beneficial use project. The quantity, quality, and timing of available on-site water supplies (see Chapters 3 and 4) then need to be assessed. The basic feasibility of beneficial use projects can be determined by comparing the available quantity with the anticipated demands, considering the possible water use applications (see Chapter 2). Major mismatches in the quantity or timing of supply and demand suggest that the project may not be appropriate or cost-effective for the site conditions. Once general feasibility is determined, system design alternatives for graywater or stormwater capture, storage, and treatment can be considered and evaluated based on their cost, other benefits provided, and the extent to which the project achieves its objectives. Unit processes within the treatment train (if applicable) would be selected based on factors such as water quality objectives, cost, maintenance requirements, and project scale. Figure 6-1 outlines the design-related steps within the broader decision framework for a project, discussed in Chapter 9.

The following sections discuss system design considerations for the beneficial use of graywater and stormwater. Unit processes for graywater and stormwater capture and use systems are discussed at the end of the chapter.

GRAYWATER SYSTEM DESIGN

Typical graywater systems are designed to supply water for irrigation and/or toilet flushing. A key factor is matching the graywater supply with demand so that the resource is managed efficiently. Approaches to estimate graywater demand are summarized in Bergdolt et al. (2011). Table 6-2 summarizes some key considerations for graywater project design across different scales. Graywater generation in multi-residential buildings may often be greater than can be effectively used for irrigation, while commercial or industrial buildings may not generate enough to meet specific needs. The following sections outline the state of practice for graywater use at the household or neighborhood scale.

Collection and Storage

Graywater collection requires separate collection of graywater sources (i.e., showers/baths, lavatory sinks, and laundry) from blackwater (i.e., toilet and kitchen water). In conventional homes, graywater sources are combined with

TABLE 6-1 Examples of System Design Objectives Relevant to On-site Use of Graywater or Stormwater

Category	Typical Design Objectives
Regulatory	Meet local, state, or federal requirements
Cost	Minimize capital, operation, and maintenance (life-cycle) costs or achieve lower unit costs than alternative water supplies
Aesthetic and public perception	Improve appearance and public perception of a site
	Contribute to positive environmental action
Maintenance	Operate within maintenance and repair schedule and requirements
	Design systems to allow for modification or expansion
Longevity	Achieve long-term functionality
Resources	Improve downstream aquatic environment
	Improve wildlife habitat
	Achieve multiple use functionality
Safety, risk, and liability	Function without significant risk or liability
	Function with minimal environmental risk downstream

SOURCE: Adapted from TRB (2006).

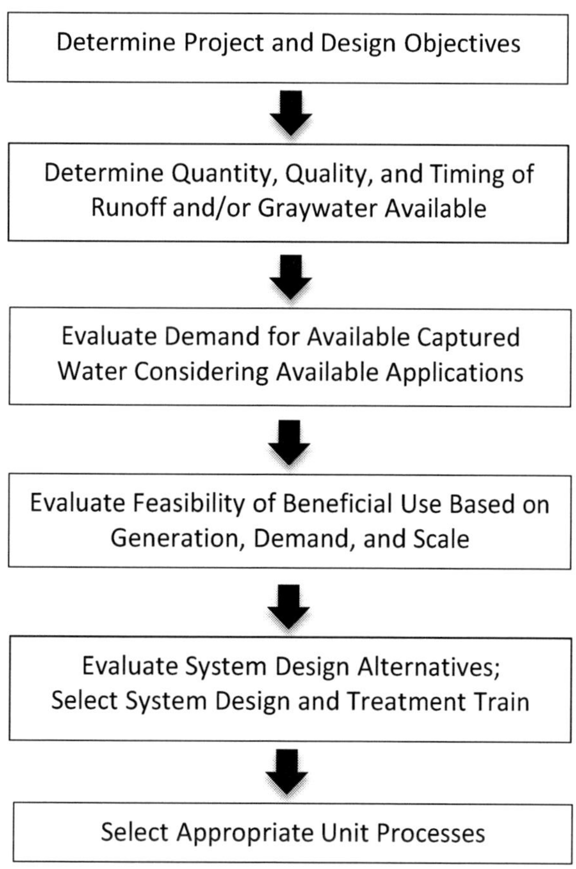

FIGURE 6-1 Typical steps to evaluate feasibility of and to design a graywater and stormwater beneficial use system.

blackwater sources very near to plumbing fixtures. Where graywater is reused, separate pipes are installed for graywater collection from fixtures to divert graywater to a treatment and/or use system (Figure 6-2). Guidance on collection and use of graywater for both irrigation and toilet flushing is included in Bergdolt et al. (2011).

In existing one-story homes with an unfinished basement or crawl space, graywater collection can be accomplished fairly simply. However, in most existing homes, retrofits for separate collection of household graywater are too costly to justify unless conducted as part of a major home remodeling effort. Simple laundry-to-landscape systems with no storage are often installed in existing development, and many homeowners can build this simple, low-cost system themselves. New development projects, however, can consider separate collection of household graywater, and graywater may even be collected and treated at a neighborhood (multi-residential) scale (see Figure 2-1). Neighborhood-scale graywater reuse removes the burden of maintenance from the homeowner and enables centralized quality assurance inspections to ensure that the system is meeting water quality requirements. The two basic graywater systems are discussed below, followed by discussions of treatment considerations.

Laundry to Landscape

A simple design for graywater irrigation systems is the laundry-to-landscape system (Figure 6-3), which is typically applied only at the household scale. This system does not require storage or filtration and has been applied at many households. Laundry water is pumped directly from the washing machine to valve boxes submerged in mulch basins. The water is then conveyed for subsurface irrigation into tree trenches or mulch basins or through subsurface irrigation to minimize human contact. Mulch basins include drip irrigation systems below ground surface covered with mulch. Many states and the 2012 Uniform Plumbing Code recommend that drip irrigation systems be covered by at least 2 inches (5 cm) of mulch to reduce human exposures (see Chapter 8, Box 8-1). Many states also provide guidance on graywater application rates based on soil type to ensure adequate infiltration of graywater and minimize build-up of salt in clay soils. Pumps internal to laundry machines provide sufficient pressure to distribute graywater through these systems. Homeowners have generally been satisfied with laundry-to-landscape systems and have not noted excessive maintenance issues (Box 6-1 and Box 2-4). In climates where freezing temperatures are observed in winter months, graywater is diverted to a wastewater collection system for those months to prevent breaks in the distribution plumbing.

Graywater Systems with Storage

Beyond the most simple laundry-to-landscape systems, graywater systems that collect water from multiple sources (e.g., laundry, shower, handwash basins, and bath tubs) need to include a storage tank (Figure 6-4). Storage tanks must be well sealed and include a vent and an overflow and drain to a wastewater collection system. The 2012 International Plumbing Code (IPC), which has been adopted by some

TABLE 6-2 Design and Feasibility Considerations Specific to Household, Neighborhood, and Commercial and Industrial Applications of Graywater Beneficial Uses

	Household	Neighborhood/Multi-residential	Commercial and Institutional
Irrigation	• Subsurface irrigation (necessary to reduce risk) sometimes not practical in grassy areas.	• At neighborhood scales, more graywater is often generated than is required for irrigation. • Treatment may be required depending on state regulations and whether there is a high likelihood for human exposure. • Human health risks from untreated graywater are higher than at the household scale, because graywater could serve as a vehicle for disease spread between households (see Chapter 5).	• Graywater generation rates vary widely among commercial and institutional settings, so graywater production rates and demands for end uses need to be evaluated on case-by-case basis (see Chapter 3). • Often graywater production is not sufficient to meet end use demands, but can be suitable when showers or laundry are on site (e.g., fitness facilities, hotels, offices with showers, aquatics centers). • Hospitals are not appropriate sites for graywater use because of contamination potential.
Toilet Flushing	• Treatment is required and systems can be complex for homeowners to maintain. • Difficult to enforce water quality requirements. • Risk for cross-connection and improper maintenance resulting in health risks.	• Graywater volume generated is often suitable for toilet flushing. • Treatment is required. • Maintenance is required (but can be provided by third-party). • Stacked plumbing can be favorable in buildings with less than 10 floors, but can be complex in high rises.	

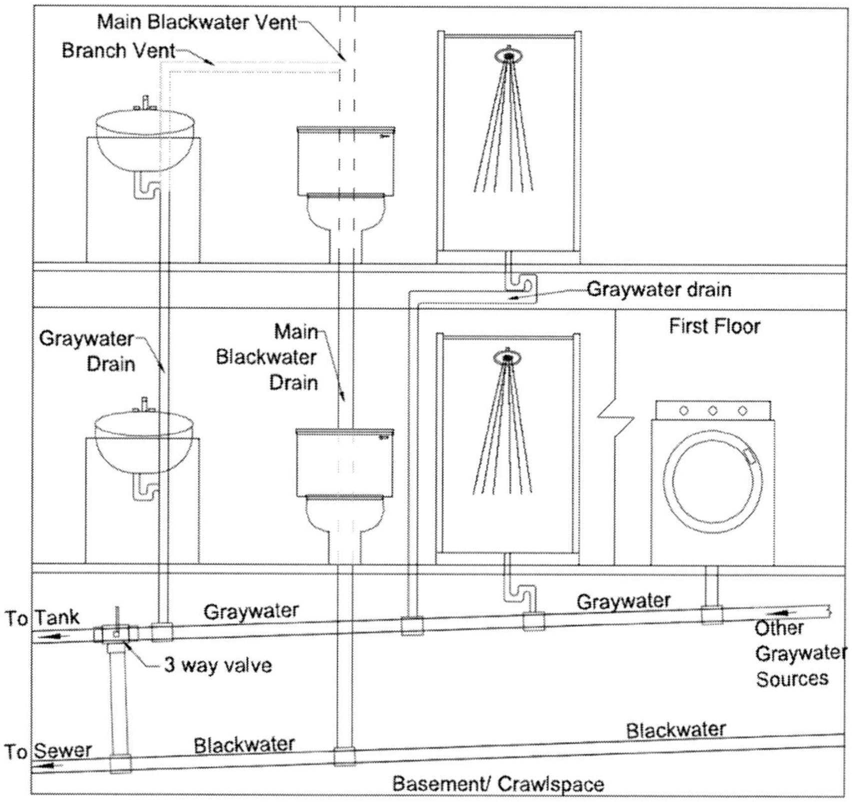

FIGURE 6-2 Plumbing for separate collection of graywater. SOURCE: Bergdolt et al. (2011).

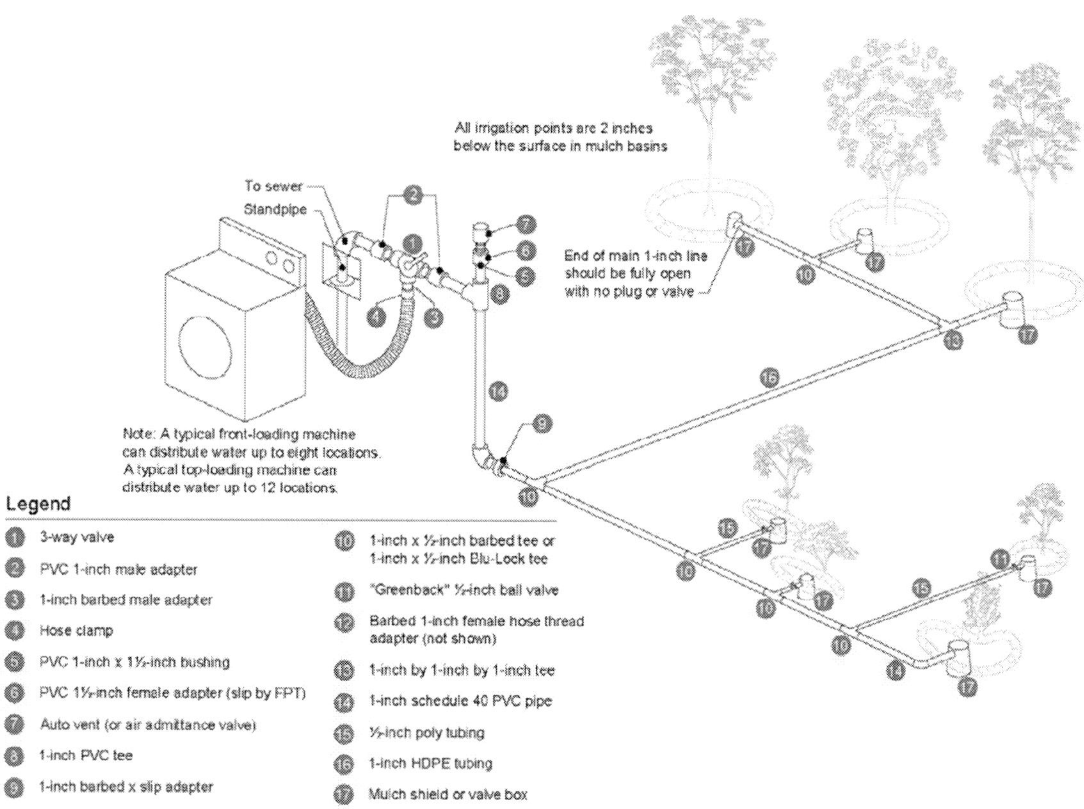

FIGURE 6-3 Laundry to landscape system. SOURCE: SFPUC (2012).

BOX 6-1 User Satisfaction with Graywater Reuse Systems

A number of studies have been conducted to assess user satisfaction with graywater reuse, and all studies focused on systems that were installed at the request of the homeowner or resident, rather than mandatory applications. A study was conducted on 25 homes in Guelph, Canada, using graywater for toilet flushing (City of Guelph, 2012). Of the 25 homes studied, the Brac Greywater Reuse System (no longer commercially available) was installed in 24. This simple treatment system includes filtration and chlorination. Organic matter and turbidity are not target constituents for removal in the system. Surveys were administered after 1 year of operation, and the majority of study participants rated their system as "good." The following quote from City of Guelph (2012) describes user feedback on the graywater reuse systems (GWRS):

> Some of the difficulties with the GWRS that were raised included: difficulties with motor controls, system operating too often and/or too noisily, overflow and flooding issues, and difficulty with access to the tank and/or filter. The frequency of filter cleaning was mentioned as a problem for some users, while others were able to adapt to this new responsibility. Most users that were interviewed indicated that the system requires a lot of diligence in cleaning. Some went as far to say that when the routine of the cleaning is established, the system functions quite well. Of those interviewed, five users indicated that they clean their filters weekly, and 5 users cleaned their filters monthly.

In summary, users were generally happy with the installed systems to flush toilets with graywater but noted some challenges in managing required maintenance.

Participants of the City of Long Beach's Laundry-to-Landscape Pilot Program were also surveyed (Gallup, City of Long Beach, personal communication, 2014). In this study, 15 of the 26 participating homeowners scored their level of satisfaction with the overall functioning of the system as 5/5, or "very satisfied." Eight reported being "satisfied" (4/5), and 3 reported being "neutral" (3/5). No participating homeowner reported being "dissatisfied" or "very dissatisfied."

In summary, the few studies done on user satisfaction with graywater reuse indicate that users were generally happy with the installed systems. Maintenance was not an issue for users of laundry-to-landscape systems, but it was for users of the system to use graywater to flush toilets.

states,[1] specifies a maximum graywater storage time of 24 hours for irrigation and 48 hours for toilet flushing. Thus, storage tanks are designed based on expected graywater generation such that graywater not used in a 24-48 hour period will overflow to a wastewater collection system. Guidance for design and sizing of storage tanks is provided in Bergdolt et al. (2011). Storage tanks act as a settling tank for large particles (e.g., lint and hair) present in graywater. Therefore, the graywater in storage tanks should not be mixed to avoid re-suspension of previously settled materials.

Graywater irrigation systems with storage also include a coarse filter and pump (Figure 6-4; Bergdolt et al., 2011). Coarse filters should be selected based on specifications provided by the manufacturer of drip supply lines and will typically be around 120-150 mesh (110-125 μm). When graywater is used for toilet flushing, a separate contact tank is recommended for disinfection. Water is drawn as needed from the storage tank and then diverted to the disinfection tank. The technology is mature, and commercially available kits to install graywater irrigation systems may increase adoption of the practice.

[1] See http://www.iccsafe.org/gr/Documents/stateadoptions.pdf.

Graywater Treatment Design Considerations

The extent and type of graywater treatment system depends on the end use for graywater. The following section outlines basic treatment design considerations for various end use graywater applications.

Subsurface Irrigation

When human exposures to graywater are minimized (e.g., subsurface irrigation), minimal treatment is required (see Chapter 5). The simple laundry-to-landscape design, where untreated graywater is supplied at least 2 inches below the ground surface, is allowed in many of the states that allow graywater use for irrigation. Among the 25 states that currently allow household-scale graywater use for irrigation (see Chapter 8), 20 states have not established enforceable water quality requirements for subsurface irrigation at the household scale. Instead, common sense best management practices are provided to guide safe application of graywater for irrigation (Sharvelle et al., 2013). In such states, laundry-to-landscape systems with no treatment (Figure 6-3)

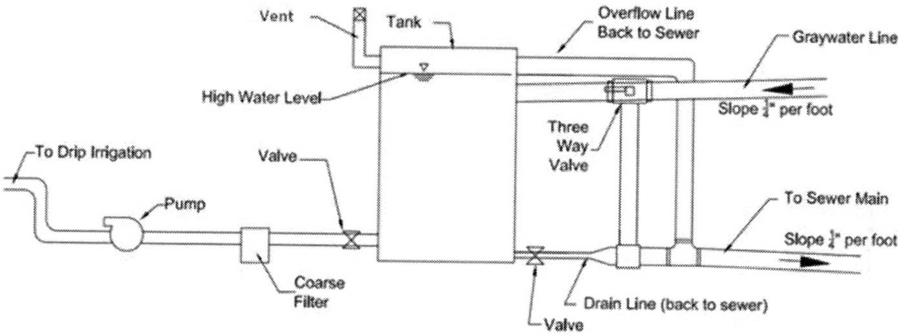

FIGURE 6-4 A storage tank for separate collection of graywater. This example shows components for irrigation end use. SOURCE: Bergdolt et al. (2011).

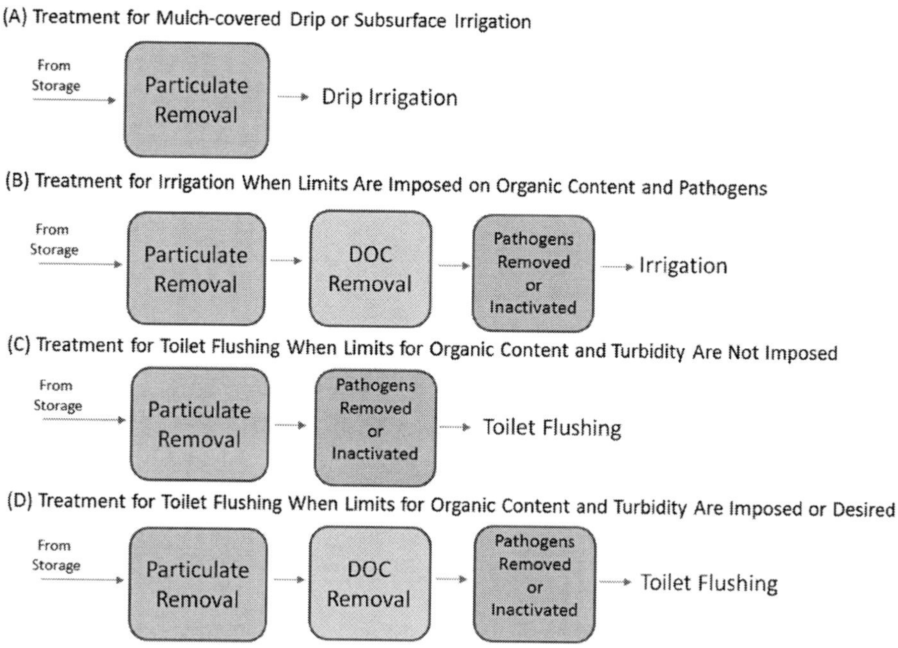

FIGURE 6-5 Recommended treatment systems for graywater for various end uses and regulatory requirements. NOTE: Storage and filtration are not required for the laundry-to-landscape system, which is not included in this figure.

or simple graywater systems with storage and particulate removal (Figures 6-4 and 6-5a) are typically used. Filtration is necessary when a conventional drip irrigation system is used for graywater to prevent clogging (Figure 6-5a). Graywater systems without disinfection are appropriate where exposures are restricted by subsurface irrigation systems. The additional risks associated with graywater applied by surface irrigation (not covered by mulch) at the household scale are less clear, because pathogen data for graywater are extremely limited. Additionally, graywater is only one of many existing vehicles of infectious disease transmission already present within the household (see Chapter 5). Potential risks, therefore, increase with exposures to untreated graywater at neighborhood and regional scales, because graywater provides a new vehicle for disease spread between people that might not otherwise be in contact.

Only four states impose water quality standards for subsurface irrigation at the household scale (Table 6-3). The states of Wisconsin and Maine have set indicator organism limits as part of their irrigation water quality requirements (Table 6-3). Based on typical graywater quality (see Chapter 4; Table 4-1), disinfection is required in those states that impose limits on indicator organisms for subsurface irrigation. Organic carbon or solids removal for single residence irrigation is required in five states that currently allow graywater irrigation (Table 6-3). In those states, requirements for organic carbon (often measured as 5-day biochemical oxygen demand, BOD_5) and total suspended solids (TSS) are well below typical amounts found in raw graywater. Thus, a treatment system would be required in these states to remove organic carbon and solids (see Figure 6-5b). Because organic carbon in graywater occurs primarily in the dissolved form (i.e., from personal care products), more extensive treatment is required to remove dissolved organic carbon (DOC) compared to stormwater, where organic carbon is often in the particulate form and can be removed through sedimentation and filtration. For graywater, biological or physicochemical treatment systems (discussed later in the chapter) are required to remove DOC if BOD_5 concentrations of less than or equal to 30 mg/L are desired. Graywater systems that employ the treatment processes identified in Figure 6-5b are most appropriate at the neighborhood scales or larger because of the complexities of system components that remove dissolved organic matter. Current regulations addressing organic matter in subsurface irrigation may be overly conservative. Research conducted by Sharvelle et al. (2012) showed that such treatment is not required for successful operation of subsurface irrigation systems that use graywater.

Some basic treatment systems are outlined in Figure 6-5, where the contaminants of concern for removal include suspended solids, pathogens, and in some cases DOC. One other water quality concern for graywater treatment is odor. Graywater can become odorous after it is stored for several hours in an anaerobic environment. Storage tanks, as shown in Figure 6-4, must be properly sealed and vented to direct odorous emissions to an appropriate location, preventing odor buildup in the building.

Toilet Flushing

Further treatment is necessary to reduce human health risks when graywater is used to flush toilets because of potential human exposures to pathogens present in untreated graywater (see Chapter 5). All states with graywater regulations that address toilet flushing require disinfection, and seven states have laws that set specific water quality requirements for toilet flushing (Table 6-4). Wide variability exists in these state regulations in terms of target concentrations

TABLE 6-3 State Water Quality Standards for Residential Graywater Irrigation Through Surface Drip and Subsurface Systems

	BOD_5 (mg/L)	TSS (mg/L)	Maximum Total Coliform (cfu/100ml)	Maximum Fecal Coliform (cfu/100ml)
Florida	25	30	-	-
Georgia	30	30	-	-
Wisconsin	30	35	-	200
Maine	30	10-30	10-100	-
State water reuse regulations for restricted access urban reuse	10-60	5-60	240[a]	23-800[a]
Reported range of typical graywater prior to treatment[a]	30-380	50-280	500-630,000	40-7,900

[a]See Table 4-1 for data sources; includes single fixture graywater and combined sources (not including the kitchen sink or dishwasher). *E. coli* reported for fecal coliform, which represents only one of the fecal coliform indicator bacteria that may be present.
NOTES: Approximately 20 states allow household-scale graywater irrigation with no water quality standards. Florida, Georgia, and Wisconsin standards apply to subsurface irrigaton only.
SOURCES: Sharvelle et al. (2013); EPA (2012a).

TABLE 6-4 State Water Quality Standards for Graywater Reuse for Toilet Flushing

	BOD_5 (mg/L)	TSS (mg/L)	Turbidity (NTU)	Total Coliform (cfu/100 ml)	Fecal Coliform (cfu/100 ml)	Disinfection
California	-	-	-	2.2[a]	-	-
New Mexico	30	30	-	-	200	-
Oregon	10	10	-	2.2[a]	-	-
Georgia	-	-	10	500[b]	100[b]	-
Texas	-	-	-	-	75[b]	-
Massachusetts	10	5	2	-	14[b]	-
Wisconsin	200	5	-	-	-	0.1-4 mg L^{-1} residual chlorine
State Water Reuse Regulations for Unrestricted Access Urban Reuse	5-60	5-60	2-10	2.2 (mean) 23-240 (max.)	2.2-14 (med. or mean) 14-200 (max.)	-
Reported range of typical Graywater prior to treatment[c]	30-380	50-280	30-240	500-630,000	40-7,900	N/A

[a]Median value.
[b]Daily or sample maximum
[c]See Table 4-1 for data sources; includes single fixture graywater and combined sources (not including the kitchen sink or dishwasher). *E. coli* reported for fecal coliform, which represents only one of the fecal coliform indicator bacteria that may be present.
SOURCES: Sharvelle et al. (2013); EPA (2012a).

for indicator organisms. The state of Wisconsin has simplified the issue by requiring a target range for residual chlorine rather than indicator organism targets (Table 6-4). Disinfection processes are also important to control odor, although treated graywater can become odorous when residual disinfectants are not maintained (Vandegrift, 2014).

Additional removal of DOC (often measured as BOD_5) and TSS or turbidity is required in some states for toilet flushing with graywater, although little research is available to clarify whether or how much DOC removal is necessary. Although a small portion of organic matter can be removed through sedimentation or other processes for particulate removal, much of the organic matter in graywater is in the form of DOC. Even extensive removal of DOC in graywater, resulting in measured BOD_5 concentrations between 5-10 mg/L, would not prevent regrowth of pathogenic organisms. Instead, disinfectant residual must be maintained to prevent regrowth of pathogens. Research has demonstrated that when adequate concentrations of residual total chlorine are achieved, regrowth of indicator organisms can be prevented for 3-4 days even when DOC is not removed from graywater during treatment (Wiles, 2013). Nevertheless, DOC or turbidity removal allows residual disinfectant concentrations to be more reliably maintained. When residual disinfectant levels are not maintained, it is possible for biofilms to form in distribution systems that could contribute to the microbial load in treated graywater. Thus, treatment with DOC removal in addition to disinfection (Figure 6-5d) is generally recommended at neighborhood or regional scales. Many commercially available systems for household-scale treatment of graywater are not designed to remove DOC, which may be acceptable as long as disinfectant concentrations are adequately maintained (Sharvelle et al., 2012). In these systems, disinfection residuals are often achieved via chlorine addition, with the potential concern associated with the formation of disinfection by-products. When treated graywater is used for toilet flushing, the ultimate fate of those disinfection by-products is a wastewater treatment system, where they may or may not be removed from the water depending on treatment practices. Many treatment technologies are available to meet the desired water quality objectives, and these unit processes are discussed later in the chapter.

Commercial graywater treatment systems for toilet flushing vary substantially in complexity and the degree of treatment achieved. Those systems that achieve a high degree of treatment are generally too expensive for current widespread adoption at the household scale, and there are reliability issues with systems that are more cost-effective. Some providers of graywater treatment systems have gone out of business in the past few years (e.g., Brac and Water Legacy), and maintenance issues have been cited as a cause for such failures. As of 2015, only two treatment systems have achieved National Sanitation Foundation (NSF) International 350 certification for toilet flushing[2] because of the extensive treatment required to meet the certification (see Box 6-2). In particular, the standards set for carbonaceous biochemical oxygen demand (CBOD) and turbidity in NSF 350 are difficult to achieve with simple treatment systems, such as sand filtration or cartridge filters (Friedler et al., 2006; Gual et al., 2008; Hodgson, 2012; Zuma et al., 2009). Even ultrafiltration membranes were not found to achieve CBOD less than 10 mg/L when applied to graywater (Ramon et al., 2004). To achieve the CBOD and turbidity requirements outlined in NSF 350, graywater must be treated through a biological reactor, membrane filtration (nanofiltration or reverse osmosis), or activated carbon, all of which are technologies that have not previously proven to be practical for residential application because of complexity and associated high costs and maintenance. Some of these technologies become more cost-effective and practical at the neighborhood scale, although one company recently launched a graywater treatment system for household use meeting NSF 350 standards (Showley, 2015).

In the committee's judgment, NSF 350 (see Box 6-2) requires more extensive treatment than necessary to provide safe water for flushing toilets, and use of this standard as a basis for toilet flushing could increase project costs, unless novel technologies are developed. Additional research is needed to better understand the impacts on risk and reliability associated with higher organic matter and turbidity levels in graywater for toilet flushing at a range of scales, because these factors significantly affect treatment costs. This research could inform the development of improved risk-based guidance that could provide the basis for standards of practice.

Spray Irrigation

If graywater is to be used for spray irrigation, then it must meet state water quality requirements for unrestricted use of reclaimed wastewater, which are summarized in Table 6-4 and detailed in EPA (2012a). Generally, the regulations require BOD_5 of less than 20-30 mg/L and very low numbers of indicator organisms (i.e., less than 23/100 mL total coliforms in California). Treatment of graywater for spray irrigation would require removal of organic matter and disinfection (Figure 6-5b), using treatment processes similar to those used for toilet flushing. Given the treatment costs and maintenance required, these systems are generally recommended at neighborhood and municipal scales, but not at the household scale.

---

[2]See http://info.nsf.org/Certified/wastewater.

BOX 6-2 NSF 350

The National Sanitation Foundation (NSF) International is a nonprofit, nongovernmental organization that develops minimum material, design, and performance standards for a variety of products so that consumers can compare systems that meet a set standard, and understand, according to defined terms, the capacity of a product. NSF has developed a set of standards for graywater use for toilet flushing and for irrigation. NSF Standard 350 was developed to establish minimum material, design, construction, and performance requirements for on-site residential and commercial water reuse treatment systems (Table 6-2-1). This particular standard applies to residential and commercial graywater systems designed for toilet flushing or outdoor unrestricted urban water use. NSF 350 makes no distinction between graywater systems that do and do not include kitchen water, which would require very different treatment approaches. To receive system certification, manufacturers must undergo a 26-week testing period where the system will be dosed with synthetic graywater. NSF constructed a second standard (NSF 350-1) that covers graywater reuse for outdoor subsurface irrigation only. This regulation covers up to 1,500 gpd (5,700 lpd) for both residential and commercial applications. The testing procedure is identical to NSF 350 in regard to duration and stress tests, but the quality requirements are less strict with CBOD and TSS values of 25 mg/L and 30 mg/L, respectively. The only other water quality requirement for NSF 350-1 is that pH is between 6 and 9.

Because of a lack of national risk-based guidance on appropriate quality for graywater use, some states (e.g., Washington and California) have used NSF 350 water quality requirements to set regulatory standards for graywater use to flush toilets. Colorado has included the NSF 350 standards in draft regulation. A risk-based approach was not applied to develop NSF 350 water quality standards. Instead, the basis for the water quality standards in NSF 350 was a review of water quality regulations in U.S. states where such standards existed and internationally, most of which applied to municipal use of treated effluent. Residential and multi-residential use of graywater differs from municipal reclaimed water systems in that storage time is less. Although E. coli, pH, and storage vessel residual chlorine limits are easily met by treatment systems (Hodgson, 2012), extensive treatment is required to meet $CBOD_5$ and turbidity limits, and no risk analysis is available to show that such strict limits for these parameters are required to ensure safe use of graywater for toilet flushing.

TABLE 6-2-1 Summary of NSF 350 Water Quality Requirements for Toilet Flushing

	Class R[a]		Class C[b]	
Measure	Test average	Single sample maximum	Test average	Single sample maximum
$CBOD_5$ (mg/L)	10	25	10	25
TSS (mg/L)	10	30	10	30
Turbidity (NTU)	5	10	2	5
E. coli (MPN/100 mL)	14	240	2.2	200
pH (SU)	6-9		6-9	
Storage vessel residual chlorine (mg/L)	≥ 0.5-≤ 2.5		≥ 0.5-≤ 2.5	

[a]Class R: Flows through graywater system are less than 400 gpd.
[b]Class C: Flows through graywater system are less than 1,500 gpd.

Consideration of Scale for Selection of Appropriate Treatment Processes

The most important drivers for selection of appropriate technology for graywater treatment processes are to achieve quality appropriate for the desired end use and to meet water quality requirements. However, another important consideration is the scale of the system.

Household Scale. Untreated graywater is commonly applied for irrigation at the household scale (Sharvelle et al., 2012). There is some level of complexity in all treatment processes applied for toilet flushing (Figures 6-5c and 6-5d). Although technologies are commercially available that implement these treatment processes at the residential scale (particularly in Australia; Sharvelle et al., 2013), they typically require substantial homeowner maintenance. Adoption of household-scale treatment systems that include the treatment steps outlined in Figures 6-5c or 6-5d has been limited in the United States because the maintenance required is generally not practical for most homeowners. The treatment processes identified in Figure 6-5c with only suspended sediment removal and disinfection are more likely to be adopted at the residential scale than are those in Figure 6-5d, where DOC is also removed from graywater. DOC and turbidity remaining in treated graywater cause some aesthetic issues that may be acceptable to individual homeowners who want to use graywater for toilet flushing.

Neighborhood and Regional Scale. Large-scale graywater systems for irrigation can be put in place in multi-residential and commercial facilities (Table 6-2). If untreated graywater is applied through subsurface irrigation systems so that human exposures are prevented, then the simplest treatment processes for graywater that include storage and coarse filtration (Figure 6-5a) are still acceptable and appropriate. However, in multi-residential and/or commercial systems, more complex treatment can be considered when staff are available to maintain the system. This allows for a higher degree of treatment and the potential to use the treated graywater for spray irrigation in addition to subsurface drip irrigation. The current standard of practice is that disinfection is conducted for neighborhood-scale graywater systems (Figure 6-5b, with or without DOC removal).

For toilet flushing at neighborhood or regional scales, DOC removal in addition to disinfection is often adopted to reduce turbidity and associated public concerns regarding water quality (Figure 6-5d). In addition, at the multi-residential scale, typically staff are available to perform routine maintenance activities (Box 6-2), rendering such treatment processes more practical than at the household scale. Multi-residential units can be a very good fit for graywater use for toilet flushing because of the reliable source of graywater and ease of collection and redistribution to toilets. In addition, economies of scale are realized and maintenance can be feasible.

Graywater System Operational Considerations

From the simplest laundry-to-landscape graywater system to the treatment processes identified in Figure 6-5, all systems require maintenance that is critical to sustain safe graywater operations. The degree of maintenance required depends on the complexity of the system to be installed. Maintenance for graywater systems typically includes activities such as changing and/or cleaning filters, replacing consumables, and replacing system components when they reach the end of their useful life. At the household scale, the burden of maintenance falls on the homeowner; thus, more success has been reported for systems that are very simple and require low maintenance. At the neighborhood scale, routine maintenance can be conducted by facilities staff or a homeowners association.

For simple graywater irrigation systems (e.g., laundry-to-landscape or the treatment system in Figure 6-5a), lack of maintenance typically results in the inability to operate the system and irrigate landscape, with limited risk to human health. However, when water quality goals are in place and the potential for human contact with graywater is high (e.g., toilet flushing or spray irrigation), lack of proper maintenance can result in human health risk, and routine monitoring is necessary to ensure safe operations. Thus, advances in online monitoring and automations would enable practical application of household graywater systems that achieve a high level of treatment (i.e., removal of organic matter and turbidity and disinfection) and thereby expand the potential use of graywater at the household scale.

Chlorine residual is an easily monitored parameter and a good indicator of the safety of treated graywater. Chlorine can be monitored online or with simple test kits. An example of online chlorine monitoring is at Aspen Residence Hall at Colorado State University (see Box 2-2) where the system automatically switches to municipal water supply when the residual chlorine concentration in treated graywater (collected directly after the treatment system) drops below 1.0 mg/L.

Energy Recovery from Graywater Systems

Heat from graywater can be captured through heat exchange systems to save energy. Graywater temperature ranges from 64 to 100°F (18-38°C; Ericksson et al., 2002), and sources of graywater tend to be heated water (e.g., showers, baths and laundry). Several studies have demonstrated the capacity to recover energy from graywater, with energy requirements of potable water heating reduced by up to 30-50 percent (Proskiw, 1998; McNabola and Shields, 2013; Smith, 1975), although the largest recoveries are possible closest to the drain point of the hot water. The heat recovered from graywater can be used for heating potable water or to improve the efficiency of absorption heat pumps.

Eliminating graywater from the wastewater stream increases concentrations of solids and organic matter compared to domestic wastewater, making it amenable to anaerobic treatment and enhanced energy recovery (Liu et al., 2004; McCarty et al., 2011). An example is a system designed by Semizentral under implementation in Qingdao, China (Semizentral, 2015), which is designed for a population of 12,000 and started in 2012 (see Box 2-1).

SYSTEM DESIGN FOR STORMWATER BENEFICIAL USE

The design of systems for beneficial uses of stormwater (Figure 6-1) results from careful consideration of available design alternatives to meet a complex web of often competing objectives. In contrast with graywater project planning, regulatory drivers are frequently the top priority when establishing stormwater-related project objectives. Often, several different regulations must be met for any given project. Stormwater regulations may dictate hydrologic, hydraulic, water quality, or design objectives. In many cases, regulations are highly prescriptive in how to design stormwater controls as well as directly govern the design process for beneficial use systems. The most common regulatory drivers that cause stakeholders to implement stormwater capture and use systems include the following:

- Combined sewer overflow regulations and associated standards for retaining certain rainfall depths on site,
- Total maximum daily load (TMDL) implementation plans,
- National Pollutant Discharge Elimination System Phase I and Phase II permit conditions,
- State regulations including water quality and design standards,
- Potable water use restrictions (e.g., during times of drought), and
- Local regulations (e.g., requiring or strongly motivating onsite use).

Stormwater beneficial use systems can be developed at many different scales, from household rain barrels and cisterns to neighborhood-scale stormwater collection to regional capture and infiltration systems to augment municipal water supplies (see Chapter 2). Regardless of scale, design for stormwater capture and use projects includes assessments of the quantity of runoff available from the source area and water quality relative to intended uses, as well as decisions on runoff conveyance, storage, treatment, and water delivery to meet the overall objectives. Many useful guidebooks exist to advise homeowners on how to design systems to capture rooftop runoff for on-site use (Carpenter et al., 2009; City of Bellingham, 2012; Despins, 2010; Jones and Hunt, 2014; Lawson et al., 2009; MPCA, 2015). The following sections describe typical approaches to storage and treatment for stormwater capture and use across a range of relevant scales, focused on two distinct strategies: (1) stormwater capture and tank storage and (2) groundwater recharge.

Stormwater Capture and Tank Storage

The capture and on-site storage of stormwater for later withdrawal and use, sometimes called "active" stormwater harvesting, can be designed across a range of scales, from households to large buildings to neighborhoods.

Stormwater Capture at Household and Single-Building Scales

Capture, Storage, and Distribution. Rooftop runoff can be readily captured at the household or large-commercial-building scale for a range of beneficial uses. Irrigation is the predominant use for household-scale projects, although large tanks could provide water for nonpotable indoor uses (e.g., toilet flushing) or commercial applications (e.g., cooling, washing, laundry [see Chapter 2 and Box 6-3]). Roof area and climate determine the quantity and timing of stormwater runoff available for beneficial use (see Figure 3-2).

Tank size determines how much of the available stormwater can be captured for beneficial use. Storage tank capacities can range from small containers, such as rain barrels that are about 35 gallons (130 liters) each, to large surface or subsurface tanks (Figures 2-1, 6-6, and 6-7). Rain barrels are most commonly used at residential properties and are a component of many public education programs. Large above-ground tanks are often made of corrugated metal, precast or cast-in-place concrete, or polyethylene plastic. Below-ground tanks are commonly made of precast concrete, plastic pipe, fiberglass reinforced plastic, or proprietary modular plastic storage systems surrounded with impermeable geotextile membranes (Figure 6-7).

Rain barrels provide a small supplemental water supply and typically only capture a small fraction of roof runoff on an annual basis, even in arid climates, while larger tanks have been used to store substantial fractions of roof runoff for beneficial uses (see Chapter 3). The committee's scenario analysis of stormwater irrigation use in Chapter 3 based on 1,500 ft^2 (140 m^2) roof areas and medium-density residential development showed that two rain barrels captured and enabled the irrigation use of between 5 and 14 percent of roof runoff in the six locations analyzed, compared to between 16 and 61 percent for large 2,200-gallon (8,300 liters) cisterns, with the magnitude of capture depending on local rainfall patterns relative to on-site water demand (see Table 3-7). The larger cisterns provided 4 to 21 percent potential potable water savings in total water use in the committee's irrigation use scenario analysis, compared to 1 to 5 percent for two rain barrels per house (see Tables 3-5 and 3-6). Table 6-5 provides an example highlighting the different levels of performance achieved, in terms of reducing stormwater runoff, by larger storage tanks for a research site in Kansas City for roofs about 1,000 ft^2 (92 m^2) in area (Pitt et al., 2014). As tank size increases, fewer rain events are likely to overfill the tank, assuming the captured water is effectively used to match the evapotranspiration demand, and a larger percentage of the on-site stormwater is also be available for use. Larger tanks are common in Australia and New Zealand at homes where they supply much of the household nonconsumptive water requirements and, in some cases, provide drinking water (Cunliffe, 2004).

Sizing appropriate storage for household- or building-scale use requires a careful assessment of the quantity and timing of on-site water demand (e.g., irrigation, toilet flushing, other uses) relative to the quantity and timing of rooftop

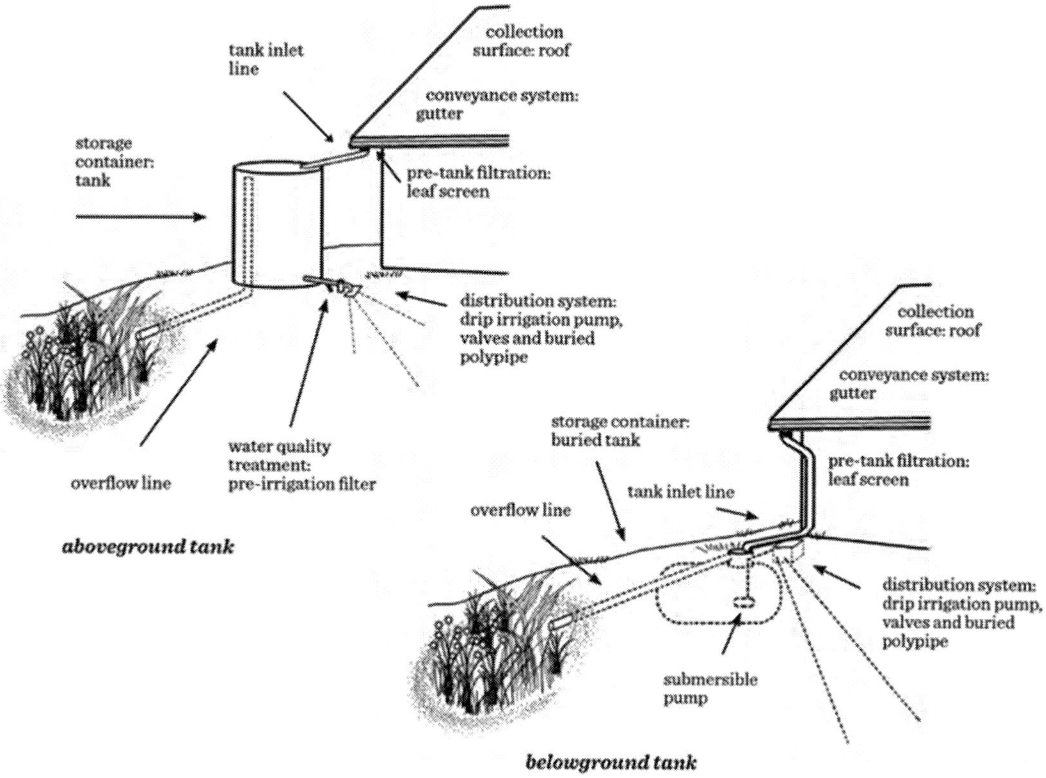

FIGURE 6-6 Surface and subsurface cisterns as part of rooftop stormwater capture systems. SOURCE: City of Bellingham (2012).

FIGURE 6-7 Subsurface stormwater storage under construction for the Darien Library (a LEED Gold facility) in Darien, Connecticut. The tank holds 83,500 gallons. SOURCE: http://www.invisiblestructures.com/project_profiles/page/2.

runoff available in the context of overall project objectives. The relative importance of water supply versus stormwater runoff control will also influence the optimal project design. Larger tanks provide larger capture and allow less stormwater runoff, but if on-site use does not keep pace with availability, even large tanks can eventually fill, which negates their hydrologic and pollution reduction benefits. Depending on precipitation rates and timing relative to on-site use, a threshold can be met where increasing tank size does not provide additional benefits. This is particularly true in areas with abundant precipitation, such as the central or eastern United States. In the Southwest, where rainfall and irrigation demand are mismatched, extremely large tanks (approximately 20,000 gallons for a 1,500 ft^2 roof in a medium-density, residential development; see Figure 3-11) would be needed to capture all of the available roof runoff, and the runoff would still not meet the total irrigation demand (see Chapter 3). Costs are also a significant factor in determining the appropriate size of storage (see Chapter 7).

The elevation of the roof area and existing gutters can typically provide passive conveyance to storage tanks (see Figure 6-6). Ideally, tanks are located at a high point on the property, so that gravity can be used to distribute the rainwater as needed, although pumps can also be added for increased water pressure.

All storage tanks need to be sealed and screened to prevent mosquito issues. Rector et al. (2014) describe additional design and maintenance suggestions to prevent mosquitoes from entering and breeding in stormwater storage tanks.

Treatment. Roof runoff usually has the best water quality of the source flows in urban areas, which reduces the need for extensive water treatment. If the captured runoff is to be used untreated, then care should be taken to minimize contamination by avoiding materials that release high levels of metals into the water (see Box 4-1) and avoiding collection from shaded roof areas, because the tree cover can serve as habitat for urban wildlife and increase bacteria loading to the

BOX 6-3 Chesapeake Bay Foundation: Rooftop Stormwater Harvesting for Nonpotable Uses

An example of a larger rainwater harvesting system is at the Chesapeake Bay Foundation's headquarters in Annapolis, Maryland (a LEED® Version 1 Platinum certified building), which was completed in 2000. The commercial office building has about 32,000 ft^2 (2,900 m^2) of floor area. Runoff from the approximately 10,000 ft^2 (930 m^2) galvanized metal roof is filtered and collected in three elevated cisterns (Figure 6-3-1). Further sand filtering treats the water before on-site use for washing, irrigation, and fire suppression. The building has only about 10 percent of the typical potable water demand compared to a conventional office building.

SOURCE: http://www.nrel.gov/docs/fy02osti/29500.pdf and http://www.cbe.berkeley.edu/mixedmode/chesapeake.html.

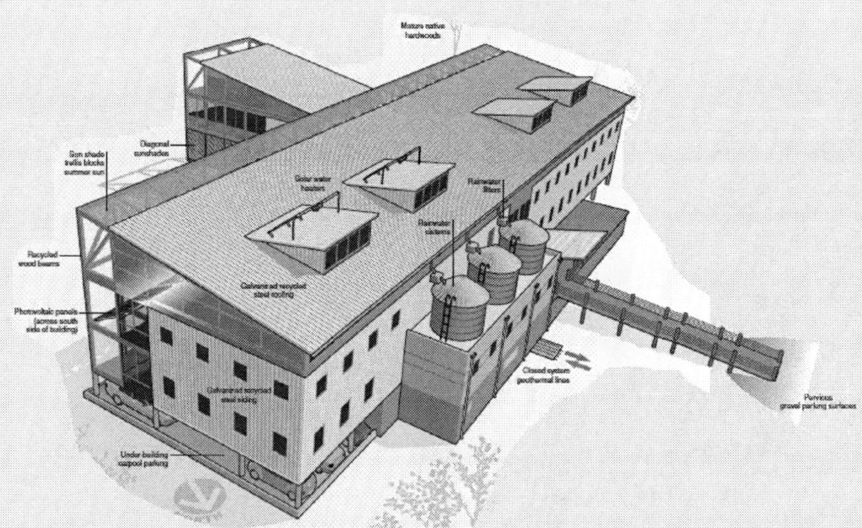

FIGURE 6-3-1 Schematic representation of the Chesapeake Bay Foundation headquarters and stormwater capture system. SOURCE: http://www.nrel.gov/docs/fy02osti/29500.pdf.

TABLE 6-5 Storage Tank Sizes for Different Ranges of Captured Runoff from a 1,000 ft^2 Residential Roof Area in Kansas City, Missouri

Storage Volume (ft^3)	Storage Volume (gal.)	Approximate Annual Roof Runoff Captured (%)	Number of 35 gal Rain Barrels Needed	Number of 5-ft-diam., 5-ft-tall tanks	Number of 10-ft-diam., 10-ft-tall tanks
10	75	20-30	2	0.1	NA
30	224	45-60	6	0.3	NA
100	748	80-90	21	1	0.1
300	2,244	98-100	64	3	0.4
500	3,740	99-100	107	5	0.7
1,000	7,480	100	214	10	1.3

NOTE: A 5-foot-diameter, 5-foot-tall tank represents 730 gallons of storage or 98 ft^3. A 10-foot-diameter, 10-foot-tall tank holds 5,900 gallons or 785 ft^3.
SOURCE: Pitt et al. (2014).

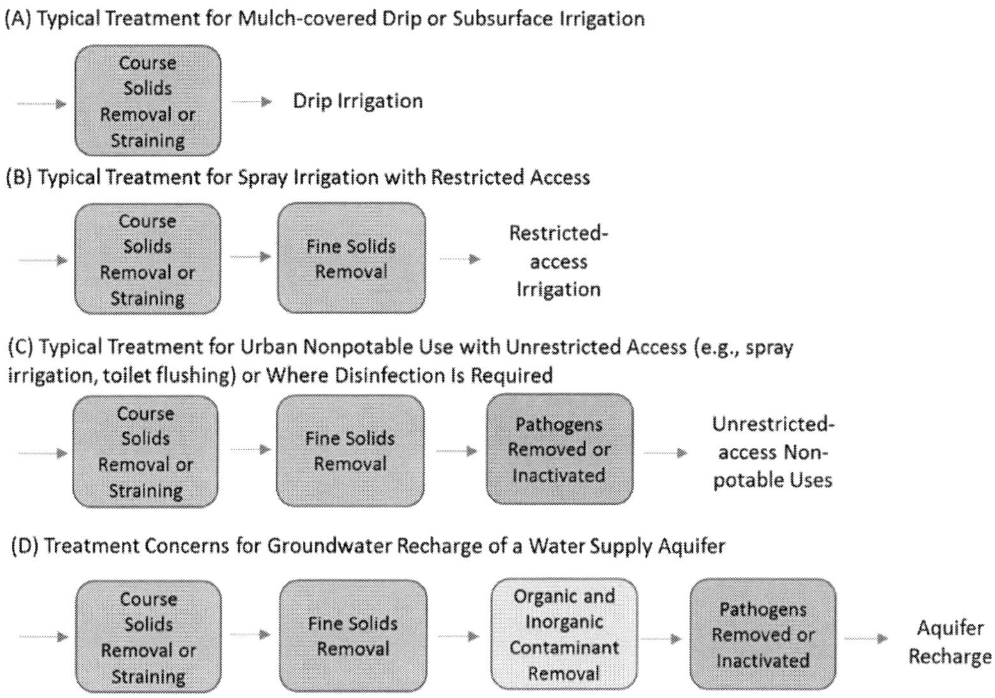

FIGURE 6-8 Recommended treatment processes for captured stormwater for various end uses and regulatory requirements.

FIGURE 6-9 Small on-demand water treatment system with ultraviolet (UV) disinfection used to treat roof runoff for vehicle washing at a Washington, DC, fire station.

BOX 6-4 Docklands Park: Neighborhood-Scale Stormwater Capture and Use in Melbourne, Australia

An example of a large stormwater storage facility is at Docklands Park in Melbourne, Australia. The 2.7-ha Docklands Park collects stormwater from downtown Melbourne and stores it in large underground tanks that supply water for park irrigation. Three wetlands in the park (Figure 6-4-1) provide the capacity to treat approximately 80 percent of the stormwater runoff in the 4.8-ha catchment. Treated stormwater is stored in underground tanks with a combined capacity of 130,000 gallons (490,000 liters), and the captured stormwater is treated using ultraviolet disinfection prior to use for irrigation.

FIGURE 6-4-1 Docklands, downtown Melbourne, Australia showing a public sculpture garden in a wetland area near very large underground tanks for stormwater storage. SOURCE: http://www.aecom.com/News/Where+We+Are/Europe/_news/Influencing+surface+water+management+direction,+policy+and+practice+at+CIWEM%E2%80%99s+conference.

roof runoff. When source control is combined with applications that minimize human exposures, such as drip irrigation or restricted access to a spray-irrigated site, no treatment is needed other than removal of suspended solids to prevent clogging (see Figures 6-8a and 6-8b; see also Chapter 5). In all cases, some type of coarse filter is usually installed in the intake line to prevent organic leaf debris from entering the storage tank. In addition, many household roof harvesting systems use "first-flush" diverters to direct the more contaminated initial runoff from the roof away from the storage tank, although there is little evidence to show the effectiveness of these first flush diverters. To reduce fine solids and minimize problems with sprinkler head clogging, a simple sand or cartridge filter can be used in-line with the pressurized line.

For applications with significant human exposures, risk-based stormwater treatment design is complicated by limited data on the occurrence of human pathogens in stormwater at a range of scales and source areas. With site-specific pathogen data (reflecting multiple samples over a single year), the risk assessment framework outlined in Chapter 5 can be used to assess whether treatment is needed. In absence of these data, a typical conservative approach bases the treatment design on known or estimated concentrations of fecal indicator bacteria; but these may be poor indicators of the presence or concentrations of human pathogens in roof runoff (see Chapter 5). Roof runoff may contain elevated fecal indicator bacteria, particularly if trees with urban wildlife (especially birds and squirrels) are close, as is common in residential areas. In some cases, the levels of indicator bacteria may decrease because of die-off during storage in the tank, but die-off is not consistent, and concentrations may even increase with storage (Pitt and Talebi, 2012b). Therefore, if bacteria reduction is necessary to ensure safe operations in areas where access is not controlled and significant exposures are possible (or to meet regulatory requirements), then disinfection will be necessary (Figures 6-8c and 6-9). An array of unit processes are available to maintain bacteriological quality (discussed later in the chapter), but historically, chlorine has been used for small-scale installations. With an adequate recirculating pump system, it may be possible to maintain bacteriological quality in the tank without high chlorine residuals.

Maintenance. Periodic maintenance of stormwater capture systems is essential to their appropriate function and safe use. Maintenance needs at the household or building scale are relatively low and mostly related to periodic removal of sediments that enter the tank as well as any routine maintenance for any treatment systems (e.g., clean or replace filters) or mechanical systems (e.g., pump, first flush diverter) used. These tasks can be performed by the homeowner or a contractor. However, utility- or community-based inspection and maintenance programs could be developed to ensure robust maintenance, cost-effective operation, and low risks to human health. Such rigorous maintenance efforts would be necessary if the household capture of roof runoff was part of a regional water supply strategy.

Stormwater Capture and Tank Storage at Neighborhood Scales

Capture, Storage, and Distribution. Neighborhood-scale stormwater capture projects may use roof runoff from several close buildings, such as at institutions or business parks, or collect mixed stormwater flows originating from several source areas (e.g., roofs, paved surfaces) located close together, allowing a larger percentage of total runoff to be captured. The most common situation is collecting at-grade surface flow from parking lots and roadways. Subsurface water storage tanks are often used in neighborhood-scale stormwater collection systems and come in a range of designs, from traditional tanks to large subsurface storage units (see Box 2-5 and Figure 6-7). In Docklands Park in Melbourne, Australia, subsurface tanks hold approximately 130,000 gallons of stormwater runoff collected from the surrounding urban area for irrigation use (see Box 6-4).

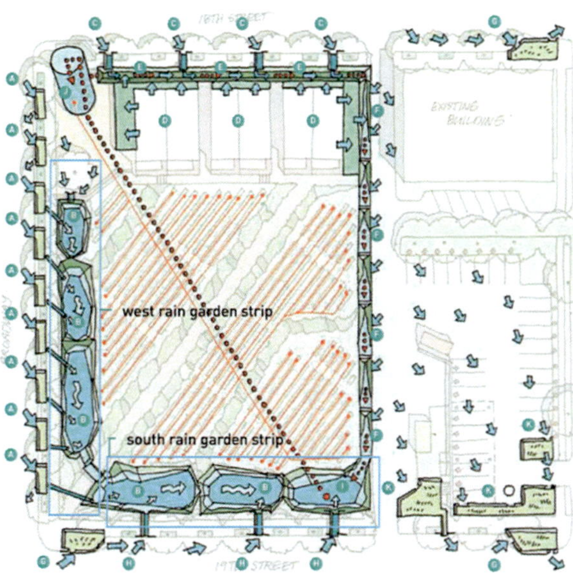

FIGURE 6-10 Kansas City 18th and Broadway downtown stormwater park showing treatment flow and underground cistern for irrigation. SOURCE: http://www.18broadway.com/water.html.

Larger neighborhood-scale systems frequently require extensive piping and pumping systems, because it is rarely the case that gravity drainage can provide adequate collection of runoff from disparate site areas to a central storage location. For these reasons it is advisable to integrate plans for the beneficial use of stormwater into site-development planning as early as possible such that long-term energy costs and infrastructure maintenance can be minimized. This can be challenging because these considerations are not typically included as driving factors in site design. When assessing the costs and benefits of such systems, one may find that some of the economies of scale derived from neighborhood systems are offset by increased energy costs and infrastructure complexity. In these cases, it is worth considering distributing storage around a site and utilizing transfer pumps to move water as needed to a central storage and treatment system. The National Park Service used this approach in retrofitting the National Mall in Washington, DC, with approximately 1 million gallons of total storage for on-site irrigation use (Box 2-5). The National Park Service designed a set of large distributed collection tanks that feed to a single central storage tank that then is used for treatment and distribution.

At the neighborhood or large-building scale, real-time, logic-controlled release of captured stormwater can be used to optimize water conservation and meet water quality objectives. These operations are usually designed as part of combined sewer overflow and stormwater management activities and discharge captured stormwater based on predictions of the amount of runoff for a coming rain. This approach keeps the maximum amount of water in the tank for beneficial uses when needed, while ensuring that sufficient storage is available to retain the water from the upcoming storm. During operations before a predicted storm, a remote-operated computer system checks the tank water volume to ensure that sufficient storage volume is available to contain the expected runoff volume for the next rain. If insufficient volume is available, then the tank is drained to allow capture of the next event. Between events, the water in the tank is available for whatever beneficial uses are suitable (WERF, 2014).

Low-cost, programmable logic controller systems are now available and can be coupled with wired and wireless internet communications and low-cost sensor systems for both new construction and retrofits of stormwater storage systems. The commercial availability of highly distributed real-time control (DRTC) technologies first became available around 2013. The use of DTRC systems in stormwater harvesting is particularly compelling because they have been shown to significantly enhance the performance of stormwater capture when internet-based forecast data are effectively integrated into control logic (WERF, 2005). Real-time systems result in lower overall water savings (based on releases for forecasted storms that do not occur), but they offer notable advantages when water quality or wet weather flow objectives are also important.

Treatment. Runoff from the non-roof areas that drain into most neighborhood-scale stormwater capture systems is often of poorer quality than roof runoff (see Chapter 4) and therefore requires significant pre-treatment prior to storage. This pre-treatment can be integrated into site design to utilize natural treatment systems such as biofiltration or bioretention to reduce sediment and other contaminant loading to a beneficial use system. For example, Figure 6-10 shows the site plan for a Kansas City neighborhood stormwater capture project that drains several blocks of the surrounding downtown urban area. Stormwater enters the peripheral zone of the site and cascades through surface swales for partial treatment. The water is then captured and pumped to a 40,000-gallon (150,000 liter) underground storage tank and subsequently withdrawn to irrigate the community gardens and other on-site landscaping. Pre-treatment prior to storage in stormwater capture projects also frequently includes the use of hydrodynamic stormwater treatment systems for gross solids and sediment removal where natural systems are infeasible to integrate into site design.

The additional treatment provided will depend on the intended use of the water and the extent of human exposures (see Figure 6-8). More sophisticated water treatment systems are usually used for neighborhood-scale systems (Figure 6-11), compared to building-scale systems. Microorganisms are the most critical constituent of concern for nonpotable applications of stormwater. Data on typical pathogen concentrations in stormwater are lacking, but disinfection is typically included when surface flows from paved or open space areas are included and significant human exposures are anticipated, because stormwater collected from land surfaces could contain pathogens from pet waste, leaking sewers, or septic tanks. ASCE (Clary et al. 2014) reported that bioretention, grass-lined detention basins, sand filters, and wet detention ponds lead to significant reductions in fecal coliform bacteria, but only disinfection consistently produced large reductions in indicator bacteria concentrations (Table 6-6). ASCE (Clary et al. 2014) reported that concrete-lined detention basins, grass filter strips, infiltration basins, manufactured hydrodynamic separators, and wetlands showed no significant reduction in fecal coliform concentrations.

Maintenance. Maintenance of neighborhood-scale stormwater capture systems includes essentially the same major items as discussed for household- or building-scale projects, although the technical sophistication and scale of the project necessitates professional maintenance to ensure the system is functioning properly. Larger scale systems can also integrate monitoring approaches that can provide real-time information to increase system reliability and performance by notifying of and/or predicting the need for maintenance.

Groundwater Recharge

An alternate strategy for the beneficial use of stormwater involves enhanced infiltration into shallow or deep aquifers, with the intent of augmenting aquifers used for water supply. The extent to which groundwater recharge actually conserves conventional water supplies depends on the degree to which the recharge area is hydrologically connected to aquifers used for water supply (or aquifers that might be used in the future) (see Figure 2-4). Thus, local hydrogeology governs the capacity for groundwater recharge of stormwater to substantially benefit public water supplies.

Groundwater recharge projects using stormwater can be designed along an array of scales from small, household-scale rain gardens to neighborhood-scale biofilters to regional-scale aquifer recharge projects. In all cases, water quality is an important consideration, because contaminants in urban stormwater may pose risks for groundwater contamination, as is discussed in Chapter 5. Unless treatment is provided to prevent groundwater contamination, source areas with the least contaminated runoff (e.g., roofs, possibly walkways and little used driveways) should be selected. Areas of automobile activity and commercial or industrial use should generally be avoided in the absence of any planned treatment. Direct injection should also be avoided, because infiltration through soils and sediment can provide additional contaminant removal (Pitt et al., 1996; discussed in more detail below). Many guidance manuals are available on the construction of stormwater infiltration structures at the household and neighborhood scales, including King County Washington's Stormwater Design Manual (King County DNRP, 2009), Denver's Urban Drainage and Flood Control's Criteria Manual (UDFCD, 2013), New York State Stormwater Design Manual (NYSDEC, 2015), Boston Water and Sewer Commission (2013), Water Environment Federation (WEF, 2014), and the EPA's website.[3] Large-scale, urban, stormwater capture and recharge systems require thoughtful design because of the flows being handled and the likelihood of contaminants increasing in number and concentration with the greater extent and types of areas contributing to runoff.

[3] See http://water.epa.gov/polwaste/npdes/stormwater.

FIGURE 6-11 Large stormwater treatment system underground of the Washington, DC, mall, designed to treat stormwater from tanks holding 1 million gallons (3.8 million liters) for spray irrigation.

TABLE 6-6 Influent and Effluent Fecal Coliform Levels for Stormwater Controls (data from the International BMP Database)

		Geometric Mean (MPN/100 mL)		Median (MPN/100 mL)	
Stormwater Control	Events Represented	Influent	Effluent	Influent	Effluent
Bioretention	27-30	3,355	886	5,000	750
Grass-lined detention basins	162-165	2,218	639	2,497	700
Disinfection systems	64-80	1,158	17	1,050	10
Sand filters	150-157	1,463	632	1,600	593
Wet detention ponds	23-24	2,930	637	3,200	1,500

NOTE: Data derived from the International Best Management Practice Database (www.bmpdatabase.org).
SOURCE: Clary et al. (2014).

Design methodologies and treatment trains for large-scale, urban, runoff recharge systems are lacking.

Household-scale Groundwater Recharge: Rain Gardens

Capture, Storage, and Infiltration. At the household scale, rain gardens are simple excavations that are 1- to 2-feet (0.3-0.6 meters) deep and sized to be about 10 to 20 percent of the roof drainage area with minimal subsurface preparation (see Figures 6-12 and 6-13). Specific site conditions including climate, rainfall intensity and frequency, and soil infiltration, will result in other production functions for other locations. As runoff enters the rain gardens, it infiltrates the underlying soil. If the entering runoff rate exceeds the infiltration rate, then the water ponds and is stored for later infiltration when the incoming runoff rate decreases. If the ponding becomes deep, then it can overflow through a surface outlet. Rain gardens are planted with vegetation selected to withstand the highly variable dry to submerged conditions.

Rain gardens are popular on-site controls that have low costs and can significantly reduce the discharges of roof runoff from homes. If designed correctly, then almost all of the roof runoff can infiltrate into the shallow groundwater with little runoff overflowing to the drainage system. Additional impervious surfaces may also drain to simple rain gardens, including walkways and driveways. Some of the considerations in the planning and design of rain gardens include the soil characteristics and depth to groundwater. The rate of infiltration limits rain garden use in poorly draining soils, as does the presence of a high groundwater table. In most cases, only a small percentage of the incoming water is lost through evapotranspiration (usually less than 10 percent) because of the large amount of water applied to a relatively small area (Pitt et al., 2014). Long-term care is required to maintain the plants and to ensure the rain garden's objectives.

Treatment. No additional treatment is typically provided for rain gardens at the household scale. Source control is used to limit stormwater collection to source areas with low contaminant levels. In cold climates, care needs to be taken to divert runoff from sidewalks or driveways that contains deicing salts from reaching the rain gardens (discussed further in the next section).

Neighborhood-Scale Groundwater Recharge: Biofilters

Capture, Storage, and Infiltration. Neighborhood-scale infiltration areas can be designed to collect and infiltrate mixed stormwater from several homes or buildings. They are more sophisticated than simple rain gardens because their engineered soils provide much greater infiltration rates over smaller sizes than rain gardens, while maintaining hydraulic conductivity over long periods. Assuming adequate underlying infiltration rates, almost complete runoff can be captured and infiltrated if sized correctly (generally about 2 percent or more of the paved drainage area). Sometimes biofilters incorporate subsurface storage tanks for beneficial use (Figure 6-14). Large facilities have also been incorporated into neighborhood parks and public gardens serving a several block area, especially when constructed in conjunction with other stormwater control elements (such as grass filtering swales and storage tanks) forming an effective treatment train. The biggest challenge when retrofitting these facilities is having adequate space to locate the biofilter in areas where the water can flow by gravity.

The performance of a biofilter is based on several key factors, including the size of the device, the inflow, the soil infiltration rate, the infiltration rate any engineered media fill used, the amount of subsurface storage (either in rock fill or structural storage), and the outlet structures. With excessive particulate solids loadings resulting in accumulations between 10 and 25 kilograms per square meter, infiltration rates can significantly decline, which results in nuisance and public health concerns associated with frequent and long-duration standing water. A biofilter with healthy plants can help sustain the infiltration rates at a desired level unless the accumulative particulate load occurs in just a few years, as can occur in biofilters that are not sized appropriately (Clark, 2000; Clark and Pitt, 2009). Pretreatment components, such as grass filters or swales, can also help reduce sediment-loading problems (Nara et al., 2006). An elevated sodium adsorption ratio, which may occur when deicing salts enter a biofilter with snowmelt, can also severely restrict infiltration capacity by causing dispersion of clays in a soil (Figure 6-15). To prevent groundwater contamination and premature clogging in biofilters with even small amounts of clay in the soil, high-salt-content water must be diverted from the infiltration facility.

Treatment. Biofilters rely upon soil treatment, sometimes amended with engineered media, to provide treatment of the percolating stormwater to decrease the groundwater contamination potential. Considerable literature exists on the capacity of biofilters, swales, and filter strips to capture contaminants

FIGURE 6-12 Household-scale rain garden. SOURCE: https://www.bae.ncsu.edu/topic/raingarden/stormwater.htm.

on soil or infiltration media for household or neighborhood-scale projects (e.g., Barraud et al., 1999; Dierkes and Geiger, 1999; Payne et al., 2015; Regnery et al., 2013; Weiss et al., 2008). Pollutants most likely to be removed during infiltration are metals or organic chemicals that sorb to soil or suspended solids, which are then removed by physical straining by the soil media, or biodegradable organic chemicals. In studies on recharge basins receiving large metal loads, the removal of most of the heavy metals occurred in either the basin sediment or the unsaturated zone (Hampson, 1986; Ku and Simmons, 1986). Metals can also be removed by surface-complexation reactions with oxide minerals such as ferric oxide (Appleyard, 1993; Dzombak and Morel, 1990). Organic compounds in urban runoff from oils and gasoline have a high affinity for organic solids and surfaces. Optimal siting conditions include deep soils containing moderate amounts of organic content for enhanced pollutant removal. The potential for contamination increases in areas with well-drained soils, typically sands with low organic content. In addition to sorption, organic compounds may also be removed through volatilization or biotransformation.

Physical straining through the soil or engineered media may remove pathogens, which are typically associated with particles, although the effectiveness will depend on the organism and the porous media properties. Straining is more effective for protozoan pathogens (greater than 3 microns), such as Giardia lamblia and Cryptosporidium parvum, and larger bacteria (approximately 1-2 microns) than for viruses, which are too small (0.02-0.08 microns) to be effectively removed by filtration through porous media, although adsorption to some types of subsurface media does occur.

Contamination potential is highest in areas where the permeability is very high or the water table is shallow. Examples from the Midwest suggest that infiltration rates should be small enough to allow for soil treatment of contaminants but high enough to ensure that the ponded water remains oxygenated. Published criteria for infiltration include rates that are greater than 0.2 in/hr (0.5 cm/hr),[4] but less than 5 in/hr (13 cm/hr).[5] States typically recommend that biofilters are sited between 2 and 5 feet (0.6-1.5 meters) above the seasonal high water table (Clark et al., 2010), although these values are generally derived from septic tank designs and may not be adequately protective of groundwater under a range of soil types and stormwater qualities. These design criteria also do not differentiate based on intended use of the groundwater, such as for drinking water supply.

Maintenance. Maintenance of biofilters includes periodic inspections to ensure proper drainage and removal of accumulated fine sediments as needed. Periodic sediment removal in the upper soil layer (50 cm or less) can also help avoid breakthrough of metals (Nieber et al., 2010). Care of plants (and replacement as needed) is important to maintain effective drainage rates.

[4] See http://www.pca.state.mn.us/index.php/water/water-types-and-programs/stormwater/stormwater-management/minnesotas-stormwater-manual.html.

[5] See http://dnr.wi.gov/topic/stormwater/standards/postconst_standards.html.

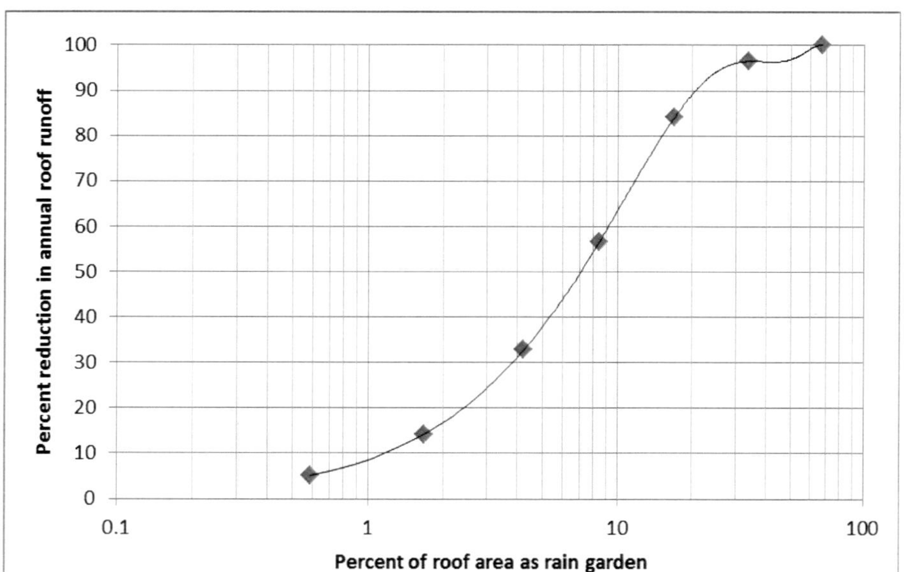

FIGURE 6-13 Percentage reduction in annual roof runoff with rain gardens for Kansas City. SOURCE: Pitt et al. (2014).

FIGURE 6-14 Biofilter/swale under construction with underground water storage vaults for irrigation use. SOURCE: Photo by Dan Bourdeau, Geosyntec, Inc.

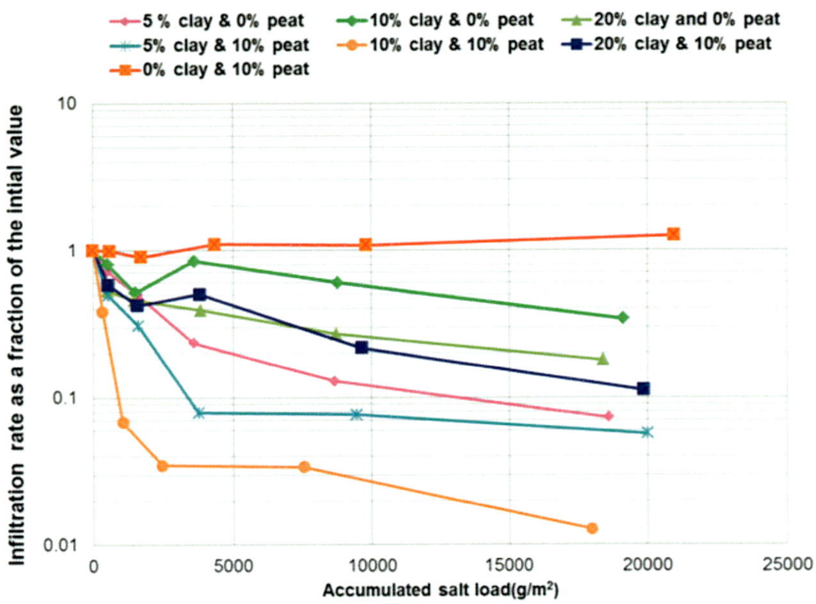

FIGURE 6-15 Decreased infiltration rates with increased salt loadings and increased amounts of clays in underlying soil. SOURCE: Kakuturu and Clark (2015).

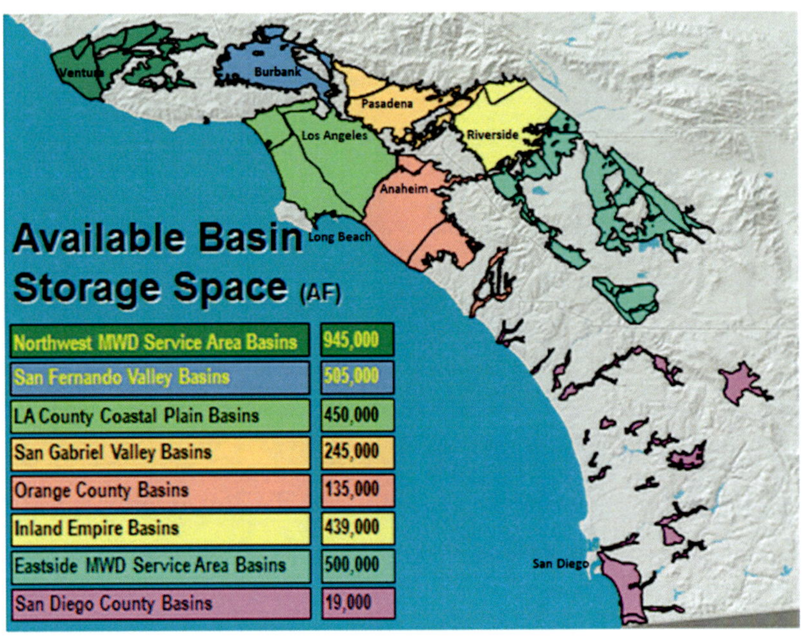

FIGURE 6-16 Available aquifer storage in Southern California (shown in acre feet). SOURCE: Metropolitan Water District of Southern California

Regional Groundwater Recharge Projects

Capture, Storage, and Infiltration. Large-scale stormwater capture projects (Figure 2-1) are designed to recharge regional aquifers for future withdrawal to augment the local water supply. This can be accomplished through riverbank infiltration systems or infiltration basins in areas that recharge aquifers developed as water supplies. It is also possible to capture the stormwater in surface impoundments that are integral components of the water supply system. Regional-scale stormwater capture facilities are common in the arid Southwest of the United States, where for decades, water suppliers and flood control agencies have been capturing floodwaters in spreading basins both to replenish groundwater basins and to manage flood risk, while monitoring stormwater for water contamination. For example, in Southern California many groundwater recharge spreading ponds and other stormwater capture facilities (e.g., inflatable rubber dams) have been built as a part of the river flood control facilities to enhance percolation of stormwater (or flood water).

The geology and soil types are critical to identifying the best sites for recharge facilities. For example, Figure 6-16 shows the variability in aquifer storage in Southern California. Near San Diego, the hydrogeological conditions (fractured crystalline and semi-consolidated sedimentary rock) create mostly brackish aquifers with limited yield and storage. Stormwater storage options are limited to alluvial aquifers in San Diego. In contrast, the Los Angeles region has several large basins with vast storage capacity, although existing water contamination in some basins creates additional water management challenges.

Regional infiltration systems are designed to achieve and sustain maximum infiltration, with periodic maintenance to remove sediment that reduces infiltration rates. Water conveyance systems within the drainage basin and recharge basins are constructed and managed to optimize infiltration (see Box 6-5). Topography to support flow to locations at the bottom of localized or regional watersheds where sufficient space for recharge is also available are key siting considerations.

Treatment. As with biofilters, discussed previously, water treatment for large stormwater infiltration systems is typically provided by soil aquifer treatment, sometimes amended with engineered media to enhance contaminant removal. In current practice, no standard design criteria exist for stormwater treatment for large groundwater infiltration projects. At the spreading basins managed by the County of Los Angeles, no additional treatment is provided beyond soil aquifer treatment, but during a storm event operators monitor total suspended solids (TSS) and allow stormwater TSS greater than 500 mg/L to bypass the facility rather than divert it into the spreading grounds. Consequently, the so-called first flush (which is expected to contain the highest pollutant levels) bypasses the spreading grounds (MWH, 2003). In addition to protecting the water quality, this practice reduces the sediment load in the basins, thereby helping to sustain high infiltration rates. In Los Angeles, a stormwater recharge project in the Sun Valley neighborhood was designed to remedy flooding and provide beneficial use of stormwater from a highly urbanized area while ensuring protection of groundwater. The project incorporates hydrodynamic (swirl-type) separators with a media filter designed to treat the first flush, thereby removing suspended solids, oil, and heavy metals (CH2M-Hill, 2006).

Hatt et al. (2006) review treatment practices employed in Australia for use of urban stormwater for nonpotable purposes. They conclude that practice is ahead of research, with no technologies designed specifically for stormwater use. Rather, standard technologies designed for general stormwater pollution control (e.g., sediment traps, swales and buffers, wetlands and ponds) are employed. The authors make the case that the absence of specific guidelines will limit stormwater use to smaller scale, less complex systems.

Maintenance and Operations. Maintenance strategies at large stormwater recharge facilities are largely concerned with maintaining high infiltration rates, which consists of managing both sediment and ponding duration in the basins. Techniques to manage stormwater infiltration basins to sustain performance are discussed in the Australian Guidelines for Water Recycling—Managed Aquifer Recharge (Australian SCEW, 2009) and in Kazner et al. (2012). Regarding sediment management, facility operators aim to minimize the accumulation of fine particulates in the surface layers of the spreading grounds, which decreases the soil's hydraulic conductivity and, consequently, infiltration rates. Maintenance activities include occasionally removing surficial soil layers and mechanically ripping the basin soil to break up any clogging layers (Chambers Group, 2014; MWH, 2003). Facility operators manage ponding duration to control groundwater mounding, which decreases the pressure differential within the vadose zone, thereby decreasing infiltration rates. Ponding management also helps avoid potential vector concerns associated with standing water. Ponding management consists of allowing basins to sufficiently dry between wetting events; to that end, when possible, operators use a battery of infiltration basins to regulate the specific placement of water at the facility to leave some basins empty long enough to properly dry (Batman, Los Angeles County Department of Public Works, personal communication, 2014).

SYSTEMS THAT INTEGRATE STORMWATER AND GRAYWATER USE

Integration of the beneficial use of stormwater and graywater can offer many advantages including achieving larger reduction in water use compared to when either approach is used alone. In particular for irrigation applications, stormwater can help to flush salts that may accumulate in graywater-irrigated soil. Below is a list of some of the possible configurations of integrated graywater and stormwater systems:

1. Graywater and stormwater are used for irrigation
2. Graywater is treated for toilet flushing, and stormwater is used for irrigation and/or supplementing indoor nonpotable uses as needed
3. Graywater is used for irrigation, and stormwater is used for indoor nonpotable use and to supplement graywater irrigation as available

There are some important design considerations for systems that integrate graywater and stormwater. Sizing of combined systems needs to account for the complexities of the variable nature of the stormwater flows, the fairly consistent supply of graywater, and the demand profile for use. Given this complexity, it may be necessary to conduct continuous simulation of these systems to determine appropriate sizing and control logic. Integrated systems may be most advantageous in combined sewer areas where potential overflow of graywater combined with stormwater does not have direct negative impacts on receiving waters. In Mediterranean climates, integrated systems could potentially provide stormwater storage during the wet season and graywater storage during dry months. To avoid septic conditions, untreated graywater should not be stored for more than 48 hours. Most stormwater storage tanks are designed to store water for long durations to maximize water supply, but raw graywater should not be combined with stormwater in storage tanks that have storage times longer than 48 hours. If graywater is treated to remove a substantial portion of organic matter, then it could be combined with stormwater for longer duration storage. Also important to consider in the design of integrated stormwater and graywater use systems is that the treatment required for each source of water is different for different uses. Therefore, to maximize treatment efficiency, graywater and stormwater may need to undergo treatment through separate processes or process trains. For example, graywater used to flush toilets may be treated through a series of filters and membranes to reduce organic content, while stormwater supplemented to graywater for toilet supply may be treated through a simple sand filter. The combined waters would then be disinfected prior to supply to toilets. In such a system, it would also be possible for stormwater to be treated through the same treatment train that is used to treat graywater, although that level of treatment provided would not be required for use in toilet flushing.

UNIT TREATMENT PROCESSES

Depending on the water quality objectives, the treatment processes may be similar for graywater and stormwater. Figures 6-5 and 6-8 present conceptual treatment systems for graywater and stormwater, respectively, for a variety of end uses and regulatory conditions. Based on the appropriate treatment system and desired water quality for the end use, unit processes can be selected. Because of a lack of national guidance for graywater and stormwater use and the wide variability of local regulations (see Table 6-4 and Chapter 8), no standard of practice shapes the design of treatment systems.

Table 6-7 summarizes applicable unit processes, organized by contaminant class that can be easily paired with treatment system designs as described in Figures 6-5 and 6-8. In some cases, unit processes that are typically applied for water and wastewater treatment but have potential to be applied to graywater or stormwater treatment are included. The most appropriate configuration of these unit processes will develop through lessons learned from current and future projects.

Unit treatment process efficiency and considerations specific to applications for graywater treatment are summarized in Table 6-8. Similar information for application of these unit processes to stormwater can be found in Strecker et al. (2005) and Clark and Pitt (2012). In addition, a summary of commercially available graywater treatment systems can be found in Sharvelle et al. (2013). Stormwater control practices are usually built as part of the drainage system, but some commercial treatment units (especially media filters and hydrodynamic separators) are available. A NRC report (2009d) summarized many stormwater control options, along with features and expected performance. Pitt et al. (2011) describe some off-the-shelf treatment systems that have been used for roof runoff capture at the household scale, along with case study descriptions of stormwater treatment systems that have been used for larger-scale stormwater beneficial use applications.

CONCLUSIONS

Graywater irrigation at the household scale can be achieved with simple systems that require little energy and maintenance. These simple systems, such as the laundry-to-landscape system or systems that include storage, coarse filtration, and pumps, typically do not include organic matter removal or disinfection, and risk is managed through a series

BOX 6-5 Stormwater Recharge in the Chino Groundwater Basin

The Inland Empire Utilities Agency (IEUA) in collaboration with the Chino Basin Watermaster (CBWM), the Chino Basin Water Conservation District (CBWCD) and the San Bernardino County Flood Control District (SBCFCD) jointly sponsor the regional groundwater recharge program. An extensive water conveyance network directs stormwater run-off, imported water, and reclaimed water to 16 recharge sites, where the water is held so that it percolates into the ground and replenishes the aquifer system (Figure 6-5-1). IEUA recharges 15,000 to 25,000 AF (18-31 million m^3) of stormwater annually in addition to 10,000 AF (12 million m^3) of reclaimed water and 40,000 to 50,000 AF (49-62 million m^3) of imported water. The objective of the program is to mitigate future water shortages by expanding the local groundwater supplies. A rigorous monitoring plan samples stormwater and local runoff quarterly in storm and non-storm events, including analysis for total dissolved solids, metals, volatile organic compounds, synthetic organic compounds, and unregulated chemicals, to ensure that the recharge water meets permit requirements.

Between 2002 and 2011, approximately $75 million in capital improvements were made to the existing flood control/stormwater capture facilities. Some of the funding was used to increase stormwater capture and recharge by approximately 12,000 acre-feet (15 million m^3) per year by enhancing percolation rates in the recharge basins and improving flood control channels to provide conveyance to off-stream stormwater retention basins. The project also developed maintenance protocols for cleaning the recharge basins without dewatering and drying them to ensure optimum percolation rates during the winter season. New monitoring wells and lysimeters were installed to monitor groundwater quality. Approximately one-half of the capital improvement program was funded by grants, and the other one-half from local water fees and taxes.

SOURCES: http//www.ieua.org/water-sources; Chino Basin Watermaster (2001); IEUA (2014).

FIGURE 6-5-1 Regional stormwater recharge in the Chino Basin. SOURCE: Rich Atwater.

of best management practices. However, there is a lack of clarity on the appropriateness of surface drip irrigation with no landscape cover, because the risk of this practice remains poorly defined at the household scale. Most guidance recommends subsurface irrigation as a conservative strategy. Neighborhood-scale systems typically provide disinfection where access-control is not feasible, which creates more system complexity and requires more energy, although the burden of maintenance is removed from individual homeowners. However, there is not yet a standard of practice for implementation of such neighborhood-scale systems, and many questions remain on who would implement, pay for, and maintain such systems.

Graywater reuse for toilet flushing requires plumbing components and treatment systems that are most appropriate in multi-residential buildings or neighborhoods. Graywater systems for toilet flushing require dual plumbing with a connection to potable water and backflow preventers that require annual inspection. Treatment systems for toilet flushing should include disinfection to reduce risk and prevent bacterial growth, and existing technologies are available. Even the simplest treatment systems require periodic maintenance that can be a burden at the household level, although such maintenance is more easily managed by contractors or on-site staff at the neighborhood/multi-residential scale. For broader adoption of graywater for toilet flushing at the house-

TABLE 6-7 Unit Processes to Achieve Removal of Target Constituents Described in Figures 6-5 and 6-8

				Applicability		Appropriate Scale		
Particulate Matter	**Capital Cost**	**Maint. Required**	**Energy**	**Graywater**	**Stormwater**	**Household**	**Neighborhood**	**Regional**
Diameter > 5-10 μm								
Sedimentation	low	low	none	✓	✓	✓	✓	✓
Diameter 1-5 μm								
Cartridge filtration	medium	medium	low	✓	✓	✓	✓	✓
Sand or soil filtration	medium	low	medium	✓	✓	✓	✓	✓
Diameter < 1 μm								
Membrane filtration	medium	medium	high	✓	✓		✓	✓
Coagulation[a]	medium	high	medium	✓				✓
DOC								
Biological aerated filters	medium	high	high	✓			✓	✓
Sequencing batch reactors	medium	medium	high	✓			✓	✓
Membrane bioreactors	high	medium	high	✓			✓	✓
Constructed wetland	medium	medium	low	✓	✓		✓	✓
Soil aquifer treatment	medium	low	low	✓	✓		✓	✓
Sand and soil filtration	medium	low	medium	✓	✓	✓	✓	✓
Membrane filtration	medium	medium	high	✓	✓		✓	✓
Activated carbon	medium	medium	medium	✓	✓	✓	✓	✓
Pathogens								
Chlorine	low	low	low	✓	✓	✓	✓	✓
UV	medium	high	high	✓	✓	✓	✓	✓
Ozonation[b]	medium	medium	medium	✓			✓	✓
Membrane filtration[b]	medium	medium	high	✓			✓	✓
Inorganic and Organic Contaminants								
Metals								
Chemically-active media filtration	high	medium	low		✓	✓	✓	✓
Sedimentation	medium	medium	low		✓	✓	✓	✓
Organic Compounds								
Air stripping (active)	medium	medium	high		✓			✓
Chemically-active filtration	high	medium	low		✓	✓	✓	✓
Sedimentation	medium	medium	low		✓	✓	✓	✓

[a]Coagulation has rarely been used with stormwater, but it can be effective.
[b]These disinfection processes are possible with stormwater, but they have been rarely used.
NOTES: These unit processes are not stand alone and should be used in treatment systems as shown in Figures 6-5 and 6-8.

hold scale, treatment systems are needed that are low maintenance and include process automation and control to ensure safe use at a reasonable cost.

Many state graywater treatment standards for toilet flushing are not risk-based or fit-for-purpose. Many graywater quality standards are based on state standards for unrestricted urban use of treated wastewater, discharge of treated wastewater to water bodies, or other standards for very different applications than use of graywater for toilet flushing. Many of the standards for graywater use for toilet flushing may be unnecessarily strict in terms of organic content and turbidity removal, resulting in requirements for technologies that are costly, energy-intensive, and require frequent maintenance. Standards also vary widely across states, resulting in an inconsistency in treatment systems that can be applied. As a result, graywater systems are more costly than they might be otherwise, because a single standard of practice has not been established across the country, and treatment system manufacturers must develop systems to meet varying objectives. Additional research is needed to determine appropriate design standards for dissolved organic carbon and turbidity that prevent aesthetic and maintenance issues while allowing proper function of disinfection systems when using graywater for toilet flushing.

New developments and future urban planning provide opportunities for rethinking the conveyance and use of various water and waste streams for maximum cost, energy, and water savings. Separation of graywater results in blackwater that is more concentrated in solids and organic matter than conventional domestic wastewater and may be amenable to produce methane biogas through anaerobic digestion. These systems can also be integrated with urine sepa-

TABLE 6-8 Summary of Unit Process Application to Graywater

	Graywater Treatment Efficiency	*Considerations Specific to Graywater Use*
Particulate Matter		
Diameter > 5-10 μm		
Sedimentation	Not well documented	No specific graywater considerations
Diameter 1-5 μm		
Cartridge filtration	Highly variable depending on filter size	Clogging noted in filters smaller than 100 μm (pre-treatment required)[a]
Sand or soil filtration	TSS: 62-85%[b,c,d] Turbidity: 46-87%[b,c]	Several successful applications, good fit technology
Diameter < 1 μm		
Membrane filtration	Turbidity: 92-97% (UF), 98% (NF)[e] TSS: 100% (NF)[e]	Potential for fouling is high, but can be reduced with extensive pretreatment
Coagulation	Turbidity: 82-91%[f]	None
DOC		
Biological aerated filters	Effluent BOD < 20 mg/L[g]	No specific graywater considerations
Sequencing batch reactors	Effluent BOD < 20 mg/L[g]	No specific graywater considerations
Membrane bioreactors	Effluent BOD < 4 mg/L[g]	No specific graywater considerations
Constructed wetland	Effluent BOD 0.7-80 mg/L[g]	No specific graywater considerations
Soil Aquifer treatment	Not Applicable	Not Applicable
Sand filtration	Effluent BOD 62 mg/L[b] COD: 37-61%[b,c,d]	Several successful applications, good fit technology
Membrane filtration	Effluent BOD 86 mg/L (UF),[h] 1.5 mg/L (RO)[h] COD: 45-96% (UF), 93% (NF), 98% (RO)[e,h]	Potential for fouling is high, but can be reduced with extensive pretreatment Due to low molecular weight of graywater DOC, NF or RO is needed for efficient removal[e]
Activated carbon	Effluent BOD 10 mg/L[h] COD: 93%[h]	No specific graywater considerations
Chemical oxidation	Not Applicable	Removal of organic matter negligible when ozone generator was sized practically for residential or multi-residential applications[i]
Photo-oxidation	Not Applicable	Graywater DOC (mostly surfactants) results in high turbidity, photo-oxidation is not effective[i]
Bacteria and Viruses		
Chlorine/Bromine	Depends on dose and initial water quality	Effective and low cost[i] Not good for plants in irrigation application
UV		Can be effective[k]
Ozonation		Safety concerns at residential scale
Hydrogen peroxide		Corrosive to toilet components[j]

[a]E. Clerico, Natural Systems Utilities, personal communication, 2014.
[b]Friedler et al. (2006).
[c]Gual et al. (2008).
[d]Zuma et al. (2009).
[e]Ramon et al. (2004).
[f]Pidou et al. (2008).
[g]Sharvelle et al. (2012).
[h]Sostar-Turk et al. (2005).
[i]Hodgson (2012).
[j]Kuru and Luettengen (2012).
[k]Wiles (2013).

ration including nutrient capture. Thus, graywater reuse can be a key element of energy-efficient urban water and resource management systems that not only minimize net water abstraction from the environment but also achieve a high level of energy and nutrient recovery.

The state of practice and development of cost effective and safe stormwater capture systems for roof runoff are hindered by the lack of data on human pathogens and the risk associated with various uses. Design and treatment standards are generally well accepted for nonpotable use of runoff collected from land surfaces, and no treatment other than coarse solids removal is needed for subsurface irrigation where human exposures are minimal. For beneficial uses of roof runoff with low to moderate exposures, additional pathogen data and risk analyses are needed to establish a consistent state of practice for on-site stormwater use. Collection, storage, and treatment technologies are mature and can be readily adapted for various scales and uses, but appropriate guidance based on risk assessments is generally lacking.

Operations and maintenance of household and neighborhood graywater and stormwater use systems is not well guided or monitored. All systems that capture graywater and stormwater for beneficial use require routine maintenance. For systems where disinfection is not required (e.g., subsurface irrigation), failure to conduct needed maintenance poses operational concerns but does not pose a significant risk for human health or environmental quality. However, for systems with disinfection processes to protect human health (i.e., systems for toilet flushing), ongoing maintenance is critical. Although many states require that installed systems meet certain water quality targets, ongoing monitoring is not required. More guidance is needed to ensure safe operations of graywater and stormwater treatment systems at household and neighborhood scales. Because frequent routine water quality analyses are expensive and impractical even at the neighborhood scale, system operational performance standards and online monitoring of surrogate parameters (e.g., residual chlorine, suspended solids, or turbidity) should be considered.

Stormwater infiltration for aquifer recharge is commonly practiced, but designs and regulations in the United States may not be adequately protective of groundwater quality. Design for large-scale stormwater infiltration projects are still emerging. For many locations, the design and performance standards for stormwater infiltration have been developed to address surface water regulatory drivers rather than the protection of groundwater quality from stormwater infiltration. Of particular concern is the infiltration of organic contaminants and salts from highly urbanized areas into water supply aquifers, although human pathogens may also be of concern depending on the infiltration site characteristics. In current practice, no standard design criteria exist for stormwater treatment for large groundwater infiltration projects. Thoughtful planning, source area selection, source control, and mechanisms to integrate treatment into the watershed could improve efficiency of these systems and reduce the amount of treatment required. Innovative treatment systems, for example, engineered wetlands and filter media, may also be needed for regional-scale systems where source control is challenging.

7

Costs and Benefits

A key issue in evaluating the merits of the potential beneficial uses of graywater and stormwater pertains to what benefits are generated, and how those benefits compare to the costs. This is not a simple exercise, and it is not possible to draw broadly generalizable inferences for the following reasons: (1) many different types of benefits and costs may be relevant, (2) the types and magnitudes of benefits and costs typically are very use- and site-specific, and (3) the benefits and costs may be borne by a wide range of different individuals and entities. Each of these key issues is discussed in turn.

Furthermore, many of the potentially important benefits are difficult to quantify and value in monetary terms (i.e., they consist of what economists refer to as "nonmarket values"). Indeed, many of the motives for pursuing beneficial use projects for graywater or stormwater include aspirational and other values that extend beyond financial returns and water supply benefits. These facts may make it challenging to provide a viable and fair comparison of the key benefits of a stormwater or graywater project to its costs: the costs are generally identified and monetized, but many key benefits may not be readily amenable to monetization.[1] This may create an unfortunate imbalance in how beneficial use projects are perceived and evaluated, unless considerable effort is applied to fully recognize, account for, and estimate the full range of important, applicable benefits.

RANGE OF POTENTIAL BENEFITS AND COSTS

The types of benefits and costs associated with graywater or stormwater use are highly diverse. Although the costs are primarily financial, the benefits may be financial, social (i.e., related to health, well-being, and quality of life),[2] and environmental. The benefits may be broadly distributed or limited to the entity implementing the project.

Financial costs refer to the total out-of-pocket expense borne by whoever is paying for a graywater or stormwater project. Financial benefits may include returns to a homeowner or municipality in terms of avoided costs (e.g., reduced purchases of potable or imported water or delayed infrastructure upgrades) or other monetary returns (such as avoided fines for stormwater violations). Environmental benefits may accrue because of potential water-related savings in energy use and related emissions of carbon dioxide and other air pollutants associated with energy generation. Additionally, stormwater capture and use can reduce harm to local watersheds from otherwise poorly managed flows. Social benefits may include providing a community with aesthetic improvements due to increased green space or reducing traffic disruptions associated with periodic urban street flooding during intense rain events. Public education and increased individual awareness of local water supply and related issues may be considered as social benefits—although ones that can be particularly difficult to quantify and monetize. Box 7-1 provides an overview of many of the types of financial, social, and environmental benefits and costs that may be derived from a given graywater or stormwater application.

Many key benefits (and costs) can be highly site specific. In locations where water is relatively scarce, tapping graywater or stormwater resources is likely to reduce the demands on potable or other supplies that may be very expensive (at the margin) or associated with environmental stressors (e.g., diminishing stream flows to levels that place

[1] Various nonmarket valuation approaches may be applied on a case-specific basis to estimate many of the potential benefits. These approaches include hedonic or other revealed preference techniques and/or survey and related stated preference methods (Adamowicz et al., 1998; Alcubilla and Lund, 2006; Cadavid and Ando, 2013; NRC, 1997; Poor et al., 2007; Van Houtven, et al. 2014). The challenge is that such methods can be complicated, data intensive, and costly to apply properly, and need to be performed on a case- and site-specific basis.

[2] This is sometimes referred to as the Triple Bottom Line (TBL) framework. In the field of economics, the term "social values" is used to include all the possible benefits (and costs), including both the internal, financial aspects borne by private individuals and entities, as well as the broader array of external impacts that may be reflected under the "social" and "environmental" categories used in this chapter, but this definition is not used in this report.

BOX 7-1 Potential Benefits and Costs Associated with Beneficial Use Options for Graywater or Stormwater

Financial [a]

Private Costs
- Lifecycle costs to install, operate, and maintain a graywater or stormwater capture and treatment system

Public Costs
- Increased wastewater management costs associated with reduced wastewater volumes, increased pollutant concentrations, or the reduced viability of centralized water reuse programs
- Reduced income from potable water sales needed to cover fixed costs

Private Benefits
- Reduced potable water purchases (for consumers)
- The value to homeowners of drought-proof landscaping (avoided replacement costs), although maintenance costs may be higher
- Possible associated increases in local property values, in properties and communities with added greening (e.g., trees) or street improvements (e.g., curbs and sidewalks)
- Potential sale of marketable stormwater retention credits (and/or reduced on-site stormwater management fees)

Public Benefits
- Reduced, avoided, or deferred utility costs related to developing new water sources, acquiring water rights, upgrading or expanding treatment plants, or replacing water or wastewater conveyance systems infrastructure
- Possible reduced energy costs related to pumping imported water and possible energy production from advanced resource recovery systems
- Reduced penalties and fines for overflow events (e.g., combined sewer overflow permit violations)
- Greater water supply reliability from stormwater capture and recharge in arid or semi-arid regions

Social [b]

Private Benefits
- On-site aesthetic enhancements
- Aspirational value to individuals and businesses for tapping local resources and promoting sustainability

Public Benefits and Costs
- Aesthetic enhancements (such as those derived from added greening of a neighborhood).
- Increased or enhanced local recreational opportunities due to added local green space, improved local riparian systems, etc.
- Public health effects (e.g., reduced heat stress from urban greening, reduced disease from combined sewer overflows, increased disease from failures in graywater or stormwater treatment or exposure control)
- Aspirational and public education value related to tapping local resources and promoting sustainability
- Increased robustness, resiliency, and reliability of local water supply portfolios, by diversifying mix of source waters
- Increase in "local control" of water resources and water supply (e.g., reducing reliance on regional or imported supplies)
- Potential increase in local job creation (e.g., to install and maintain stormwater or graywater infrastructure), if such jobs engage otherwise under-employed labor resources (i.e., not a transfer of jobs from one region or sector to another)

Environmental

Public Benefits
- Reduced hydromodification of urban streams
- Reduced carbon footprint and improved air quality (where energy consumption is reduced)
- Reduced loadings of stormwater pollutants to surface waters
- Reduced environmental impacts associated with developing new water supplies
- Potential reduction in downstream water availability if projects expand irrigation and evapotranspiration (see Chapter 5)

[a] These are cash flow impacts on households, businesses, utilities, and other institutions that directly engage in beneficial use applications of graywater or stormwater.
[b] This includes monetary and nonmarket benefits and costs for entities that may not be directly engaged in beneficial use projects (e.g., external benefits and costs).

special status fish species at risk, creating a large carbon footprint from seawater desalination). In such settings, the benefits of graywater or stormwater use may include both financial benefits and an array of benefits to the broader human and natural community. These same benefits would not be experienced in a location with abundant conventional water supplies.

Benefits and costs are also highly dependent on the specific type of application (i.e., end use) and the scale of the application. Graywater is typically reused through simple systems at the household scale, but large multi-residential properties and even regional systems are possible (see Table 2-1 and Box 2-1) and involve more costly treatment and dual-distribution systems. Stormwater applications can also vary widely in scale, ranging from household to neighborhood to regional levels. The scale will impact relative costs (e.g., because of economies of scale) and also the level and types of benefits generated.

When examining the costs and benefits of graywater and stormwater use projects, who bears the costs and who realizes the benefits are additional important considerations. For example, when a project generates significant external benefits (such as environmental benefits) that are not fully captured by those who bear the costs, how should the costs be equitably shared? In some cases, subsidies may provide a financial incentive for providing broader benefits. The distribution of the benefits and costs must address concerns about affordability, social equity (fairness), and environmental justice.

FINANCIAL COSTS

This section presents what is known about the financial costs of graywater and stormwater projects. Ideally, financial costs should account for the full costs during the life cycle of the project, including initial installation (e.g., equipment, construction, permitting), annual operation and maintenance (O&M) over the full course of the project's effective operating lifetime, any capital replacement required during the asset's effective life, and any decommissioning and subsequent replacement costs at the end of useful life. These costs can be presented as total annualized costs (with capital expenses converted to equivalent amortized annual values, and added to annual O&M expenses) or as a total "present value" of all project costs in current dollars. However, there is a dearth of well-documented data on the full costs of such projects, particularly their life-cycle costs, because most are new with minimal data about long-term maintenance costs and performance effectiveness.

It is important to include all costs, regardless of who bears them, and to keep track of who bears which costs. For example, a household may bear much of the expense of installing a graywater system that taps its laundry machine and diverts the effluent to a subsurface drip system irrigating its garden. However, the local municipality and/or a local nongovernmental organization may provide subsidized support for these activities, and these expenses must be included in a full accounting of financial costs.

Factors That Affect Financial Cost

Several key factors impact the costs of graywater and stormwater beneficial use projects. These factors include whether treatment is provided, the type and level of treatment, and the scale of the application (including size of storage). For simple, household-scale, laundry-to-landscape systems (Box 2-4) where untreated graywater is conveyed without additional pumps to a subsurface irrigation system, capital costs are fairly minimal (primarily labor), with little if any O&M costs. In a multi-residential apartment building, graywater reuse for toilet flushing requires treatment (see Chapter 6) as well as some mechanical pumping and more elaborate piping and storage investments. Treatment costs may vary depending on the water quality objectives, which may be governed by regulations that vary from state to state or even by locality (see Chapter 8). Costs will also be influenced by whether the application is part of a new development or a major renovation effort, or a retrofit into an existing structure or established neighborhood. Retrofit on-site capture and use projects are typically considerably more expensive than installations in new developments.

Stormwater projects will also vary widely in their financial costs, depending on the scale, location, and the type and extent of conveyance and treatment (if any) that may be necessary. When stormwater is collected and infiltrated into local groundwater basins in the same general proximity in which it is collected, the costs will be less, all else being equal, than when stormwater is harvested at one location and pumped a considerable distance to another location where it can be more suitably put to direct beneficial use. Stormwater infiltration projects may often require the purchase of land, which varies widely in price. The capture and use of stormwater in some states may also require the purchase of water rights from downstream users.

Summary of Cost Data

This section summarizes the limited available data on the financial costs of graywater reuse and stormwater capture and use at the household and neighborhood scales. Most available data are focused on the household scale. Some data on regional-scale stormwater capture for groundwater recharge are also included, although such project costs would be highly site specific. Monetary values reported in this chapter have been adjusted to 2014 U.S. dollars, unless stated otherwise.[3]

Household-scale Graywater Reuse

Graywater system costs at the household scale vary widely depending on the size and complexity of the system. Information on installing do-it-yourself laundry-to-landscape systems is widely available online, and some water utilities provide subsidies for material costs. San Francisco Public Utilities Commission (SFPUC, 2012) reports that laundry-to-landscape systems cost a few hundred dollars for a self-installed system and $1,000-$2,000 for a professionally installed system. SFPUC has offered a $112 subsidy toward a $117 graywater kit that can be installed by a homeowner without plumbing skills and without a permit (P. Kehoe, SFPUC, personal communication, 2014). Costs increase if a pump or filtration system is also needed and with professional installation (see Box 7-2). Nonmonetary costs arise when homeowners devote their own time to installing the system (which may be valued in monetary terms based on a prevailing wage rate).

The financial costs of household-scale graywater systems increase with complexity, especially if they are professionally installed and/or must be professionally maintained. Some systems that are more complex than laundry-to-landscape kits can still be installed by homeowners with basic plumbing and electrical skills, reducing out-of-pocket installation costs but requiring investment of the homeowner's time. SFPUC (2012) reports that a simple whole-house graywater system with a tank and pump, but no filtration costs between $500 and several thousand dollars, depending on the extent of plumbing work and whether the system is professionally installed. Adding a sand filter system (but no disinfection) increases per household costs into the range of $5,000 to $15,000 (SFPUC, 2012).[4] Although cost data are limited, one whole-house graywater collection system for irrigation with professionally installed dual plumbing (installed during construction) and homeowner-installed tank and pump costs $2,300 for a five-bedroom house, with the bulk of that cost ($1,700) being associated with the dual plumbing system (L. Roesner, Colorado State University, personal communication, 2015). A smaller home would have lower plumbing installation costs. The costs of retrofitting dual plumbing in an existing home would be substantially higher.

Neighborhood and Regional Graywater Reuse

The costs associated with graywater systems naturally increase as they are scaled up to neighborhoods and regions, although economies of scale may occur. In multi-residential settings, graywater is commonly used for both landscape irrigation and toilet flushing. The costs for graywater reuse at the neighborhood scale depend on several factors, including the type and scale of graywater treatment, the extent of dual-plumbing required (including whether dual collection for irrigation or dual collection and distribution for toilet flushing is used), and whether the plumbing is installed as part of a new construction project or as part of a retrofit into an existing building. Extremely limited cost data are available on neighborhood-scale graywater projects.

Hodgson (2012) breaks down capital and O&M costs for 14 different graywater system sizes (50-5,000 gpd; 190-190,000 lpd) and for various disinfectant schemes (see Table 7-1). For example, for a 1,000 gpd (3,800 lpd) system (capable of treating enough graywater to provide toilet flushing for about 150 to 160 people in a dormitory-type setting),[5] total storage and treatment costs (plumbing not included) amount to $4.39 per 1,000 gallons (Kgal). As shown in Table 7-1, initial and annual operating costs increase with capacity, but units over 100 to 300 gpd (380 to 1,100 lpd) achieve economies of scale that significantly reduce the cost per unit of water (Hodgson, 2012). These costs are comparable to or lower than potable water rates in many cities,[6] and they are significantly lower if combined wastewater charges are considered. However, the plumbing cost, which is omitted but may be the most expensive component, especially for a retrofit application, would also need to be taken into account. An analysis of costs of graywater use in a mixed-use development ranging from 40,000 to 500,000 ft^2 (3,700 to 46,000 m^2) revealed that installation of dual collection and distribution plumbing during building construction represented 77 to 93 percent of the capital costs for the project (RMC Water and Environment, 2012), although Cordery-Cotter and Sharvelle (2014) reported lower plumbing costs for a small dormitory-scale project (see Box 7-3).

An additional potential effect of wide-scale graywater use (if not accompanied by population growth) within existing development is the reduction of wastewater discharges to sewer systems, which may increase costs for wastewater utilities if reduced flows impede the in-sewer transport of wastes to the wastewater treatment facility and cause other problems in the sewer network. Reuse of graywater not only reduces the combined wastewater flow but also increases concentrations because most of the wastewater constituents must still be conveyed for treatment (see Box 3-3). Reduced flow decreases flow velocity in wastewater collection systems, which results in increased residence time, potentially

[3] See http://data.bls.gov/cgi-bin/cpicalc.pl.

[4] See additional cost estimates at http://www.graywateraction.org.

[5] Based on data provided in Cordery-Cotter and Sharvelle (2014).

[6] See http://www.circleofblue.org/waternews/wp-content/uploads/2014/05/WaterPricing2014GraphsInteractive1.pdf.

BOX 7-2 Benefit-Cost Analysis of a Single-Family Graywater Irrigation System

Costs. The costs of a laundry-to-landscape system range from approximately $120 for a simple do-it-yourself (DIY) system with no treatment (see Box 2-4; P. Kehoe, SFPUC, personal communication, 2014) to $1,250 with professional installation (J. Gallup, City of Long Beach, personal communication, 2014). A self-installed, whole-house, graywater collection and irrigation system with dual plumbing professionally installed during home construction and a self-installed 150-gallon (570-liter) tank and water pump has been reported to cost $2,300 (L. Roesner, Colorado State University, personal communication, 2015), although professionally installed whole-house systems cost between $5,000 and $15,000.

A hidden cost to each household might occur if widespread graywater implementation in a community leads to significant reductions in potable water use, because it is possible that the water utility would need to raise its rates to continue to meet its fixed-cost obligations.

Financial Benefits. Potable water savings from single-family graywater systems depend on the behavior of the homeowner and how the irrigation rates change with installation of the graywater system. In laundry-to-landscape graywater pilot programs by the SFPUC and the City of Long Beach, the majority of households with newly installed laundry-to-landscape systems showed increased water demands (Gallup, personal communication, 2014; Kehoe, personal communication, 2014), possibly because of an expanded irrigated landscape or higher irrigation rates. A household only realizes a financial benefit from water savings to the extent that graywater reduces potable water costs.

For a consistent comparison, the committee used the potential water savings calculated for a typical medium-density, residential development in six locations in the United States (see Table 7-2-1 and Chapter 3) to roughly estimate potential financial benefits for laundry-to-landscape and household (i.e., shower, laundry, bathroom sinks) graywater systems. These estimates are based on conservation irrigation to meet the evapotranspiration deficit for turfgrass, so actual savings could be more or less depending on the irrigation rates and extent and types of vegetation to be irrigated. Savings could also be reduced if the household water use increases with the availability of a no-cost water source. Additionally, these calculations are based on average graywater produced (see Box 3-1), but high-efficiency washing machines will generate less water. At a potable water charge of $10 per 1,000 gallons (including associated wastewater charges), potential water supply cost savings for a household might amount to approximately $23 to $53 per year for a laundry-to-landscape system and $50 to $130 per year (on average) for a whole-house graywater system (based on the assumed water savings). A graywater system may also sustain valuable plants during drought restrictions, but such benefits are not included in this calculation.

Social and Environmental Benefits (and Costs). Various social and environmental benefits may arise, including:

- Public education and water awareness benefits, as households become more directly engaged with their own water use and enjoy aspirational values by reusing water on-site.
- Neighborhood aesthetics and potential property value enhancements may apply, to the extent that the graywater system leads to improved landscaping conditions, particularly during an extended drought.

Benefit-Cost Comparison. Based on the committee's scenario assumptions (see Chapter 3), do-it-yourself, laundry-to-landscape systems can potentially offer low payback periods of 2.5 to 6.0 years at a 5 percent discount rate, assuming that total potable water use is actually reduced by the installation of the graywater system and no value is assigned to the homeowner's labor. Payback periods are more than 50 years with the professionally installed laundry-to-landscape system, even with high-end water savings, and 44 years or longer for the whole-house system, considering water savings only. Payback periods would be shorter in households with more people and a higher existing irrigation demand. Payback periods would also be shorter in areas with near-year-round irrigation demand and longer in areas with irrigation demand limited to the summer. If water use increases substantially with installation of the reuse system, then no savings may occur. Nonfinancial benefits may also be important for households and communities.

TABLE 7-2-1 Summary of Scenario Analysis of Potential Water Savings from Conservation Irrigation with Laundry-to-Landscape and Household Graywater in Six U.S. Regions

	Self-installed Laundry-to-Landscape		Self-installed Household Graywater	
	Average potential household water savings (gallons/year)	Estimated potential annual cost savings ($/year)	Average potential household water savings (gallons/year)	Estimated potential annual cost savings ($/year)
Los Angeles	4,800	48	13,000	130
Seattle	2,800	28	7,800	78
Lincoln	5,300	53	13,000	130
Madison	2,600	26	6,200	62
Birmingham	4,200	42	9,000	90
Newark	2,300	23	5,000	50

NOTE: See Chapter 3 for details of the analysis. Scenarios assume medium-density, residential development with three persons per household and irrigation to meet the evapotranspiration deficit for turfgrass. Landscaped area and residential density determined by location-specific data (see Appendix A).

TABLE 7-1 Cost of Graywater Treatment and Storage Systems of Varying Scale, Including a Coarse Filter and Chlorine Disinfection System (Plumbing Excluded)

System Size (gpd)	Capital	Annual Chemical and Energy	Annual Maintenance	$/Kgal
50	$2,136	$18	$253	$24.33
75	$2,136	$26	$253	$16.50
85	$2,294	$29	$253	$15.17
100	$2,294	$35	$505	$18.91
150	$2,376	$56	$505	$13.08
300	$2,369	$112	$505	$6.97
500	$3,185	$186	$758	$6.16
750	$3,767	$279	$758	$4.61
900	$4,218	$335	$758	$4.13
1000	$4,218	$372	$1,010	$4.39
2000	$5,270	$744	$2,021	$3.96
3000	$6,906	$1,116	$3,031	$3.86
4000	$8,773	$1,487	$4,041	$3.83
5000	$12,435	$1,859	$5,052	$3.92

NOTES: Costs have been adjusted to 2014 dollars. The costs of dual plumbing and distribution plumbing during new construction projects (omitted in this table) have been reported to represent 77 to 93 percent of the cost for a large commercial graywater project (RMC Water and Environment, 2012), although the committee was unable to find detailed data on plumbing costs across a range of project types. Life-cycle costs are reported by Hodgson based on a 3 percent discount rate and a 10-year effective lifetime.
SOURCE: Hodgson (2012).

causing solids sedimentation, anaerobic conditions (septicity), odor and corrosion problems, and increased frequency of sewer blockages.

These effects on wastewater flows are already resulting from increased indoor water conservation from water saving devices. By comparison, the impact of graywater reuse within existing development is anticipated to be minor because of the challenge of retrofitting buildings for indoor nonpotable reuse. In new development, however, wastewater collection systems can be designed to accommodate reduced flows where significant graywater reclamation and reuse is planned. A full accounting of costs would consider increased wastewater system costs associated with widespread graywater implementation, although such an assessment has not been conducted. Widespread potable water use reductions, if they arise across the community, might result in the need for utilities to increase their water and wastewater rates to cover their fixed costs, potentially offsetting financial savings enjoyed by residents.

Harvesting Roof Runoff

Household roof runoff harvesting systems typically consist of rain barrel or cistern collection systems (see Chapter 6). The cost of a residential-scale rain barrel is estimated to be in the range of $60 to $100. Some of this cost may be provided through utility subsidies. Rain barrels vary in size and material, with a typical barrel holding up to 35 gallons (130 liters) of roof runoff (Pitt et al., 2011). O&M costs are negligible.

For larger cistern-based household systems, data from Pitt et al. (2011) for two homeowner-constructed rainwater harvest systems indicate an upfront cost of about $1,500 to $1,600, although costs vary with the extent of new collection infrastructure required and the inclusion/exclusion of treatment and pumps. One such system is located in Montana and entails a 2,500-gallon (9,500 liter) storage tank drawing water collected from a 925-square-foot (86 m^2) roof, including a first flush diverter and a pump. The storage tank cost $900 and the gutters cost $380, comprising the majority of the $1,500 initial cost. The second system, located in Portland, Oregon, and approved by the city for all household water uses, includes a 1,500-gallon (5,700 liter) tank with filters ($900), ultraviolet (UV) disinfection ($380), backflow prevention device ($130), and other components, for a total cost of $1,600. Labor costs for installation (either by homeowner or contractor) are not included.

BOX 7-3 Benefit-Cost Analysis of a Dormitory Graywater System for Toilet Flushing

Costs. Installation costs of a graywater toilet-flushing system for a dormitory with 28 residents (described in Box 2-2) were $66,500. Plumbing retrofitting for this system represented about 30 percent of the installation costs. O&M costs were $5,000 per year (Cordery-Cotter and Sharvelle, 2014).

Financial Benefits. Water savings were estimated at 65,000 gallons (250,000 liters) per year. Using local utility rates of $6 per 1,000 gallons, the water supply cost savings are estimated to be $400 per year (Cordery-Cotter and Sharvelle, 2014).

Social and Environmental Benefits (and Costs). Various social and environmental benefits may arise from household-level applications of graywater for toilet flushing, including water awareness benefits and the aspirational value associated with contributing to water conservation. Reductions in potable water demands may also ease environmental and hydrological stresses on source waters.

Benefit-Cost Comparison. The financial benefits are considerably less than the financial costs in this illustration (and less than annual O&M costs). Relatively high costs are associated with retrofitting an existing building. However, the pilot project application included monitoring and other cost elements that are likely to be higher than those for a more routine installation and operation. Economies of scale may be feasible for larger scaled residential complexes (see Boxes 1-1 and 2-1), although the committee could not obtain detailed cost data for these systems. Incentives and other nonfinancial benefits may be important in decision making.

Several commercial enterprises provide roof runoff capture systems for residential buildings. No published costs are available, but these systems generally cost up to about 10 times the cost of the above homeowner-built systems (Pitt et al., 2011). The advantages of the commercial systems are the vendor's relationships with local regulators and knowledge of the regulations, professional design and installation, and greater confidence concerning safety issues. Because local and regional regulations pertaining to rainwater harvesting and its use vary greatly throughout the country, the extra service provided by the commercial suppliers of these systems may be beneficial.

Rooftop stormwater harvesting may be applied in larger contexts, such as commercial buildings or building complexes. For example, the Frankfurt Airport in Germany installed six 26,000-gallon (100-m^3) cisterns to capture rooftop runoff for toilet flushing and outdoor irrigation. The system cost $109,000 and saves 26 million gallons (98 million liters) per year for an effective cost of $4.19/Kgal (Pitt et al., 2011). O&M costs are not available.

Los Angeles Department of Water and Power (2015) reports that the highest costs per unit of water captured are associated with "on-site direct use" (Figure 7-1). Reported costs ranged from approximately $3,200/acre foot (AF) to nearly $14,000/AF, with a mean of more than $7,000/AF (or $21/Kgal) captured.

Neighborhood Stormwater Capture and Use

Cost estimates for a wide range of stormwater capture or groundwater recharge projects at household and neighborhood scales are provided by Los Angeles Department of Water and Power (LADWP) in Figure 7-1. LADWP (2015) reports the cost thresholds for projects identified for investments (less than $1,100/AF [$3.40/Kgal] for infiltration or less than $1,550/AF [$4.80/Kgal] for direct use), which are generally competitive with other new water supply options available in southern California (see Figure 7-2). The cost threshold for infiltration is lower because it does not reflect an analysis of how much of the stormwater is recoverable, and the actual cost per acre foot of recovered water is likely to be higher.

Of the alternatives, infiltration projects were significantly lower in costs than the stormwater tank capture and use projects, and neighborhood ("subregional") scale infiltration projects are reported to be the most economical of the alternatives (approximately $1,000/AF [$3.10/Kgal] average). Neighborhood-scale tank storage and use projects ("subregional direct use") were considerably more costly, ranging from approximately $1,200 to nearly $7,000/AF captured, with a median of $3,300/AF ($10/Kgal) considering life-cycle costs (LADWP, 2015). By comparison, imported water through the State Water Project costs in the range of $450 to $1,300/AF (or $1.40 to $4.00/Kgal) depending on pricing "tier" and treatment. The costs of imported waters, however, are increasing at a rate greater than general inflation, and the water is not always available for communities wishing to acquire additional supplies. Other new supply options such as desalination or water reuse typically cost between $950 and $2,200/AF ($2.90 to $6.75/Kgal) or more, as shown in Figure 7-2 (Raucher and Tchobanoglous, 2014; Sunding, 2013).

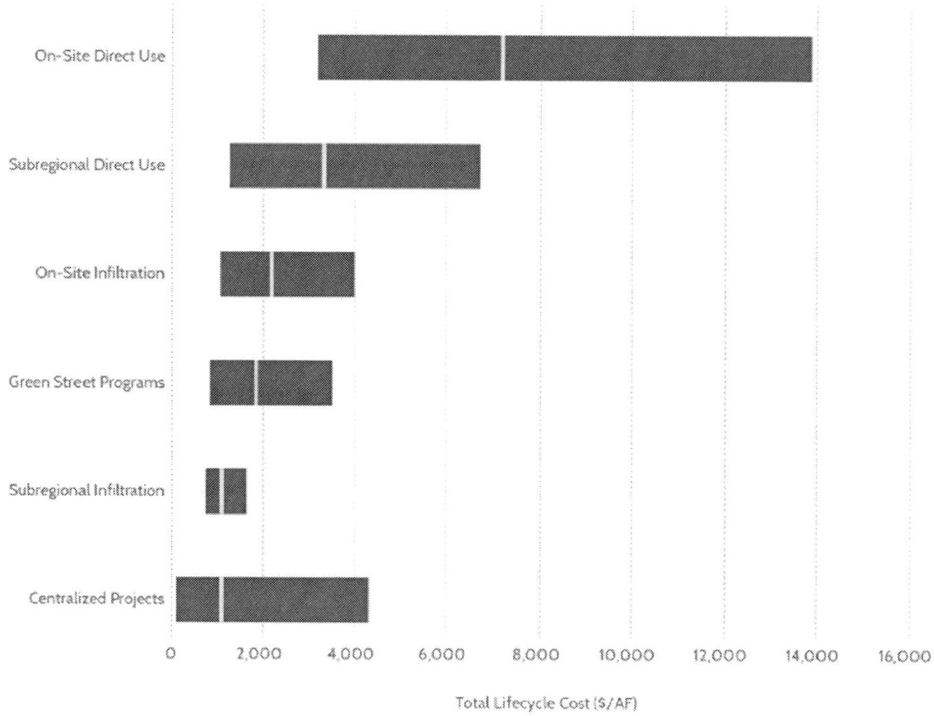

FIGURE 7-1 Life-cycle costs per captured volume, although infiltrated water may not be fully recoverable and may understate the true cost of supply. SOURCE: LADWP (2015).

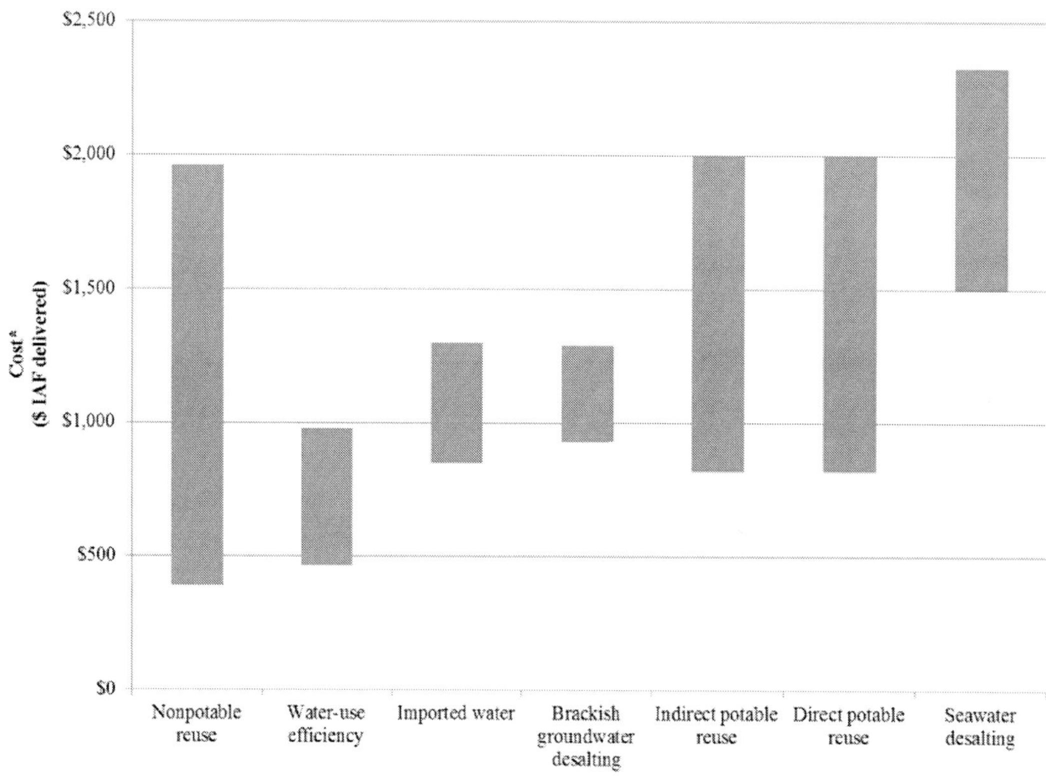

FIGURE 7-2 Comparative lifetime unit costs of alternative water supply options, based on typical costs for California water utilities. NOTE: Costs may be considerably higher than shown for some options and locations, depending on site- and project-specific factors and other considerations. The displayed water use efficiency (i.e., conservation program) cost range reflects water utility-borne costs only (e.g., rebates for water efficient appliances), and do not include costs borne by customers. SOURCE: Raucher and Tchobanoglous (2014).

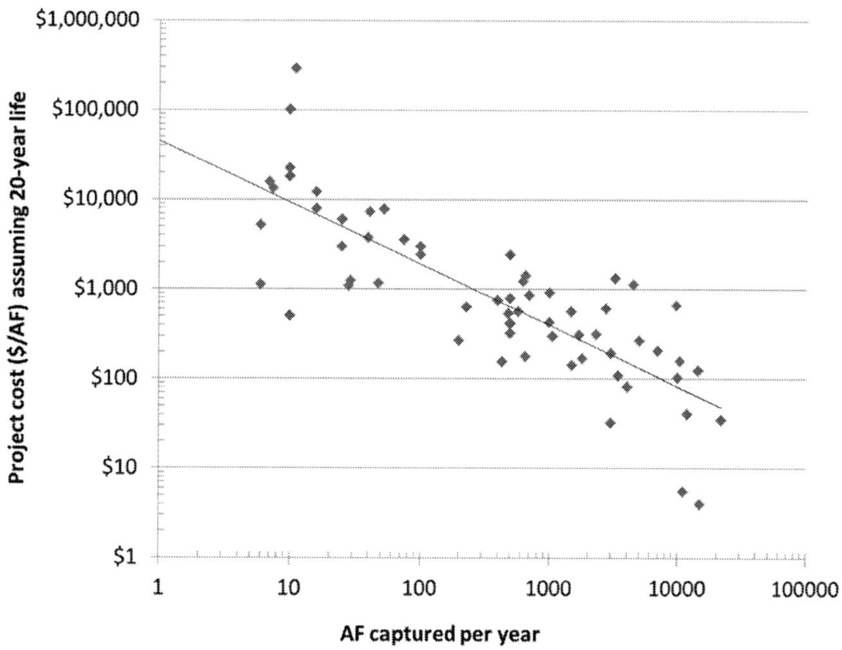

FIGURE 7-3 Cost of stormwater capture and use projects by project size. SOURCE: Adapted from West et al. (2014).

Regional Stormwater Capture

The costs of large centralized stormwater capture projects are highly location specific, and land acquisition requirements and the suitability of existing flood water capture and conveyance systems for a stormwater capture project significantly affect overall costs. Regional-scale stormwater capture and recharge projects may be more cost-effective where existing flood management basins and related facilities can be upgraded rather than developing new facilities with associated land acquisition costs.

Existing publicly available cost data for regional stormwater capture are limited to southern California, so the full range of costs in other locations may not be represented. In an analysis of 118 stormwater projects in southern California, West et al. (2014) observed clear economies of scale among the larger stormwater capture projects (Figure 7-3). LADWP (2010) provides details on 11 large-scale stormwater capture and recharge projects, which were designed to provide an estimated 26,000 AF/year (32,000 m^3/year) of additional groundwater recharge (essentially doubling the region's prior stormwater capture and recharge levels), at a combined capital cost of $251.9 million (assumed to be reported in 2010 dollars). LADWP (2010) estimates life-cycle costs at $60-$300/AF ($0.18-$0.92/Kgal) for these projects. By comparison, LADWP (2015) analyzed 35 large centralized stormwater capture projects and reported a range of life-cycle costs from approximately $100/AF to $4,200/AF, with an average of approximately $1,000/AF (or $3.10/Kgal). Compared to other new water alternatives in Figure 7-2, these costs are very reasonable. A more case-specific benefit-cost comparison for enhancement of a single, large-scale, regional stormwater capture and recharge project is provided in Box 7-4. The neighborhood-scale example in Chapter 2, Box 2-6, is illustrative of how community amenities and other nonmonetized co-benefits influence decision making for a major investment in a new stormwater capture system.

FINANCIAL BENEFITS

Financial benefits include the monetary (i.e., cash flow) savings or earnings that accrue to those that implement or otherwise benefit from a beneficial use of graywater or stormwater. There may also be monetary and nonmonetary benefits and costs that accrue to parties other than those that implement the project, and these impacts also need to be properly reflected in a comprehensive assessment of total benefits and costs. Environmental and social benefits are described later in the chapter.

Financial Benefits from Household-level Applications

For a typical household-level application such as the use of graywater or on-site captured stormwater for landscape irrigation, the financial benefits primarily include the savings the user realizes if these practices lower their potable

BOX 7-4 Benefit-Cost Analysis of a Regional-scale Los Angeles Stormwater Recharge Enhancement Project

The Dominguez Gap Spreading Grounds West Basin Percolation Enhancements Project is intended to expand and rehabilitate an existing flood control and stormwater management basin managed by the Los Angeles County Flood Control District (LACFCD). When implemented, the project is estimated to increase local water supplies by an average of 1,000 AF/year (1.2 million m^3/year) for the Central Groundwater Basin. This project secures local water supplies in a water-scarce region for long-term water supply reliability and will allow the region to reduce the amount of water imported from the Sacramento River Bay Delta.

Monetized benefits from the project include avoided water import costs and reduced net carbon emissions from importing water. For 2013, LADWP paid $860/AF (in 2014 U.S. dollars) of treated water imported via the State Water Project, and this cost is expected to increase steadily over coming years. Reduced reliance on imported water will avoid the extensive energy requirements associated with transporting water from the Bay Delta and avoid 958 metric tons of carbon dioxide (CO_2) equivalents per year in greenhouse gas emissions associated with the production of this energy.[a] The social cost of carbon is estimated as the aggregate net economic value of damages from climate change across the globe and is expressed in terms of future net benefits and costs that are discounted to the present (IPCC, 2007), based on guidance for regulatory benefit-cost analyses from the Interagency Working Group on Social Cost of Carbon (2010; $23.23/MT in 2014 U.S. dollars). Over the 50-year project period, the social cost of carbon is escalated by 2.4 percent per year because CO_2 will produce larger incremental damages as physical and economic systems become more stressed in responding to greater climate change. Nonmonetizable benefits include social health and safety and water quality benefits.

Several costs are also associated with this increased local water supply. Primarily, the costs are associated with pumping the recharged groundwater from the aquifer at the point of extraction. In this case, that cost is $76/AF, for a total of more than $1 million (in 2014 dollars) over the life of the project (MWDSC, 2007).

A summary of all benefits and costs of the project are provided in Table 7-4-1. The monetizable benefits far outweigh the present value costs of this stormwater recharge project. Additionally, the project provides nonmonetized benefits such as improved water quality and reduced demand for net diversions from the Delta.

[a] This value was calculated by applying a factor of 0.724 pounds of CO_2 equivalents per kWh and converting to total tons of CO_2 equivalents, based on the California Action Registry, General Reporting Protocol. See http://www.climateregistry.org/tools/protocols/general-reporting-protocol.html.

TABLE 7-4-1 Benefit-Cost Analysis Overview of Dominguez Gap Spreading Grounds Improvement Project Over the 50-year Lifespan (present values, in 2014 U.S. dollars, 6 percent discount rate)

	Present Value
Costs—Total Capital and O&M	$4,486,000
Monetizable Benefits	
Increased local water supply (1,000 AF/year imports avoided)	$16,687,000
958 metric tons per year of CO_2 equivalents avoided (includes 2.4%/year escalation)	$469,000
Local groundwater pumping costs (cost)	($1,010,000)
Total Monetizable Benefits	$16,146,000
Qualitative Benefit or Cost	**Effect on Net Benefits**
Social health and safety	[a]
Other social benefits	[a]
Improve water quality	[a]
Improve long-term management of California groundwater resources	[a]
Reduce demand for net diversions from the Delta	[b]
Provide a long-term solution	[a]
Improve water supply reliability	[a]

[a]Likely to increase net benefits relative to quantified estimates.
[b]Likely to increase net benefits significantly.

water use (assuming there is not an offsetting water utility rate increase). Where wastewater charges are linked to metered potable use, lower potable use may also translate into savings on wastewater charges as well. If the household is on a private well or septic system, they may save some of the operating and maintenance expense associated with running their own on-site supply and wastewater systems, although the effect of diverting graywater on the life of septic systems remains unclear.

Calculating the potable water savings from graywater or stormwater use requires information on end uses (irrigation or additional uses, such as toilet flushing), seasonal irrigation requirements of landscape, and graywater or stormwater availability. The discussion in Chapter 3 presents a scenario analysis of potential water savings based on conservation irrigation of turfgrass in a medium-density development in six U.S. regions, but many other scenarios exist.

Graywater

The committee's analysis showed potential for 9 to 19 percent reduction in water use from whole-house graywater for conservation irrigation and 13 to 26 percent potable water savings where graywater is used for both toilet flushing and irrigation (see Chapter 3). This analysis assumes that the availability of a low- or no-cost water supply does not change household water use—an assumption that remains untested (see Box 3-2 for discussion of other assumptions and uncertainties of the analysis). Water savings are lower for laundry-to-landscape irrigation systems, with only 4 to 8 percent water savings in the committee's analysis (see Table 3-2). The amount of graywater from washing machines depends considerably on the volume of water used per load, with a typical front-loading washer using 20 gallons (SFPUC, 2012). Water savings will also depend on the fill volume settings that the user selects, the number of loads per week, and the number of uses where the drain water is diverted to the sewer or septic system (e.g., when diapers have been washed). Systems that incorporate bathtub, shower, and bathroom sink graywater may provide water savings up to 19,000 gallons (72,000 liters) per household per year (assuming 2.5 persons per household).

Actual financial benefits are determined by the extent to which graywater actually offsets potable water irrigation demand. In locations where irrigation is only needed during one-half of the year, financial benefits would only accrue during periods when irrigation is needed. Therefore, the highest financial benefits from graywater systems are associated with year-round (or near-year-round) water demand, such as for toilet flushing and laundry or for irrigation in the Southwest or central United States (see Chapter 3). Example analyses of financial benefits from household-scale graywater use for irrigation and dormitory use for toilet flushing are outlined in Boxes 7-2 and 7-3.

Additional financial benefits may be derived from using graywater as a drought-proof irrigation source. For example, using graywater for landscape irrigation may enable households to maintain plantings in drought periods when water use curtailments would otherwise preclude outdoor watering. Savings may be realized by avoiding the need to replace landscape vegetation after droughts.

Stormwater

Water savings from stormwater capture projects will depend on local climate conditions, the size of the capture tank(s) (see Chapter 3), and the degree to which the beneficial use of stormwater offsets potable water demand. When used for conservation irrigation, two 35-gallon (130-liter) rain barrels generated potable water savings of 1 to 5 percent in the committee's scenario analysis, and a 2,200-gallon (8,300 liter) tank for irrigation resulted in savings of 4 to 21 percent, depending on the region of the country. When both irrigation and toilet flushing were considered, the larger tank resulted in potential potable water savings from 5 to 28 percent. In all cases, financial benefits depend on the user actually reducing potable water demands. In addition, financial benefits may be decreased if potable water rate adjustments are needed because of reduced demand to cover fixed utility costs. Financial benefits from rain barrels and cisterns are described in Boxes 7-5 and 7-6.

Financial Benefits of Neighborhood- and Regional-scale Applications

Water savings from neighborhood- and regional-scale graywater reuse projects depend heavily on the capacity of the system to offset potable water demand, in addition to local and imported water rates. Larger building-wide graywater systems often see economies of scale over smaller systems. Hodgson (2012) compared graywater reuse in multi-residential buildings against local water rates across eight major U.S. cities and found that in the case of a 1,000 gpd (3,800 lpd) system, graywater reuse provided water savings that would cover treatment and storage costs in 0.7 to 4.6 years. However, Hodgson (2012) does not include plumbing costs, which would extend payback periods (perhaps considerably), given that retrofit plumbing costs may represent 80 to 90 percent of project costs (RMC Water and Environment, 2012).

In multi-family dwellings where individual households pay the water bill, the financial benefits of graywater or stormwater use may not accrue to those who incur the costs of installation and maintenance (unless these costs are somehow conveyed to the households). If the benefits and costs accrue to different parties, then there may be less incentive to pursue such investments.

Neighborhood-scale stormwater groundwater replenishment projects may be associated with significant net cost savings to water utilities compared with the cost of other new water sources (Figures 7-1 and 7-2). Utilities can then rely on a greater volume of extractable local groundwater in lieu of far more expensive imported surface waters.

At a more regional scale, water supply cost savings may arise at the utility and community levels in a number of ways, including avoided or deferred large-scale capital investments in developing new (or expanding existing) water supply and/or water treatment facilities. These benefits may arise where graywater or stormwater use offsets sufficient quantities of potable water demands such that expensive, large-scale investments in existing conventional water infrastructure can

BOX 7-5 Benefit-Cost Analysis of Rain Barrels for Irrigation

Costs. The cost of a residential-scale rain barrel is estimated to be in the range of $60 to $100. Some of this cost may be offset by utility subsidies. Rain barrels vary in size and material, with a typical barrel holding up to 35 gallons of roof runoff (Pitt et al., 2011). Operating costs tend to be negligible when households divert the stored water to outdoor landscape irrigation (assuming no pumping or treatment is applied). The barrel lifetimes are assumed to be 20 years.

Financial Benefits. Potential potable water savings for a basic rain barrel system—consisting of two 35-gallon rain barrels—used for irrigation are calculated based on the committee's scenario analysis in Chapter 3, which assumed medium-density, residential development (with housing density that varies somewhat by region) with 1,500 square feet of roof surface area per home and irrigation to meet the evapotranspiration deficit of turfgrass. Modeled average potential household potable water savings range from 1,100 gallons/year in Los Angeles to 3,400 gallons/year in Lincoln, Nebraska (see Table 7-2). Implicit in these savings estimates are the routine filling of the rain barrel from periodic rain events, the emptying of these tanks for irrigation during dry periods between precipitation events, and an assumption that the harvested water fully displaces potable supply (e.g., that potable water irrigation decreases fully by the amount that the rain barrels provide). Using a potable water charge of $10 per 1,000 gallons (including a typically linked wastewater charge), the water supply cost savings for a household might amount to approximately $11 to $34 per year.

Social and Environmental Benefits (and Costs). Various social and environmental benefits may arise from household-level applications of rainwater harvesting barrels to supply landscape irrigation. These benefits might potentially include the following:

- Public education and water awareness benefits, as households become more directly engaged with their own water use and the local precipitation patterns and enjoy aspirational values from helping to conserve water by harvesting their own supply.
- Surface water quality improvements associated with reduced pollutant discharges.

Benefit-Cost Comparison. Based on a two-barrel system costing an average of $160, assumed water supply savings imply a potential payback period between 5 and 26 years (applying a 5 percent discount rate), depending on regional conditions. These payback periods are for equipment only and do not include costs associated with installation or maintenance. The short payback period corresponds to Lincoln where precipitation timing and near-year-round irrigation needs are better aligned for rain barrel use (although the scenario analysis may overestimate irrigation demand during winter months (see Box 3-2). The longer payback period reflects Los Angeles conditions, where irrigation needs do not coordinate well with time periods when rainfall is available. However, nonfinancial benefits, including water quality improvement and public awareness, may be important for some households and communities, regardless of anticipated financial paybacks. Indeed, the popularity of rain barrel programs despite the relatively long fiscal pay-off periods in some regions may be indicative that cost-effectiveness is not the prime motivation for households choosing to adopt these measures.

be postponed, downsized, or avoided (e.g., see Box 2-1). Savings in O&M expenses can also be realized based on the potential offset in potable water demands.

An example of water supply cost savings arises in the context of southern California, where developing local water resources (including graywater and stormwater) can offset the amount of costly water imports via the State Water Project. Box 7-4 provides a benefit-cost analysis of one large-scale regional stormwater recharge project in the Los Angeles region, in which the $4 million in present value costs to improve an existing facility is offset by $16 million in present value benefits in the form of avoided water imports, as well as a reduced carbon footprint, and several important nonquantified social and environmental benefits.

Similar types of financial savings may be realized when stormwater or graywater management and beneficial use reduce wastewater system costs. For example, reduced volumes of graywater or stormwater inflow into combined sewer systems may defer investments in expanding wastewater conveyance or treatment plant capacity. These costs would be highly site-specific but could amount to large cost savings if major infrastructure upgrades could be avoided. A potentially significant cost savings may be realized when stormwater capture and use enables a community to cost-effectively address combined sewer overflow (CSO) or municipal separate storm sewer system (MS4) compliance issues. Many municipal CSO-related consent decrees are associated with compliance costs of $1 billion or more, not including legal fees. Therefore, opportunities exist for significant potential cost savings if beneficial use of stormwater can reduce CSO and other stormwater-related compliance issues and costs.

Another financial benefit of stormwater capture and use is associated with stormwater control credit trading. The District of Columbia has implemented a market-based stormwater control credit trading system,[7] in which developers can buy credits as needed to meet local stormwater management

[7] See http://green.dc.gov/src.

BOX 7-6 Benefit-Cost Analysis of Rooftop Stormwater Capture Using 2,200-gallon Cisterns for Irrigation

Costs. The cost of stormwater capture using storage tanks depends on several factors, including size and placement, the need for new collection equipment, and the extent to which treatment, pumping, and/or plumbing is required. This illustration considers a storage tank with a capacity of 2,200 gallons is placed above ground adjacent to the home (requiring about 50 square feet of surface area and about 6 feet of height). As noted elsewhere in this chapter, costs for tank capture systems are about $1,500-$1,600, for do-it-yourself homeowner installation and operation (the value of homeowner labor is not included). More elaborate, contractor-supplied and -installed units may cost up to $16,000 (based on Pitt et al., 2011).

Financial Benefits. Potable water savings associated with a 2,200-gallon-capacity stormwater capture tank depend on the region (and associated seasonal precipitation levels and patterns) and intended water uses (irrigation alone, toilet flushing alone, or a combination of both if stored water volumes suffice). This benefit-cost analysis considers outdoor landscape irrigation only, and potential water savings calculated from the committee's scenario analysis (see Chapter 3) are estimated to range between 4,600 gallons per household per year in Los Angeles and 14,000 gallons per household per year in Lincoln, Nebraska (see Table 7-2). Using a potable water charge of $10 per 1,000 gallons (including a linked wastewater charge), the water supply cost savings for a household might amount to approximately $46 to $140 per year (region-dependent).

Social and Environmental Benefits (and Costs). Various social and environmental benefits may arise from household-level applications of stormwater capture, similar to those described in Box 7-5. Large household tanks could result in surface water quality improvements associated with reduced pollutant discharges and public health benefits associated with reduced combined sewer overflows.

Benefit-Cost Comparison. In regions where precipitation patterns are reasonably well coordinated with outdoor irrigation needs, potable water savings may offer sufficient financial benefit to offset the costs for a relatively simple do-it-yourself, household-level rainwater harvest storage tank system with a capacity on the order of 2,200 gallons (not including labor costs). Applying a cost of $1,500 (representing a simple homeowner-installed system) and a 5 percent discount rate, the payback period is 16 years in Lincoln and greater than 50 years in Los Angeles. Payback periods are longer if homeowner labor is considered, and with professional installation, accrued financial benefits never exceed the costs (assuming a 5 percent interest rate). Under scenarios with larger irrigated areas or vegetation requiring larger irrigation rates, the benefits would be greater and the payback period shorter. It is possible that other, nonfinancial benefits may be very important for some households and communities, thereby incentivizing broader use of such stormwater capture systems.

mandates, or sell credits when stormwater control systems are implemented that exceed regulatory requirements. The availability of stormwater credits will provide fiscal gains for those parties who can manage stormwater in a less costly manner through decentralized projects.

SOCIAL BENEFITS AND COSTS

Beyond the range of financial costs and benefits, there is a broad array of potential social benefits that may arise from the beneficial use of graywater or stormwater, as listed in Box 7-1. Not all of these benefits and costs will apply in every application or location; thus, care should be taken to identify when and where such benefits and costs apply, and whether they may be of appreciable value in that site- and scale-specific context.

Households may opt to engage in on-site graywater or stormwater use because of other values and goals that are important to them, such as the aesthetic value enjoyed from any additional or improved landscaping or garden areas attributable to the graywater- or stormwater-supplied irrigation. These nonfinancial benefits to the implementing households may be quite important and may be sufficiently valuable to the individuals to motivate their engaging in these activities regardless of the potential for no direct financial net gain.

Widespread beneficial use of graywater or stormwater use across a community could collectively yield sufficient water savings to provide a more reliable (e.g., drought-resistant) community-wide water supply, reduce energy demands by the local water utilities, and provide more aesthetically pleasing neighborhoods, which would improve public health, enhance property values, and offer recreational opportunities (see Philadelphia case study provided in Box 7-7). Additionally, the direct hands-on involvement of citizens in water-related activities (such as managing an on-site rain barrel) may provide a higher level of awareness of the water cycle that may foster constructive engagement for local water and other resource management issues, although no research could be found that documents these changes. Regional-scale applications can yield significant levels of social beneficial value to a community, by providing a more diverse and resilient water portfolio and expanding local supplies, making these communities less vulnerable to limitations on imported water as well as droughts and climate change impacts.

BOX 7-7 Estimating the Value of Social Benefits of Philadelphia's Green Infrastructure Initiatives

The Philadelphia Water Department (PWD) is pursuing an integrated approach for managing stormwater runoff and associated combined sewer overflow (CSO) control. This approach includes a range of green infrastructure strategies that use natural and engineered approaches to capture and infiltrate stormwater on site (e.g., green roofs, rain gardens, permeable pavement). Although Philadelphia designed these strategies to address water quality concerns (rather than water supply shortages), lessons can be learned from assessments of the social benefits that extend beyond improvements in receiving stream water quality, such as improved public health, increased recreation opportunities, and increased property values (Raucher et al., 2009). Understanding the full social costs and benefits of the program has helped justify the program to ratepayers, as well as regulators.

Raucher et al. (2009) assessed the quantitative and monetary value of a range of benefits and costs, applying standard techniques used in environmental and natural resource economics and related disciplines. Social benefits associated with a green infrastructure approach to stormwater management included streamside recreational opportunities from stream restoration and riparian buffer improvements, increased community aesthetics and property values with increased tree cover and other vegetation, and reduced heat-stress. Social benefits also included local green jobs that can be filled by local unskilled or otherwise unemployed laborers and reduced construction-associated disruption. Environmental benefits included aquatic ecosystem improvements, wetland creation and enhancement, energy savings, and improved air quality. Table 7-7-1 summarizes the monetary value of the benefits for a scenario where runoff from 50 percent of impervious surface in Philadelphia is managed through green infrastructure, compared to the traditional engineered approach using a 30-foot-diameter tunnel. These options were chosen to demonstrate the difference in net benefits between green and traditional infrastructure.

TABLE 7-7-1 Philadelphia-wide Benefits Through 2049 of Two Combined Sewer Overflow Options

Benefit Categories	50% Green Infrastructure Option	Traditional 30-ft-diam. Tunnel Option
Increased recreational opportunities	$579	
Improved aesthetics/property value	$634	
Reduction in heat stress mortality	$1,167	
Water quality/aquatic habitat enhancement	$371	$209
Wetland services	$1.8	
Social costs avoided by green collar jobs for under-employed locals	$138	
Air quality improvements from trees	$145	
Energy savings/usage	$37	$(2.8)
Reduced (increased) damage from air pollution	$51	$(50)
Reduced (increased) damage from CO_2 emissions	$23	$(6.5)
Disruption costs from construction and maintenance	$(6.2)	$(15)
Total	$3,141	$135

SOURCE: Raucher et al. (2009).

TABLE 7-2 Summary of Scenario Analysis of Potential Water Savings from Conservation Irrigation with Rain Barrels and Cisterns in Six U.S. Locations

	Two 35-gallon Rain Barrels		2,200-gallon Cistern	
	Average potential household water savings (gallons/year)	Estimated potential annual cost savings ($/year)	Average potential household water savings (gallons/year)	Estimated potential annual cost savings ($/year)
Los Angeles	1,100	11	4,600	46
Seattle	2,200	22	5,600	56
Lincoln	3,400	34	14,000	140
Madison	2,100	21	7,300	73
Birmingham	2,300	23	9,400	94
Newark	1,900	19	6,900	69

NOTE: See Chapter 3 for details of the analysis. Scenarios assume medium-density, residential development, 1,500 ft^2 rooftops, and irrigation to meet the evapotranspiration deficit for turfgrass, and results are based on 1995-1999 precipitation data. Landscaped area and residential density determined by location-specific data (see Appendix A).

ENVIRONMENTAL BENEFITS AND COSTS

Environmental benefits and costs of graywater and stormwater projects are also important to consider in the overall assessment of project costs and benefits. Environmental benefits of graywater and stormwater could include greenhouse gas reductions if on-site water use results in significant energy savings. Beneficial use of stormwater can result in significant environmental benefits related to improved water quality and hydrology, although adverse environmental impacts of graywater and stormwater are also possible.

Water Quantity and Quality Impacts

Potential environmental benefits of widespread graywater use for toilet flushing or other nonconsumptive uses would be decreased stresses on existing surface water sources, which may enhance conditions for aquatic life. As discussed in Chapter 5, the potential environmental impacts of graywater use for irrigation are limited if best management practices that prevent surface ponding are followed. Some impact to soil properties or plant health may occur depending on the graywater quality, particularly with elevated levels of sodium or boron, depending on local soil and climatic conditions (see Chapter 5). If graywater use for irrigation is implemented at a large scale in a way that uses more water for irrigation than was previously used with only potable water, then downstream flows could be impacted, which could have a negative impact on aquatic life (see Box 3-1).

Stormwater capture for beneficial use can provide significant environmental benefits and some environmental risks. Increased infiltration of stormwater for the purpose of groundwater recharge and pollution management offers major environmental benefits to surface waters, particularly in areas with combined sewer systems. Stormwater capture or infiltration can reduce and delay peak surface water flows and reduce hydromodification of urban streams (see Chapter 1, Figure 1-7). There are typically reduced deleterious effects on stream biota because of lower stream velocities and lower erosion rates. Stormwater infiltration can help reestablish natural groundwater levels and increase base flow in streams across long reaches in a watershed. An example of changing hydrology based on enhanced stormwater infiltration is shown in Box 7-8. In coastal areas, increased groundwater levels can provide enhanced groundwater discharge and reduce salt-water intrusion. Reduced stormwater flows reduce loads of phosphorus, sediment, metals, bacteria, and organic contaminants to inland and coastal waters. For combined sewer systems, reduction of stormwater flowing to the combined system can reduce combined sewer overflows and therefore reduce contamination of receiving water bodies. In some locations, these benefits are so large that they are often the driving objectives for stormwater management projects (see Chapter 1). These benefits can be quantified to assess the magnitude of contaminant reductions by various stormwater capture strategies (see Figure 7-4).

If stormwater is captured and used in a way that consumptive use of stormwater exceeds prior use of potable

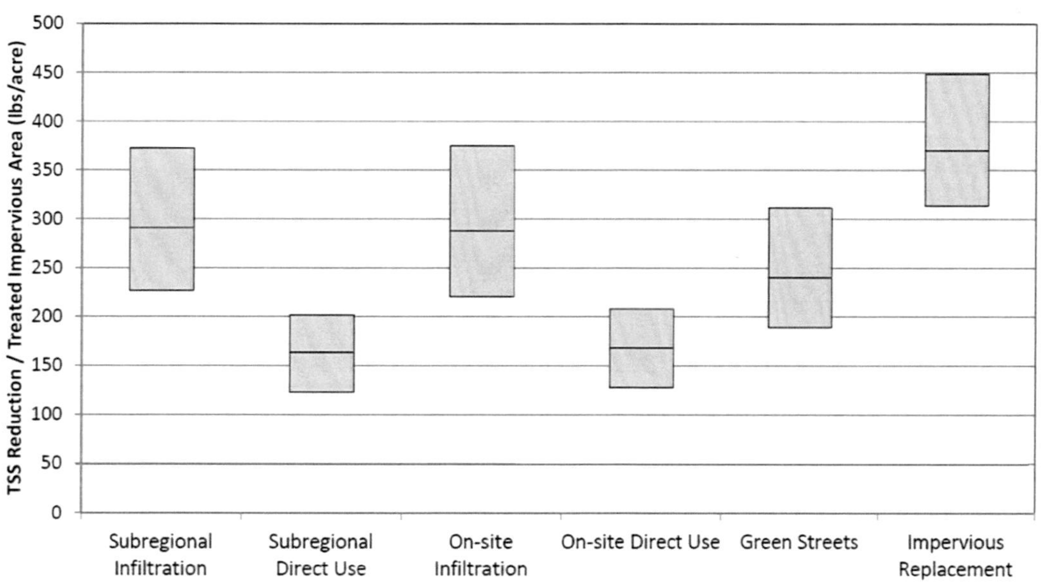

FIGURE 7-4 Total suspended sediment reductions vary with different stormwater capture or management strategies. SOURCE: Modified from LADWP (2015).

BOX 7-8 Effects of Distributed Infiltration Practices on City-scale Hydrology

In 2012, the City of Philadelphia entered into an agreement with EPA to reduce the city's combined sewer overflows (CSOs) by investing $2 billion in stormwater infiltration facilities over a 25- year period (EPA, 2012b). The Philadelphia Water Department (PWD) aims to replace up to 40 percent of impervious surface area with green infrastructure that includes swales, pervious pavement, rain gardens, and tree trenches to promote stormwater infiltration. Concerns have been raised about the potential for such a massive introduction of stormwater into the subsurface to cause flooding of building basements in areas of high building density.

In an effort to assess the potential changes to long-term water table elevations resulting from increased infiltration practices, PWD and its contractor, CDM-Smith, constructed a three-dimensional finite element groundwater model of Philadelphia's 166 km^2 CSO area (Maimone et al., 2011). The land area is underlain by Piedmont fractured rock to the west and the Atlantic Coastal Plain to the east, which is composed of layered sediments comprising aquifers and aquitards. Inputs to the model included estimates of recharge in areas without green infrastructure (45 cm/yr) as well as estimates of recharge in greened areas of the city (63 cm/yr). The model was run in the steady-state mode for a scenario of no green infrastructure and a scenario of a greened city, to evaluate predicted changes in water table elevation.

Results showed that the predicted rise in water table was spatially variable. Increases in water table elevation were less than 0.5 m in the Atlantic Coastal Plain and up to 1.8 m in small areas of the Piedmont (Figure 7-8-1). However, areas of greatest increase in water table elevation in the Piedmont were located in areas with existing greatest depth to groundwater (~9 m). The study concluded that the increased recharge due to green infrastructure infiltration practices is not likely to cause basement flooding.

This study is unique in its attempt to quantify city-wide impacts on groundwater levels from implementation of distributed green infrastructure.

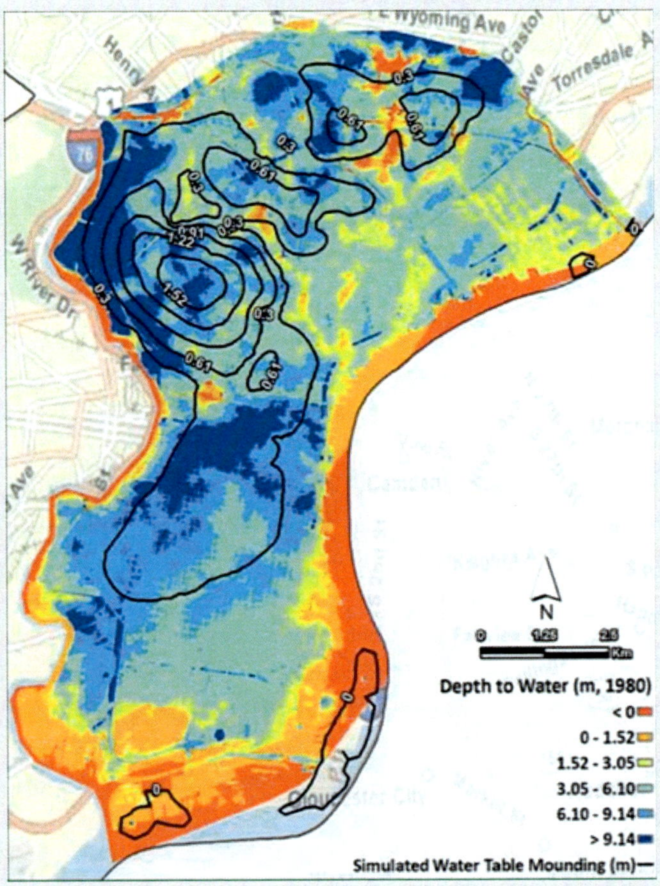

FIGURE 7-8-1 Simulated water table mounding in response to infiltration from green infrastructure in the combined sewer region of Philadelphia, Pennsylvania. NOTES: Color code indicates depth to the water table at baseline conditions with no green infrastructure. Black contour lines indicate predicted increases in water table elevation due to installation distributed stormwater facilities. SOURCE: Maimone et al. (2011).

water (see Chapter 3), then there may be impacts to downstream flows, which could, in turn, impact aquatic organisms in streams and estuaries. There is also the potential risk that stormwater contaminants may find their way into local groundwater, which could threaten local water supplies that are dependent on groundwater withdrawal (see Chapter 5). In addition, enhanced infiltration has the potential to cause geotechnical problems such as flooding basements (see Box 7-8) or altering movement of groundwater contaminant plumes.

Energy Footprint of Stormwater and Graywater Compared to Traditional Water Systems

Energy could be saved in the process of substituting conventional sources of water with alternative local or onsite sources. Energy production results in air and water pollution and generation of waste; therefore, energy saved results in reduced pollution and waste. Depending on the area of the United States, saving electricity can translate to either more or less avoided emissions. For example, based on the amount of fossil fuels burned in electricity generation (EPA, 2011), a MWh of electricity is responsible for about 100 kg of CO_2 emissions for the electricity mix in Washington state and about 800 kg in Ohio. Furthermore, because energy production requires water (e.g., for cooling in electric power plants), energy savings can also result in water savings.

For decision making, it is important to quantify the energy intensities and environmental impacts of these alternatives and compare them to conventional supply of water. The following summarizes what is known about the energy needs of on-site graywater and stormwater use based on the currently available literature.

Energy Intensity of Water from Conventional Systems

The energy intensity of water provided to customers depends on the water source(s), pumping needs, treatment processes, and storage options. A range for embedded energy data for delivering water to customers is available from the published literature. However, the numbers may differ depending on the comprehensiveness of the energy assessment (e.g., if supply-chain effects were also considered). The energy intensity of water supplied to U.S. cities is typically around 0.9 kWh/m^3 (Copeland, 2014), but it can be five times higher, as reported in Table 7-4. Systems report higher energy use when they rely upon a large percentage of imported water (pumped from a long distance), use brackish or seawater desalination, or require groundwater pumping. For example, conveying water from northern to southern California via the California State Water Project is estimated to have an energy requirement of 2.1-4.4 kWh/m^3 (Spencer, 2013). Groundwater pumping in southern California requires an additional 0.3-0.8 kWh/m^3 above energy needs for treatment and distribution.

Overall, pumping and treatment tend to be the largest energy uses for water systems, but other life-cycle phases and supply chains (e.g., treatment chemicals and equipment, pipes, pumps, buildings, and utility vehicle fleets) may have significant embedded energy that should be accounted for in a comprehensive assessment of the energy intensity of water. Life-cycle assessment (LCA) is a methodology that can provide comprehensive environmental analyses (e.g., Stokes and Horvath, 2009). Table 7-4 includes three examples of life-cycle energy intensities using LCA. In those examples, most of the energy is expended in conveying the water from the source to the water treatment plant and then to the customers (Stokes et al., 2014).

Energy Intensity of Stormwater Capture Systems

Seventy-five percent of residential U.S. rooftop stormwater capture systems are utilized for irrigation purposes (Thomas et al., 2014). By replacing potable water with stormwater, rooftop systems offer opportunities to conserve local water sources while avoiding the required material and energy inputs needed to supply potable water for irrigation the traditional way. Theoretically, energy costs for pumping and treatment could be notably reduced. However, life-cycle energy demands can vary significantly based on the design and scale of the stormwater capture system and end uses. In the existing literature, empirical analyses of rooftop runoff capture systems report a greater energy demand (median is 1.4 kWh/m^3) than theoretical studies (0.2 kWh/m^3) (Vieira et al., 2014). For Australian conditions, 1.4-1.8 kWh/m^3 median energy consumption was reported for rainwater supply to urban dwellings (Sharma et al., 2015), but depending on the pump design, values range between about 0.4 kWh/m^3 and 5 kWh/m^3. The energy intensity for household systems is characterized by material inputs (58 percent of total energy demand) and for agricultural systems is dominated by pumping needs (95 percent) (Ghimire et al., 2014).

In combined sewer systems, stormwater use for irrigation can reduce energy requirements associated with downstream wastewater treatment of stormwater. These savings were reported as 0.3 kWh/m^3 in one study (Blackhurst et al., 2010), while another study, focusing on wastewater treatment, calculated the energy cost to be 0.6 to 1.2 kWh/m^3 (Stokes and Horvath, 2010). Ultimately, these wastewater energy benefits would depend on the volume of stormwater diverted from the combined sewer system, which is likely to be less than the amount of stormwater used for on-site irrigation.

TABLE 7-4 Energy Requirements of Water from Several U.S. Cities or Regions

City/Utility	Energy Requirement (kWh/m^3)	Life-cycle Assessment-based Number?
New York City	0.7	yes
Austin, Texas	1.3	yes
Santa Clara Valley Water District (northern California)—current water mix	1.3	no
Brackish water Santa Clara Valley Water District—brackish water desalination	2.6	no
Small utility (northern California, serving 40,000 people; 100% groundwater)	1.9	yes
Medium utility (southern California, serving 180,000 people; 83% imported water, 17% brackish groundwater desalination, <1% recycled)	4.9	yes
Large utility (northern California, serving 1.3 million people; 95% imported water, 5% local runoff collected in reservoirs, <1% recycled)	1.7	yes

SOURCES: Copeland (2014); SCVWD (2011); Stokes et al. (2014).

Energy Intensity of Graywater Reuse Systems

The energy footprint of graywater reuse depends on the system design (e.g., pipes, pumps, and other components), pumping needs, type and level of treatment, and the end use of water. For a graywater irrigation system that includes only storage and filtration, the primary energy demand is pumping (beyond the embedded energy in system materials). The laundry-to-landscape system typically requires no energy beyond that embedded in system components. Systems that provide extensive graywater treatment, such as membrane filters and biological treatment, will have a high energy demand.

The current literature offers limited data on graywater energy use, particularly at the household scale. In the United Kingdom, the nonpotable water needs (met from graywater) of 500 households were calculated to require the following amounts of energy by three different treatment types: 9,200 kWh for a reed bed, 197,000 kWh for membrane bioreactor (MBR), and 238,000 kWh for membrane chemical reactor (Memon et al., 2007). An economic feasibility study of MBRs treating graywater estimated energy use at 1.0-1.5 kWh/m^3 (Friedler and Hadari, 2006). A California study assessed a sand-filtration decentralized treatment system that reused graywater for irrigation purposes, including ultraviolet treatment of effluents. The study found the life-cycle energy consumption to be 10.3 kWh/m^3 (Shehabi et al., 2012). For comparison, Hendrickson et al. (2015) estimated the energy consumption of an office building-scale wetland system that recycles wastewater (not just graywater) for nonpotable use (toilet flushing and irrigation in a neighboring park) to be about 5.5 kWh/m^3 throughout the system's life cycle. There may also be opportunities to capture energy from blackwater when graywater is source-separated for reuse, as described in Chapter 6.

INCENTIVES AND DECISION MAKING

One of the challenges in promoting sound graywater and stormwater beneficial use projects is the lack of direct financial incentives for those who bear the implementation costs. For example, at the household level, a family that invests in stormwater capture and beneficial use via large cisterns may not save enough money on its water utility bills to justify the expense (i.e., the payback period may be longer than desired). As the cost of potable water increases, the financial incentives may become larger. For multi-residential projects, the building owner or developer who makes the infrastructure investments to install a graywater or stormwater use system may not directly benefit, because the cost savings from reduced water use accrue to the residents. Nonfinancial motives such as the enjoyment associated with an irrigated garden and related environmental and social values (e.g., the desire to use local water resources wisely) also create positive incentives for projects that tap graywater or stormwater for beneficial uses, and these benefits may extend to the larger community.

Ultimately, aligning the proper fiscal incentives boils down to understanding and communicating where there are important external benefits to a beneficial use project. Once the externalities are recognized, such as through a comprehensive benefit-cost analysis that embodies full social and environmental accounting, the challenge is finding and implementing mechanisms that better align who pays with who benefits (e.g., by cost-share agreements or public-sector subsidies).

For regional-scale projects, with relatively large budget implications, incentives may be more fiscally constrained, with altruistic values playing a lesser role. In such cases, it is important that the utility—and the broader regional community—recognize the important fiscal, social, and environmental values that can be derived from a beneficial use project.[8] It is important as well to account for the positive and negative externalities associated with these projects. For example, other communities may derive benefits when water savings or local water resource development relieves pressure on limited water resources shared by the broader area (e.g., where stormwater capture and recharge reduces regional demands for scarce and expensive imported waters). In such instances, some form of cost sharing across all beneficiaries will help incentivize projects that have a large number of beneficiaries outside of the immediate utility service area (i.e., subsidies may help internalize the positive impacts of such projects). This may take the form of suitable subsidies for projects that provide such external benefits and is evident in grants and cost-share arrangements such as those provided by the Bureau of Reclamation (e.g., Title XVI grants), the State of California (e.g., grants funded by Proposition 84), and the Metropolitan Water District of Southern California (via its local resources program).

CONCLUSIONS

It is important to recognize the full suite of benefits—as well as the full costs—of graywater and stormwater projects, although it may be empirically challenging to do so. A wide array of potential benefits may arise from projects that use graywater or stormwater. Some of these benefits are financial and can be readily estimated and portrayed in monetary terms, such as the value of water savings or the avoided cost of obtaining water from an alternative supply. In addition, important societal and environmental benefits may apply. These benefits may be difficult to quantify or monetize and are typically highly site- and scale-specific and dependent on the type of application. Costs for graywater and stormwater projects are also highly dependent on scale, system design, and plumbing requirements, and they are generally better understood than the benefits. Yet there is a lack of well-documented and complete cost information for many of the possible applications.

Simple household-scale graywater reuse or roof runoff capture systems can offer reasonable financial payback periods under certain water use scenarios and appropriate climate conditions. For example, considering the committee's scenario analysis of potential water savings in medium-density, residential development, simple laundry-to-landscape graywater systems can offer payback periods as low as 2.5-6 years (not accounting for the cost of labor), with the shortest payback periods in the Southwest and central United States. These estimates assume graywater for irrigation actually offsets potable use—an assumption that remains to be demonstrated. Longer payback periods were estimated for rain barrels (5-26 years) and cisterns (14 to more than 50 years, not accounting for labor) used for conservation irrigation. The longer payback periods reflect locations where distinct wet and dry seasons do not coordinate well with irrigation demands, as in the arid Southwest. Shorter payback periods may be possible in more humid climates for households with large irrigated areas. The cost of installation (whether by contracting with a paid professional, or valuing homeowner-provided labor) greatly extends the payback period, as do water uses in which additional plumbing and treatment are required. However, in household-scale applications, it may well be the non-financial benefits that motivate households to adopt beneficial use—such as a sense of conserving resources or having outdoor irrigation water reliably available to support landscaping during times of drought.

Some neighborhood- or regional-scale stormwater capture and use projects provide financial benefits that exceed costs, sometimes by a wide margin, in addition to other social and environmental benefits, and economies of scale are evident for both tank capture and infiltration projects. The regional stormwater capture and recharge projects in southern California, for example, can pay back large dividends to the community in the form of significant water supply enhancements (i.e., avoiding the cost of expensive imported water). Based on available unit cost data, stormwater tank capture and use at the neighborhood scale tend to be much more costly than alternatives designed to recharge groundwater. Larger scale projects may also reduce stormwater-related regulatory compliance costs and provide a wide variety of highly valued societal and environmental benefits, including enhanced aesthetics, property values, and recreational opportunities.

Published cost data from larger-scale graywater projects are extremely limited, and therefore the financial benefits and cost options are difficult to assess beyond the pilot scale. Some efficiencies of scale would be associated with graywater toilet flushing systems in large, new, multi-residential developments (particularly compared to smaller retrofits), and additional incentives may be possible if such investments defer water and wastewater infrastructure expansion in densely populated urban areas.

Depending on the stormwater or graywater system design, energy savings are possible compared with conventional water supplies, but data for a sound assessment

[8] One potentially useful approach is provided in Marsden (2013), which provides a water recycling evaluation framework that was applied as the basis for evaluating alternative stormwater harvesting systems in an Australian community by Dandy et al. (2014). Dandy's work accounts for financial costs and greenhouse gas emissions, and addresses the different benefits perceived by different stakeholders.

are lacking. For decision making, it is important to quantify the life-cycle energy intensities of alternative water supplies and compare them to conventional supply of water. Energy saved could also result in reduced greenhouse gas and other pollutant emissions associated with energy production. The current literature contains little energy data for conventional and alternative systems. Conventional water systems in the United States are reported to provide water to customers at an energy cost of less than 1 kWh/m^3 to almost 5 kWh/m^3, depending mostly on pumping costs for conveying the water from the source to the water treatment plant. Roof runoff capture systems have been reported in a very limited number of studies to have a greater energy demand (median is 1.4 kWh/m^3) in practice than in theoretical studies (0.2 kWh/m^3), and thus may not be less energy intensive than conventional drinking water systems. Where stormwater is diverted into combined sewer systems, additional energy savings of between 0.3 and 1.2 kWh/m^3 may be obtained. Many potential variables (e.g., scale, pumping, treatment, material inputs) will drastically affect the life-cycle energy demands of these systems, and the effects of these variables in practice remain poorly understood.

8

Legal and Regulatory Issues

This chapter outlines legal and regulatory factors that impact the capacity to use graywater or stormwater for beneficial purposes. Water rights can be a barrier in some states, and local regulations related to water quality may limit potential uses. However, the legal and regulatory framework for the beneficial use of on-site water sources appears to be rapidly evolving.

POTENTIAL LEGAL BARRIERS IN THE WATER QUANTITY CONTEXT

Legal barriers related to the water quantity impacts of graywater or stormwater use primarily concern water rights laws, which determine the rights to the use of water. Other laws protecting threatened and endangered species may also have relevance to large diversions of water from streams.

State Water Rights Laws

Laws concerning the allocation of water, particularly surface water, determine the rights to stormwater and graywater and thus can significantly influence the harvesting and use of that water. Water allocation is regulated primarily at the state level. Although each state has a unique set of laws, nearly every state uses one of two doctrines—prior appropriation or riparian rights—and in some cases both, as the foundation for its water allocation (see Figure 8-1).

Prior Appropriation

The prior appropriation doctrine is the sole or predominant system of water allocation used by most states in the western half of the United States (Getches, 2009). Under this doctrine, the right to use water is allocated based upon the historical order in which rights to it were acquired, hence the adage "first in time, first in right." In essence, the oldest right is fulfilled, then the next oldest, then the next oldest, and so forth until there is no water left to disperse or all rights are fulfilled. Traditionally, a key tenet of the prior appropriation doctrine is the prohibition against impairing other water rights. Therefore, appropriative water rights are generally well protected in law, as well as in politics and cultural norms.

When downstream water right holders exist, the effect, or even the potential effect, of stormwater and graywater harvesting and use on water rights can pose a challenge to the application of these technologies in prior appropriation states. Large-scale stormwater capture could result in less water eventually making its way to the stream and, as a result, cause a water right that otherwise would have been fulfilled to not to be. The same possibly could be said for small-scale stormwater capture, such as rain barrels and cisterns, when widely used in existing developments. Concern about this potential impact could lead, and in some cases has led, to opposition by water right holders and caution by state regulators with regard to the development of such infrastructure. Because urban development increases stormwater runoff and reduces evapotranspiration,[1] on-site stormwater capture systems (depending on their design) may have minimal impact on downstream flows compared to pre-development conditions. However, each western state has unique regulatory frameworks for governing water rights, which are not necessarily allocated based only on pre-development conditions.

Colorado is currently the only prior appropriation state with regulations that restrict the beneficial use of stormwater,[2] although a pilot program is under way to gather data in new developments regarding the feasibility of stormwater capture for water conservation without injuring the water rights of others.[3] Six prior appropriation states have developed regu-

[1] See http://www.coastal.ca.gov/nps/watercyclefacts.pdf.

[2] An exception is allowed for rural residential property owners whose water is supplied by certain wells.

[3] New developments that qualify as one of a limited number of pilot projects may harvest rainwater from impervious surfaces for nonpotable uses as long as the entire amount harvested is replaced to the stream by some other source. If, after a time, the development can calculate the amount of harvested rainwater that would have

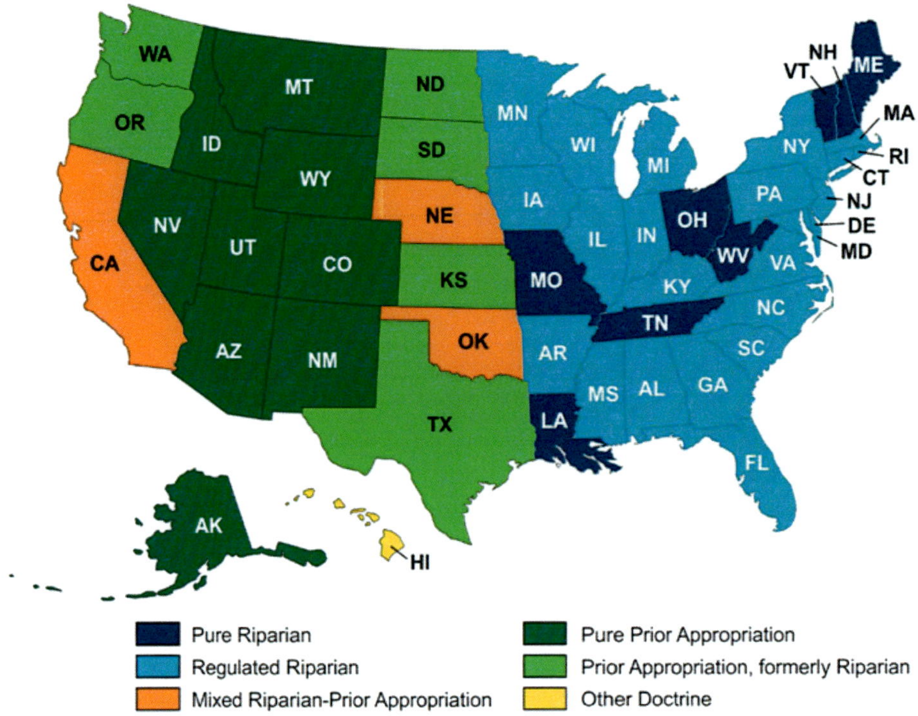

FIGURE 8-1 U.S. water rights systems by state. SOURCE: DOE (2014).

lations to allow stormwater capture and use in some form, providing specific exemptions from water rights permitting, mostly for smaller-scale systems (see Table 8-1; supporting detail is provided in Table B-1 in Appendix B). Four of these six are coastal states, suggesting perhaps that states discharging stormwater directly to the ocean or Gulf of Mexico may be more willing to consider regulatory exemptions to water rights permitting for stormwater capture and use if the benefits within the state are judged to outweigh the negative impacts. Eleven prior appropriation states (all located inland) have not yet set regulations regarding stormwater use. Lack of specific regulations does not necessarily prohibit its use—three of these states have issued statements or tax credits that encourage the capture and use of roof runoff (Table 8-1). New Mexico specifically states "the collection of water harvested in this manner should not reduce the amount of runoff that would have occurred from the site in its natural, pre-development state."[4] However, it is possible that water right holders would sue owners of stormwater or graywater infrastructure for impairing their water rights, and the laws of the jurisdiction would heavily influence the success of such a suit.

Explicit laws concerning these matters are still relatively rare, and judicial interpretation of water appropriation laws for these purposes is in its infancy. Although the precise rights of water right holders to runoff and return flows are, on the whole, unresolved at this point, it is not unprecedented for courts to hold in favor of technological improvement and perceived efficiency over the water rights of others.[5] The answer also may depend on whether the individual capturing and using the stormwater or graywater has a pre-existing water right.

Riparian Rights

The riparian rights doctrine is the sole system of water allocation used by 29 states (Getches, 2009). Borrowed from England, this doctrine grants the owners of land that abuts a natural stream, river, lake, or pond rights to that water. Most riparian rights states allow riparian landowners unlimited use of water for drinking, washing, and modest animal and garden needs and limit all other uses to "reasonable use." Traditionally, if the reasonable needs of all riparian landowners cannot be met with the available water, then usage is reduced proportionately. Many states using the riparian rights

---

been consumed by evapotranspiration from native vegetation, the development can ask the water court for permission to not replace that amount.

[4] See http://www.rmwea.org/reuse/NewMexico.html.

[5] See, e.g., Montana v. Wyoming, 131 S. Ct. 1765 (2011).

TABLE 8-1 Examples of State Regulation of Stormwater Use

Prior Appropriation States Without On-site Stormwater Capture and Use Regulation[a]	States with Regulations That Specifically Allow On-site Capture and Use	States with Regulations That Limit or Prohibit On-site Capture and Use
Alaska	California (rooftop capture exempted from water rights permitting)	Colorado
Arizona[b]	Georgia	
Idaho[c]	Kansas (domestic, <15 AF/yr)	
Montana	Maryland	
Nebraska	North Carolina	
Nevada	Ohio	
New Mexico[d]	Oregon	
North Dakota	Texas	
Oklahoma	Utah (<2,500 gallon storage with permit or <200 gallon without)	
South Dakota	Washington (<360 gallon storage without permit and for single family dwellings)	
Wyoming		

[a]Lack of regulation should not be interpreted as a prohibition on such activities.
[b]Although there is no formal regulation, Arizona has provided tax credits for rainwater harvesting systems.
[c]Although there is no formal regulation to this effect, the Idaho Deputy Attorney General issued an opinion that on-site rainwater may be captured for beneficial use as long as there is no injury to the water rights of others.
[d]Although there is no formal regulation, the New Mexico Office of the State Engineer encourages the capture and on-site use of roof runoff.
NOTE: See Appendix B-1 for details.

doctrine now have some form of "regulated riparianism," effectively a permitting system for the use of large volumes of surface water (see Figure 8-1; Getches, 2009).

The right to unlimited use of surface water by riparian landowners for drinking, washing, and modest animal and garden needs should put stormwater and graywater use for these purposes squarely within the rights of riparian landowners. The use of that water for other purposes, to the extent that the total volume is less than what requires a permit, and perhaps even if more, would likely also fall within riparian landowner rights, although it might need to be "reasonable use" and might be subject to reduction in times of scarcity. Stormwater and graywater capture and use may be less secure for non-riparian landowners, if the harvested water otherwise would have flowed to a stream or river. However, rights to water before it reaches a natural watercourse, whether from runoff or discharge, may be even more tenuous in this allocation system than under prior appropriation. The concept of equity that underlies the riparian rights doctrine, coupled with the limited accounting of volumes used, results in rights to water that are more difficult to quantify, potentially dissuading lawsuits and political objections to stormwater and graywater harvesting and use.

Determining a Violation of Downstream Water Rights

With either allocation system, there is the possibility, however small, that a court finds state legislation allowing stormwater or graywater capture and use to result in a taking of a property right in water. The first step of that analysis would involve determining whether "property" is at issue and if "background principles of the State's law of property and nuisance" already limited the property right.[6] This involves asking at what point in the water's migration does the right begin—for example, a roof, the ground, or its entrance into a natural watercourse. The answer to this question will depend on the laws of the jurisdiction and, in many states, has not been resolved. If the water right extends to the falling of rain and snow, then the remainder of the complex takings analysis must be conducted, the application of which to surface water rights neither has proven simple, nor is yet clear (Echeverria, 2010; Lock, 2000). Thus, additional court challenges may be necessary to fully clarify the legal framework for stormwater and graywater, particularly in prior-appropriation states.

Minimum Flows to Protect Aquatic Life

Reductions in streamflow that may result from stormwater and graywater capture and consumptive use have the potential to adversely affect not only water rights, but also the riparian ecosystems on which protected species rely. The U.S. Environmental Protection Agency (EPA), in its *2012 Guidelines for Water Reuse*, highlighted the concern: "[t]he most significant constraint affecting use of reclaimed water is the need to assure instream flows sufficient to protect aquatic habitat. This is especially necessary in locations where instream flows are necessary to protect the habitat of threatened and endangered species" (EPA, 2012a). The federal Endangered Species Act, for example, has influenced many water allocation and use decisions, because of the potential to violate explicit protection measures for species, such as minimum instream flows, or the potential impact an action may have on a population of threat-

[6] Lucas v. South Carolina Coastal Council, 505 U.S. 1003, 1029 (1992).

ened or endangered species themselves (Craig, 2008; EPA, 2012a; Getches, 2001). Although the committee is unaware of any specific examples where graywater or stormwater use is currently threatening instream flows to support endangered species, many states have established instream flow laws to protect species in over-allocated basins (Bonham, 2006), and concerns may arise in the context of extensive consumptive graywater and stormwater use in areas with the potential to affect important habitats.

POTENTIAL LEGAL ISSUES IN THE WATER QUALITY CONTEXT

Although water allocation systems are generally a matter of state law, federal, state, and local laws all influence the framework for water quality regulation as it relates to the beneficial use of stormwater and graywater. Although this section focuses on existing U.S. federal, state, and local laws, regulatory frameworks or water quality guidance from other countries (e.g., NRMMC, et al., 2009a,b) illuminate alternative strategies for managing water quality concerns.

Stormwater Capture and Use

Federal Laws

Federal laws do not specifically govern the use of captured stormwater, but federal water quality laws may influence stormwater capture projects through the Clean Water Act (CWA) or the Underground Injection Control (UIC) program.

The Clean Water Act and National Pollutant Discharge Elimination System Permits. The Federal Water Pollution Control Act, commonly known as the Clean Water Act (CWA), serves as the foundation for water quality regulation by the federal government and the states. The CWA sets out broad water quality restoration and maintenance objectives and establishes a two-pronged approach to achieve them, combining effluent limitations with ambient water quality standards. Under the water quality standards program, states are directed to adopt water quality criteria for contaminants, based on the water body's designated use (e.g., public water supplies, recreation, protection of fish and wildlife), which must be approved by the EPA as adequately protective of the public health and welfare (33 U.S.C. § 1313). The primary tool in the CWA for meeting these criteria is the prohibition against discharging pollutants from any "point source" into U.S. waters without a permit issued under the National Pollutant Discharge Elimination System (NPDES) (33 U.S.C. § 1342).

Most stormwater runoff discharges, such as those from certain construction sites and many categories of industrial facilities as well as municipal separate storm sewer systems (MS4s), fall under the definition of "point source," and thus require a NPDES permit (EPA, 2012c). The permits, which are issued by authorized states[7] or the EPA, can include requirements for site-level stormwater management plans or programs. Traditionally, stormwater management practices have relied mainly on "end-of-pipe" treatment best management practices (BMPs; e.g., filters) to trap or remove pollutants shortly before the water is discharged to reduce the quantity of pollutants that reach surface waters. However, an emerging trend favoring low impact development (LID) encourages implementation of "green infrastructure" to mimic an area's natural hydrology by retaining and managing stormwater on-site (see Chapters 1 and 6; EPA, 2014a). LID includes various measures to reduce impervious surfaces and thus allow infiltration of water into the soil and the capture and storage of stormwater for use or timed release.

Despite increasing recognition of the environmental, social, and economic benefits of green infrastructure (see Chapter 7), many local stormwater permits, administrative orders, and other enforceable documents that hold permittees to their stormwater management obligations under the NPDES program have not been updated to include these alternative approaches (Stoner and Giles, 2011). Existing legal requirements with a historical preference for end-of-pipe BMPs are likely to act as a deterrent to some facilities' and municipalities' adoption of stormwater capture and use, unless permittees can be confident they will receive legal credit (EPA, 2010). Thus, even though EPA has developed the *Green Infrastructure Action Strategy* to encourage the incorporation of LID into stormwater management planning (EPA, 2008a, 2013a), current regulatory and enforcement programs leave many permittees unable or unwilling to implement alternative BMPs (Stoner and Giles, 2011). EPA continues to work to address this issue (EPA, 2013a) and has developed extensive guidance for state and EPA NPDES permitting and enforcement staff on integrating green infrastructure approaches into permits, control plans, and consent decrees.[8]

Numerous states and municipalities are adopting policies and tools to grant LID practices legal credit and "put it on a

[7] Forty-six states are authorized to issue NPDES permits. The EPA currently implements the NPDES permit program for Idaho, New Mexico, New Hampshire, Massachusetts, the District of Columbia, and tribal lands. See http://water.epa.gov/polwaste/npdes/basics/NPDES-State-Program-Status.cfm.

[8] See http://water.epa.gov/infrastructure/greeninfrastructure/gi_regulatory.cfm#permittingseries.

level playing field with other BMPs" (EPA, 2013b). States such as Virginia and North Carolina are developing stormwater regulations that acknowledge the water quality benefits related to reductions in stormwater volume (EPA, 2013b). In southern California, several recently adopted MS4 stormwater permits require the on-site retention of a large percentage of expected runoff from rain events using a set of LID practices that includes rainwater capture and use (Strecker and Poresky, 2010). At the federal level, the 2007 Energy Independence and Security Act set green infrastructure requirements in federal development and redevelopment projects, calling for "strategies for the property to maintain or restore, to the maximum extent technically feasible, the predevelopment hydrology of the property," which has been interpreted by the EPA to require on-site retention of the 95th percentile storm event (42 U.S.C. § 17094; EPA , 2009).

Underground Injection Control. Federal or state UIC Class V regulations govern projects that use shallow injection systems (including wells, dry wells, and manufactured infiltration chambers) to speed infiltration of captured stormwater into the subsurface. These UIC regulations are designed to protect potential sources of drinking water from contamination. Most stormwater infiltration practices (e.g., infiltration trenches, bioretention basins) do not meet the EPA Class V well definition ("any bored, drilled, driven shaft, or dug hole that is deeper than its widest surface dimension") and can be installed without regulatory oversight, provided they meet any set state criteria protecting groundwater supplies (EPA, 2008b). Existing stormwater UIC regulations vary in structure and approach from state to state. For example, in Florida, Class V wells must be constructed so that they do not violate water quality standards at the point of groundwater discharge. California is more stringent in its regulations and prohibits any degradation in water quality while stored, a rule that can impose costly pretreatment requirements (NRC, 2009a). Because of concerns over potential groundwater contamination by direct injection or the costs of pretreatment, many projects instead use surface infiltration, which can be installed without federal oversight. Those that meet the Class V criteria can be operated without an individual permit, provided the injection does not endanger an underground source of drinking water and the operator submits basic inventory information to the permitting authority (EPA, 2013c).

State and Local Regulations Relevant to Stormwater Use

Only a few states have regulations that specifically address the use of stormwater as an on-site water supply (see Table 8-1). The legality of stormwater capture and use in the remaining states is, in many cases, less clear, but the lack of formal regulation should not be interpreted as a prohibition on such systems. In fact, some cities or states without formal regulation have promoted and issued detailed guidance for installing on-site stormwater capture and use systems (e.g., City of Tucson, 2005; Minnesota's Stormwater Manual [MPCA, 2015]). State and local regulations on stormwater use for on-site water supply are evolving quickly as the practices are becoming more widespread. Yet, even in states where stormwater capture and use is allowed and encouraged through specific regulations (see Appendix B) or guidance, state and local public health codes may limit potential beneficial uses or system designs. Underground injection control regulations may also impact how states are permitted to infiltrate stormwater in larger scale projects.

Regulations on Stormwater Water Quality Criteria or System Design. Responding to citizen interest and concerns about water conservation and stormwater management, at least 10 states have developed specific regulations on stormwater capture and use (Table 8-1). These regulations vary widely in complexity and consideration of possible uses (see Table B-1) and may be written from a range of perspectives, including public health, environmental management, or tax crediting. Many provide basic requirements for the design and permitting of stormwater capture systems, such as that described in the 2012 Uniform Plumbing Code (UPC; see Box 8-1), which as of December 2014 had been adopted (at least in part) by several states, including California, Hawaii, Iowa, and Washington. Other states have developed detailed, but not legally enforceable, guidance for developing stormwater capture and use systems (e.g., Carpenter et al., 2009). Many state regulations (and the 2012 UPC) limit water collection to roof runoff, although California allows collection from surface runoff, including paved roads and parking areas, as long as the water is used exclusively for subsurface irrigation or treated to applicable local water quality requirements (or National Sanitation Foundation [NSF] 350 in absence of such requirements; see Box 6-2). NSF-350 was developed for wastewater reuse systems and does not provide risk-based guidance for stormwater capture and use (see Chapter 6). Many of the state regulations (e.g., Utah, Washington) also provide permitting exemptions or reduced requirements for small rooftop harvesting systems used only for outdoor irrigation.

As of December 2014, only one state—California—had established specific water quality criteria for stormwater use projects. Box 8-2 summarizes water quality guidance or requirements for four states, which vary substantially. Additionally, local governments may establish their own water quality criteria. For example, Los Angeles County Department of Public Health established extensive tiered guidelines for outdoor uses of stormwater, ranging from no water quality criteria for rain barrel collection systems to rigorous permitting and monitoring requirements for systems with stormwater draining from industrial, high-traffic, or agricultural areas (see Table 8-2). However, Los Angeles County does not yet have water quality guidelines for indoor uses, such as toilet flushing.[9] The lack of a consistent, authoritative source for water quality criteria that could be adopted by state or local governments serves as a major impediment to expanding the use of stormwater for indoor or large-volume

[9] In early 2016, after the release of this report in prepublication form, the Los Angeles Department of Public Health released new guidelines for both indoor and outdoor nonpotable uses of graywater and stormwater. See http://publichealth.lacounty.gov/eh/docs/ep_cross_con_AltWaterSourcesGuideline.pdf.

BOX 8-1 Model Plumbing Codes

A plumbing code is the section of a building/housing code that requires plumbing facilities in buildings intended for human occupancy to be safe, sanitary, and efficient. The U.S. federal government does not promulgate national building codes; instead, various professional organizations have developed model codes (sometimes referred to as national or international codes), which are then adopted by state or local governments at their discretion. Model codes and standards become enforceable only through the state or local government's adoption process, at which point they become law. Model codes are often amended to fit the circumstances as well as the other laws and policies of the adopting jurisdiction. Although these provisions must be interpreted in the context of other state laws, they may have the effect of authorizing graywater (or stormwater) use.

The two most prevalent model plumbing codes are the International Plumbing Code (IPC) and the Uniform Plumbing Code (UPC), which address the design and installation of plumbing systems using prescriptive and performance-related provisions. Updated every few years, the IPC and UPC play a significant role in the widespread acceptance and implementation of innovative plumbing technologies and practices. The most current editions (ICC 2012) of both codes include provisions governing the on-site collection and reuse of graywater. The 2012 UPC also considers the use of rainwater and "treated non-potable water."

Graywater in the IPC

The IPC, updated triennially by the International Code Council, has provided for some form of graywater use since 1997. In the 2012 edition, Chapter 13 regulates the design and installation of graywater collection and disposal systems for use in subsurface landscape irrigation and toilet flushing. Under the IPC, all graywater must be filtered to some degree and must then enter a collection reservoir. Percolation tests are required for subsurface irrigation. Graywater for toilet flushing must be disinfected and dyed blue or green prior to use. No specific water quality criteria are prescribed. The 2012 IPC does not address the beneficial use of stormwater.

Stormwater and Graywater in the UPC

The UPC is issued by the membership-based International Association of Plumbing and Mechanical Officials. Chapters 16 and 17 of the 2012 edition govern reuse of roof runoff ("rainwater"), untreated graywater, and "treated nonpotable water" and outlines system and permitting requirements for various applications, including subsurface or surface drip irrigation and toilet flushing. Significantly revised from the 2009 UPC, the 2012 UPC added rainwater as an on-site water source. For untreated graywater, the code provides for subsurface irrigation (with supply piping covered by at least 2 inches of mulch) as long as there is no surface ponding. Food crop irrigation is not permitted. Rainwater runoff from rooftops may be used for subsurface or surface drip irrigation, and small systems (less than 360 gallons on-site storage) are exempt from the permitting process. The 2012 UPC also describes treatment requirements for toilet flushing. On-site, nonpotable water must be filtered and disinfected for use in toilet flushing. Disinfection is not specifically required for roof runoff, but all water for toilet flushing must meet applicable water quality requirements set by the "public health authority having jurisdiction." In absence of local water quality standards, the 2012 UPC directs local agencies to the EPA Water Reuse Guidelines (EPA, 2004b) for guidance that might assist local development of such standards. As discussed in NRC (2012a), these nonpotable guidelines are not risk based, and for toilet flushing, are quite conservative compared to other state regulations in existence (see Table 6-4).

The existence of graywater provisions in the IPC and UPC and rooftop rainwater collection in the UPC effectively legalizes the installation of these systems in the states that have adopted the relevant sections of those codes. However, a state's adoption of the IPC or UPC may not be sufficient to guarantee that such systems are legal in that state. Both the UPC and IPC note that permits are to be obtained by the appropriate authority before installation of new plumbing systems (although the UPC exempts small rainwater catchment systems), and building permits are issued by a local agency, which must determine the legality of on-site water systems in light of local laws. Moreover, internal inconsistencies between the adopted plumbing code and other areas of the state code (e.g., public health or sewage disposal laws) occur in a number of states.

outdoor uses. Even when water quality guidelines are established, communities may lack mechanisms for regulatory oversight of these requirements.

Regulations to Prevent Health Hazards from Standing Water. Large quantities of nonpotable standing water may create public health hazards associated with mosquito breeding, use of contaminated water, and algae growth. Therefore, some states with stormwater use regulations or guidelines have included express requirements aimed at public health protection, such as requiring opaque tanks to inhibit algal growth and screens to limit mosquito breeding (e.g., WVDEP, 2012; see also EPA, 2013b). However, in jurisdictions where the capture of roof runoff is not formally regulated or ex-

BOX 8-2 State Water Quality Criteria for Stormwater Use

Only a few states require or recommend specific water quality for end uses, and these criteria vary in terms of the considered end uses and contaminants as well as the recommended values. This box briefly summarizes California's water quality requirements as established in the 2013 state plumbing code (Title 24 CCR Part 5) and the water quality guidelines established by three states (Texas, Georgia, and Minnesota). Guidelines are not legally enforceable unless adopted by local regulators. The District of Columbia has established a tiered risk framework (see Figure 4-1 and Table 4-2) for setting stormwater quality criteria based on the extent of anticipated exposure.

California

California's revisions to the 2012 UPC set minimum water quality requirements for several stormwater use applications. For toilet flushing, ornamental fountains, spray irrigation (more than 360 gallon storage), and cooling tower make up, California requires:

- E. coli <100 cfu/100 ml and
- Turbidity <10 NTU,

in addition to minimum treatment technology requirements for filtration. No water quality criteria are provided for car washing and surface, subsurface, or drip irrigation, provided that the designs for these applications include a debris screen and a 100-μm filter.

Texas

In response to a mandate from the Texas legislature, the Texas Water Development Board (TWDB, 2006) recommended criteria for nonpotable indoor use (including toilet flushing) as

- total coliforms <500 cfu /100 ml and
- fecal coliforms <100 cfu/100 ml.

No water quality criteria are recommended for outdoor use. As of early 2015, Texas has not formally adopted these criteria into law.

Georgia

The Georgia Rainwater Harvesting Guidelines (Carpenter et al., 2009) include the same water quality criteria published in TWDB (2006).

Minnesota

Minnesota has developed detailed guidelines for public access irrigation systems, using stormwater, considering public health, plant health, and system function (Table 8-2-1). Additional water quality guidance is anticipated at a later date for irrigation of food crops and irrigation of areas with restricted access.[a]

[a] See http://stormwater.pca.state.mn.us/index.php/Stormwater_re-use_and_rainwater_harvesting.

TABLE 8-2-1 Minnesota Water Quality Guidelines for Stormwater Capture and Use for Irrigation

Water Quality Parameter	Impact of Parameter	Water Quality Guideline—Public Access Areas
E. coli	Public health	126 *E. coli*/100mL
Turbidity	Irrigation system function	2-3 NTU
Total suspended solids (TSS)	Irrigation system function	5 mg/L
pH	Plant health	6-9
Chloride	Plant heath, corrosion of metals	500 mg/L
Zinc	Plant health	2 mg/L (long-term use) 10 mg/L (short-term use)
Copper	Plant health	0.2 mg/L (long-term use) 5 mg/L (short-term use)
Temperature	Public health	Guidance to be determined at a future date

SOURCE: MPCA (2015).

TABLE 8-2 Summary of Los Angeles Tiered Guidelines for Stormwater Capture and Outdoor Use

	Requirement	Uses	Water Quality Standard	Treatment
Tier 1 Roof runoff collected in rain barrels with on-site use (*Household scale*)	• Labeling and design requirements specified	• Irrigation • Car washing	None	None
Tier 2 Roof runoff collected in cisterns with on-site use (*Large household scale; no source water from agricultural, manufacturing, or industrial land uses*)	• Department of Public Health review required • Screens to prevent vector intrusion	• Drip or subsurface irrigation	None	• Pre-screening
		• Spray irrigation[a] • Non-interactive outdoor water feature	• Total coliforms < 10,000 MPN/100 ml • Fecal coliforms < 400 MPN/100 ml • Enterococcus < 104 MPN/100 ml	• Pre-screening • Disinfection
Tier 3 On-site or off-site stormwater collection in cisterns; off-site or onsite use (*No high transportation corridors or agricultural, manufacturing, or industrial land uses*)	• Reviews by Department of Public Health and building and safety department required • System design requirements	• Drip or subsurface irrigation	None	• Pre-screening
		• Spray irrigation[a] • Non-interactive outdoor water feature • Street sweeping • Dust control	Same as Tier 2 spray irrigation	• Pre-screening • Disinfection • Retention/ sedimentation (for street sweeping only)
Tier 4 On-site or off-site stormwater collection in cisterns; off-site or onsite use (*Includes high transportation corridors or agricultural, manufacturing, or industrial land uses*)	• Reviews by Department of Public Health and building and safety department required • System design requirements • Extensive monitoring requirements, including metals, volatile organic chemicals (VOCs) and semi-VOCs	• On-site drip or subsurface irrigation	None	• Pre-screening
		• Spray irrigation[a] • Non-interactive outdoor water feature • Street sweeping • Dust control	Same as Tier 2 spray irrigation • Must meet Calif. Maximum Contaminant Levels and Calif. Toxics Rule Standards	• Pre-screening • Disinfection • Retention/ sedimentation (for street sweeping only)

[a]Spray irrigation is only allowed when there is negligible human exposure, such as after sunset and before sunrise.
SOURCE: Los Angeles County Department of Public Health (2011).

emptions have not been established, increased amounts of stagnant water (in cisterns or infiltration basins) may violate state or local public health laws that require private property owners to prevent conditions that contribute to vector harborage, which is typically considered a public nuisance. In Connecticut, for instance, the state environmental commissioner is empowered to issue orders aimed at eliminating mosquito-breeding places (Connecticut Gen. Stat. § 22a-45b), while local health authorities are required to investigate reports that rain barrels or other receptacles near human habitations are breeding mosquitoes and cause any such breeding places to be abolished, screened, or treated (Connecticut Gen. Stat. § 19a-213). An ordinance adopted by the city of Petersburg, Virginia, allows its health director to order occupants of private property to drain standing water that is detrimental to the health, comfort, or general welfare of any of the inhabitants of the city (Petersburg, Va. Code of Ordinances § 50-64). In extreme cases, adverse effects on a neighbor's use and enjoyment of property caused by vectors or odors also could leave a stormwater harvester vulnerable to private nuisance liability.

Graywater Reuse

No federal laws directly govern on-site management of graywater, leaving to the states policy decisions about whether and how to regulate on-site graywater reuse.

State Graywater Reuse Laws

Although some states have recognized graywater as legally distinct from wastewater for many years and have even encouraged segregated plumbing and reuse, a relatively recent surge of interest in graywater use as a conservation alternative has prompted a flurry of new legislation. As of December 2014, at least 26 states have laws allowing segregation and reuse of graywater under less stringent treatment standards than those applied to reclaimed wastewater (see Table 8-3 and Appendix B, Table B-2).

State regulations are widely variable with respect to allowable graywater reuse applications and treatment requirements (Yu et al., 2013). As a threshold matter, states differ in what sources of household water are included in the defini-

TABLE 8-3 Summary of State Regulation of Graywater Reuse

States Without Formal Graywater Regulations			States with Regulations Allowing Graywater Reuse		
States allowing wastewater reclamation that define graywater as wastewater	States not defining graywater	States treating graywater as septic	States permitting graywater using a tiered approach	States regulating graywater reuse without a tiered approach	States allowing graywater for irrigation uses only
Alabama	Illinois	Connecticut	Arizona	Colorado	Hawaii
Alaska	North Dakota	Kentucky	California	Florida	Idaho
Arkansas	South Carolina	Maryland	New Mexico	Georgia	Kansas
Delaware	Tennessee	Michigan	Oregon	Indiana	Maine
Iowa		Minnesota	Washington	Massachusetts	Nevada
Louisiana		Nebraska		Montana	Ohio
Mississippi		New Hampshire		New York (non-residential)	Oklahoma
Missouri		New Jersey		North Carolina	Utah
Pennsylvania		West Virginia		South Dakota	
Rhode Island				Texas	
Vermont				Virginia	
				Wisconsin	
				Wyoming	

NOTE: For details on states with regulations allowing graywater reuse see Appendix B, Table B-2.
SOURCE: Updated from Sharvelle et al. (2013).

tion of "graywater"—generally faucets and showers, sometimes laundry, and sometimes the kitchen sink or dishwasher. In addition, states differ on where within their statutory and regulatory codes they cover the topic: plumbing or building codes, sewage disposal regulations, water pollution control regulations, health and safety codes, water and wastewater regulations, or a distinct section of code dedicated to graywater reuse requirements (Yu et al., 2013). The location and thoroughness of a state's graywater provisions may be indicative of its general priorities and objectives with respect to graywater reuse.

States in arid regions, where ever-growing pressure on municipal water supplies has prompted widespread interest in reuse, tend to have the most comprehensive graywater reuse regulations or guidance (Martinez, 2013). For example, California—the first state to adopt legislation promoting graywater reuse in 1992—has updated its graywater regulations several times to establish a workable framework for regulating residential and nonresidential graywater systems (Snodgrass, 2010). Arizona's tiered permitting system is widely considered a model of effective regulation for graywater irrigation (see Table B-3 in Appendix B).

In tiered regulatory frameworks (see Table 8-3), the requirements for permitting increase with the size of the system and the expected human exposures. For projects with large volumes captured for beneficial use, permitting may require design review, site inspections, and monitoring to ensure adequate public health protection. However, many tiered frameworks and some nontiered regulations allow the on-site reuse of small volumes of graywater without a formal permit. Some policy analysts have noted that when a permitting process is too burdensome and is perceived by potential small-volume users as an added cost, it may have the effect of discouraging installation of graywater systems or incentivizing unauthorized reuse (Snodgrass, 2010; Yu et al., 2013). In fact, states where on-site graywater reuse is considered to be comparatively widespread also tend to be states whose regulatory schemes allow small volumes of graywater to be collected and reused without obtaining a formal permit—Arizona, California, New Mexico, Montana, Texas, and Wyoming.

Among states with graywater regulations, there are numerous differences with respect to allowable uses, permissible equipment, and treatment/water quality standards. For example, Table 6-4 highlights varying graywater quality standards for toilet flushing in seven states, and Table 8-3 summarizes the different regulatory strategies and end use limitations used by states (see Table B-2 for more detail). Yu et al. (2013) suggest that this inconsistent regulatory landscape may be a barrier to graywater implementation. Specifically, the varied scales for which on-site systems are allowed and to what purposes the water may be put can keep the cost of available infrastructure products on the market high because the manufacturing industry cannot reach optimum scales of production. Additionally, lack of consistent, authoritative risk-based guidance may deter local or state governments from establishing their own regulations where they do not already exist or may lead local or state health departments to question the safety of these applications.

Where water quality standards have been established for household- or neighborhood-scale systems, enforcing those standards and developing cost-effective regulatory or oversight mechanisms to ensure the systems are appropriately maintained is a challenge. As noted in Chapter 6, additional maintenance and monitoring guidance is needed, with clear performance standards and possible online monitoring of surrogate parameters for neighborhood- or large-building-scale systems. As part of San Francisco's Non-potable Water Program, the San Francisco Department of Public Health oversees water quality and monitoring requirements for graywater and stormwater use at multi-residential and commercial sites (see Box 8-3).

Graywater Provisions in Plumbing Codes

Even in states without comprehensive graywater regulations, special provisions for graywater systems can sometimes be found in plumbing codes (see Box 8-1). In a number of states, the plumbing code includes separate standards for the design and operation of on-site graywater reuse systems. The UPC and IPC, each of which has been adopted by many states, are the two most prevalent model plumbing codes. These model codes identify permissible graywater reuse procedures and contain treatment specifications for graywater according to categories of use (see Box 8-1). These model plumbing codes generally restrict legal use of graywater to use that limits human exposure to pathogens—namely, subsurface irrigation and toilet flushing—and specify any treatment necessary (Yu et al., 2013). Provisions in a plumbing code also address cross-connections, backflow valves, air gaps, and other aspects of plumbing configuration, which may determine not only whether a certain system is permissible, but also how the system is installed and how efficiently and economically it operates (Sharvelle et al., 2013). For example, the 2012 IPC requires a graywater storage tank for subsurface irrigation, which would prohibit the more economical, household-scale, laundry-to-landscape systems described in Chapter 6.

On-site Wastewater Disposal Laws

Even in states where the plumbing code accepts graywater for a limited class of nonpotable uses, the legality of reuse may ultimately be determined by wastewater disposal laws designed to protect drinking water supplies and regulate the entry of polluted water into the environment. On-site wastewater disposal laws affect the application of graywater to soil, which occurs in the course of the most common outdoor uses of graywater (e.g., subsurface irrigation). On-site wastewater disposal systems are permissible in many states under certain conditions. Some states only allow on-site disposal systems where public sewer connections are impossible or overly burdensome, but many are now also recognizing "innovative" or alternative on-site systems. In a number of states, graywater can be applied to subsurface irrigation under these laws as part of the general domestic wastewater stream.[10] However, some on-site wastewater disposal codes allow segregation and application of graywater at lower treatment standards than water that may contain sewage. In Maine, for example, the on-site wastewater disposal code permits "primitive" or "limited" graywater disposal to water plants with untreated graywater by hand (Code of Maine Regs. § 10-144-241).

Despite increasing recognition of graywater systems as attractive disposal alternatives, in some states the inconsistencies between plumbing code provisions and state or municipal on-site sewage and health codes may yet constrain a property owner's ability to legally implement a graywater reuse system (Snodgrass, 2010). In extreme cases, as in South Carolina and Maryland, a state may have regulations for graywater systems in its plumbing code, yet have public health or sewage disposal laws requiring all domestic wastewater to be discharged to the sewer system (Sharvelle et al., 2013). In other cases, it is unclear whether a law allowing subsurface application of graywater as a means of "disposal" has the effect of authorizing the application of graywater for subsurface "irrigation." These contradictions and ambiguities within a state's own code can create direct legal obstacles to implementation. To make matters still more complicated, the agency responsible for enforcing building codes and issuing permits for indoor plumbing installations may differ from the public health or environmental agency charged with regulating any storage or application of graywater that occurs outside the building and/or the agency that regulates beneficial use of water resources.

Potential Procedural Barriers

State and federal environmental impact analysis (EIA) laws provide varying levels of procedural and, on occasion, substantive protections of the public's interests. EIA laws seek to identify risks in advance of a project; balance environmental and socioeconomic values; ensure opportunities for stakeholder participation; and demand informed, reasonable decision making. Since EIA's emergence, litigation has played an important role in achieving these objectives (Taylor, 1984; Wathern, 1988). Federal projects are subject to the environmental review requirements of the National Environmental Policy Act (NEPA) (42 U.S.C. § 4331 et seq.), and many states have enacted "little NEPAs"—state EIA statutes that apply the same general principles to state

[10] For example, see http://extension.psu.edu/natural-resources/water/septic-systems/on-site-wastewater-treatment-system-options/drip-irrigation-on-lot-sewage-disposal-system.

and municipal projects (Bender, 2014). Some of these state laws, notably the California Environmental Quality Act, define "project" and "environmental impact" quite broadly so that many government actions—from large-scale projects to permitting decisions and even policy changes—are subject to the environmental review requirements (California Public Resources Code § 21000 et seq.).

Although few of the state EIA laws include substantive provisions—and only a small percentage of the legal challenges result in the court prohibiting a project from ultimately moving forward—opponents of proposed actions who can credibly argue that a project's potential environmental impacts will injure them often can influence the how and when of a project through EIA litigation (Riccardi, 2011; Wathern, 1988). Particularly in states where EIA laws are robust and litigated often, for example California, actions that alter downstream flows may face the additional hurdle of litigating claims brought under EIA laws.

POLICY IMPLICATIONS

Significant future expansion of graywater or stormwater to meet water supply needs raises a number of policy issues to be addressed by local, state, or national governments. These issues involve water quality standards, source control, downstream impacts, and decentralized infrastructure.

As discussed in this chapter, few states have water quality regulations or guidance for graywater and stormwater use, and those that exist are widely variable (see Box 8-2 and Table 6-2). National guidance is lacking in the United States, although Australia has developed extensive guidelines on the use of stormwater and graywater (NRMMC, et al., 2009a,b). The 2012 EPA Water Reuse Guidelines are often cited as a reference, although its categories of recommended water quality criteria are broad and lump spray irrigation, fire protection, and toilet flushing—categories with far different exposures—into a single category of "urban uses," resulting in suggested water quality criteria that may be more protective than needed to protect public health. EPA (2012c) may be more useful for large-scale, groundwater infiltration projects or industrial applications. However, the wastewater sources considered in the development of these recommendations are significantly more contaminated with microorganisms and may contain lower concentrations of other contaminants, such as metals and organic contaminants found in urban, stormwater-draining, paved areas. Without risk-based guidance, local projects may face resistance from health departments or the public, who question the safety and reliability of these practices, or may lead to unnecessary treatment (and associated cost) to meet overly restrictive guidelines that were developed for different risk scenarios (or may not be risk-based at all). Therefore, additional risk-based water quality guidance is needed that can serve as the basis for developing standards of practice for typical stormwater or graywater applications. Fit-for-purpose guidelines will require substantial public education so that on-site waters are used appropriately with the necessary treatment.

Increased use of urban stormwater, particularly at large scales, leads to water quality concerns with runoff from roadways, parking lots, and industrial areas (see Chapter 4). Treatment is an important component of large stormwater systems, but policy issues emerge about the role of enhanced source control in areas that rely upon urban stormwater to recharge potable aquifers. Limits on roofing materials, road salt application, or even tire or brake pad composition could lead to significant improvements in stormwater quality (see Chapter 4), but policy makers should weigh the costs and benefits of additional source control restrictions.

As the on-site water capture and use movement grows, particularly for consumptive uses such as irrigation, ecosystems and communities downstream will be impacted if the total volume of water used for irrigation exceeds that under prior conditions with only potable water use (see Chapter 3 and Box 3-3). Thus, policy makers should evaluate the benefits of on-site water use for various applications compared to the risks of harm to those downstream. In some prior appropriation states, the legal framework restricts the use of on-site graywater or stormwater out of concern for impacts to downstream water rights holders. However, impacts to downstream ecosystems, including streams and estuaries, also need to be considered. Although enhanced stormwater infiltration projects should increase base flow to streams, stormwater capture projects for expanded irrigation could decrease stream flows, and such effects should be carefully assessed in advance (see Chapter 3). Graywater and stormwater projects for nonconsumptive uses, such as toilet flushing, do not impact the quantity of water delivered downstream and are, therefore, ideal applications in inland communities.

Fully embracing the use of on-site water sources at household, neighborhood, and municipal scales requires a shift to decentralized treatment systems that ultimately supplement larger centralized facilities. This shift may have important policy implications in the way these decentralized facilities are permitted, managed, monitored, and maintained, particularly for communities with entirely centralized water and wastewater systems. Additionally, the intersection of green building practices, stormwater management, and water supply brings together a diverse array of local government entities with a role in managing these decentralized projects. Thus, neighborhood- or municipal-scale implementation requires the involvement of agencies responsible for city planning, water supply, wastewater, stormwater, building safety, and public health, possibly necessitating new strategies for government collaboration (see Box 8-3).

Finally, advances in potable water conservation (including but certainly not limited to graywater and stormwater capture and use) reduce the incomes of water and wastewater utilities, whose costs do not necessarily decline proportionally with water use (Beecher and Chesnutt, 2012). In growing cities, conservation allows utilities to serve more customers without the need to invest in expensive new water sources. However, when conservation causes total utility incomes to decline, increased rates, supplemental fees, or new rate structures may be necessary to maintain water utility operations (Donnelly and Christian-Smith, 2013).

CONCLUSIONS

As technologies and strategies continue to advance, graywater and stormwater use is being incorporated into law in a variety of respects at the federal, state, and local levels. A number of new laws at the state and local levels promote or regulate stormwater and graywater capture and use, and model plumbing codes have been updated to include provisions for these practices. Additionally, green infrastructure practices, including stormwater capture and use, are increasingly being incorporated into NPDES permits. However, as is often the case with innovative technologies, the law has not evolved quickly enough to keep up with advances in the technology and its use. Several legal and regulatory constraints remain that hinder the capacity for graywater and stormwater to significantly expand the nation's water supplies.

In most western states, acquisition of water rights is a requirement for large-scale stormwater capture and use projects, and water rights may limit widespread implementation of smaller-scale stormwater and graywater projects for consumptive uses. The use of graywater and stormwater for consumptive applications, such as irrigation or cooling, may impact the water available to downstream users if total water use (including potable and nonpotable) for those applications exceeds previous potable water use (see Chapter 3). Thus, unless water rights can be acquired or legislative solutions developed, opportunities for large-scale stormwater capture projects to expand existing water supplies could largely be limited to coastal regions with no downstream users or to nonconsumptive uses (e.g., toilet flushing). Several states (e.g., California, Kansas, Oregon, Utah, and Washington) have established regulations that allow small-scale roof runoff capture projects to proceed with-

BOX 8-3 San Francisco Non-potable Water Program

In dense urban areas, the use of on-site "alternate water sources" is an important strategy for conserving potable water. In San Francisco, the reuse of graywater and stormwater are addressed under the city's Non-potable Water Program. Established by ordinance in September 2012, this program has created a streamlined approval process for new commercial, multi-family, and mixed-use developments to collect, treat, and reuse alternate water sources for toilet flushing, irrigation, and other nonpotable uses. The program developed a guidebook (City and County of San Francisco, 2014) for developers interested in installing nonpotable water systems in buildings. The guidebook includes information on alternate water sources, permissible use applications, water quality parameters, design and construction basics, and ongoing operation of on-site systems.

Aimed at promoting reuse of nonpotable water while ensuring appropriate water quality standards, the program facilitates coordination between three municipal agencies involved in regulating on-site water reuse systems: the San Francisco Public Utilities Commission (SFPUC), the San Francisco Department of Building Inspection (SFDBI), and the San Francisco Department of Public Health (SFDPH). SFPUC reviews project water budgets, serves as a technical resource, and provides financial incentives for customers who are interested in nonpotable water use. SFDBI oversees the design and construction of nonpotable water systems by administering permits, conducting inspections, and issuing certificates of occupancy. Finally, SFDPH regulates the water quality and monitoring requirements for on-site treatment systems, issuing operating permits and establishing reporting requirements for on-site treatment systems.

As of May 2014, less than 2 years after the program was implemented, six developments in San Francisco were operating a nonpotable water system or were in the process of installing one, including the Bill Sorro Community affordable housing project and the Market Street Place retail center (SFPUC, 2014a). The program was originally voluntary but was amended in July 2015 to require that all new buildings of at least 250,000 square feet to be constructed, operated, and maintained using available alternate water sources for urinal and toilet flushing and irrigation. New buildings located inside the city's designated recycled water use area are required to meet this requirement beginning November 1, 2015, and new buildings within San Francisco city and county located outside of the designated area are required to meet this requirement by November 1, 2016.

SOURCE: SFPUC (2015).

out water rights permits, and only one state (Colorado) has strict limits on stormwater capture and use out of concern for water rights impacts. However, the right to stormwater and graywater use in most prior-appropriation states has not been firmly resolved through judicial decisions, leaving an unclear outlook for projects that have not acquired water rights, because they could be vulnerable to legal challenges. New scientific analyses of the impacts to return flows of various on-site water uses in different regions would help clarify these concerns, but additional legal research and guidance could better facilitate the use of on-site water supplies, considering potential legal challenges.

There is substantial variation in on-site graywater and stormwater regulations at the state level with respect to design and water quality, which leads to varying exposures and risk. As one example, there is lack of consistency among states on whether outdoor graywater use is limited to subsurface irrigation. At least three states allow drip irrigation without landscape cover, which could lead to higher pathogen exposures. In addition, states vary on their regulation of untreated graywater irrigation of food crops. Whether such exposures would lead to unacceptable risks at various scales has not been definitively resolved, but higher risks are likely with increased exposures. Regulations affecting large-scale graywater and stormwater use where public access is not controlled tend to include conservative public health protections, such as disinfection, and meeting state maximum contaminant levels, but household-scale protections are more variable.

The lack of authoritative, risk-based guidelines for the design and potential applications of graywater and stormwater in the United States is a major impediment to their expanded use. The wide variability in existing regulations and absence of federal guidance leaves stakeholders and local decision makers uncertain about the safety of these practices and the appropriate level of treatment necessary for particular uses. Development of rigorous, risk-based guidelines for graywater and stormwater across a range of possible uses and exposures could improve safety, build public confidence in the practices, reduce expenditures on unnecessary treatment, and assist communities that lack an existing regulatory framework for on-site water supplies. Such guidelines could be developed by the EPA, a collaboration of states, or a collaboration of U.S. water organizations, including the Water Research Foundation, WateReuse, and the Water Environment Research Foundation working with the EPA. The Australian Guidelines for Water Recycling provide a useful example of such an effort. This guidance can then serve as a basis for developing standards of practice for on-site, nonpotable water use. Oversight and enforcement of water quality standards for applications with significant exposures is also important but challenging, and local enforcement agencies would benefit from additional guidance on appropriate, cost-effective maintenance, monitoring, and reporting strategies.

Inconsistencies often exist between plumbing codes and public health or on-site disposal laws within the same state, especially in the case of graywater, that need to be resolved to ease project implementation. Increased use of graywater and stormwater will require enhanced collaboration among agencies with jurisdiction over different elements of on-site water systems, including wastewater disposal, water supply, public health, pollution prevention, building safety, and city planning. As regulators continue to update laws to reflect increasing acceptance of new water reuse systems, legal barriers such as inconsistent or conflicting regulations are likely to be resolved.

9

Graywater and Stormwater in the Context of Integrated Water Supply Planning

Much information has been provided in this report about the quality and quantity of stormwater and graywater available in different locations, possible end uses, known risks, reported costs and benefits, and legal and regulatory constraints. Decision makers should understand these factors to determine the potential risks, costs, and benefits of investments in graywater and stormwater capture and use systems at a range of scales at the local level. For small-scale systems, home and business owners may want to determine whether their investments are better served by graywater or stormwater capture for their given geographic and building circumstances, water supply needs, and personal objectives. At a regional scale, graywater and stormwater use affects many aspects of water, wastewater, and stormwater management, and decision making is best served by a holistic view of costs and benefits. This chapter attempts to synthesize this information within a water supply planning framework to help local decision makers, at the household, neighborhood, or regional scale, consider key information in assessing the potential role of stormwater and/or graywater as alternative local supplies to meet water needs.

Over the long term, with increasing urban population growth and the potential for more climate variability, there will be an increasing trend to maximize water conservation. Efforts will also continue to address stormwater pollution through retrofits and new construction designs. The potential role for graywater and stormwater within this future will play out with different priorities and urgencies, depending on which drivers described in Chapter 1 are most relevant to local decision makers. Opportunities for graywater use will increase with (re)development and growth of urban populations residing in multiple dwelling units. As a practical matter, beneficial use of stormwater will be driven primarily by water scarcity and pollution regulations. The next 20 to 30 years is likely to see continued evolution in the nation's approach to water management, but from today's viewpoint, no clear pathway or single technology is evident. Political and geographical realities will affect decision making region by region.

DECISION FRAMEWORK

Figure 9-1 presents broad decision steps for those considering stormwater and/or graywater capture and use. The major steps include defining objectives, identifying opportunities and constraints, characterizing sites, identifying candidate strategies, selecting the system design, implementing the system, and engaging stakeholder involvement throughout the process. Each of these will be discussed below in the context of the major findings of this report.

Stakeholder Engagement

Successful implementation of alternative water systems requires the effective engagement of a broad range of groups and individuals, typically referred to as stakeholders. A common definition of a stakeholder is any individual or group who can affect implementation of the subject project or program. Stakeholders are defined based on their legitimate interests in the matter at hand. Stakeholders may oppose or support the subject project. They may be internal or external to the responsible organization. They are not only those who will benefit from or be impacted by a project but also those who are involved in implementing and operating the required infrastructure and/or are concerned about this practice. Therefore, stakeholders are relevant for projects across a range of scales and may include individual residents, frequent visitors, business owners, employees, and organizations. An effective stakeholder engagement process will identify the relevant stakeholders and involve them proactively in the decision process so that opponents' concerns are fully considered and negative impacts are mitigated, if feasible, and supporters are fully informed.

For larger stormwater or graywater beneficial use projects, stakeholders can be identified by determining the issues relevant to the set of decisions to be made. These issues may relate to, for example, public health, environmental impacts, implementation, and costs in the context of other available

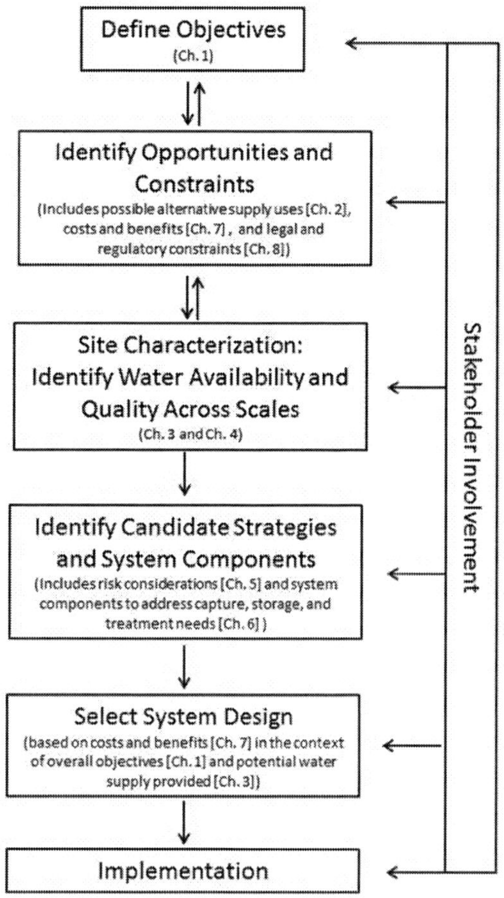

FIGURE 9-1 Steps in a general decision framework applicable to graywater and/or stormwater across multiple scales.

water supply options. A stakeholder analysis (Table 9-1) is used to identify the groups and individuals that could be affected by these issues and their credible representatives (e.g., governmental and nongovernmental organizations). The significance of the impact of each identified issue on the decisions to be made is also characterized. The next step is to reach out and engage the identified stakeholders in the decision process.

Stakeholder engagement should begin relatively early in the decision process so that the relevant issues are appropriately considered throughout the process. For large projects, it is certainly possible that different issues become more relevant at different times in the decision process. If this is the case, then the extent of involvement of specific stakeholders may vary during the process. However, engagement of all relevant stakeholders from the beginning will reduce the tendency for newly involved stakeholders to want to bring the process "back to square one." Consistent and early involvement helps to build a base of understanding and commitment to reach the necessary decisions by all relevant stakeholders.

A rich literature is available on effective decision processes (Lockie and Rockloff, 2005; NRC, 2005, 2008b, 2009c, 2012b; World Bank, 2012), and it is beyond the committee's charge to explore this in detail. In general, the process should be structured to involve three groups of participants—stakeholders, subject matter experts, and facilitators—whose relevant roles and responsibilities are clearly defined. Stakeholders define the issues and considerations that should be addressed in the decision process, and they bring a set of values that in an effective process are used to prioritize the relative importance of the identified issues and considerations. The subject matter experts bring technical knowledge that provides a factual basis for decision making. Facilitators structure the overall process to enable the various participants to function efficiently and effectively in the context of their legitimate roles, responsibilities, and interests. As a practical matter, effective stakeholder engagement works best among groups that fundamentally trust one another and see the benefits of working together, even though strongly held opinions may vary. An example

TABLE 9-1 Example Framework for Stakeholder Analysis

Issue	Stakeholder Group	Key People and Representative Groups	Type of Impact	Significance of Impact	Significance of Group Impact on Decision
Issue 1					
Issue 2					
Issue 3					

is the Watershed Management Groups in greater Los Angeles, which consists of agencies, cities, and nongovernmental organizations that focus on water resource management and future funding. Agencies and environmental groups (including TreePeople and Green LA Coalition) are responsible for developing stormwater capture projects among stakeholders in Los Angeles (Luthy and Sedlak, 2015). However, stakeholder groups comprising disparate interests and intractable adversaries are unlikely to resolve conflicts or achieve implementable projects. The CALFED Bay-Delta Program has been criticized for such failures (LAO, 2006).

Defining Objectives

A critical early step of any alternative water supply project is to determine the objectives of the project. This step is sometimes overlooked, to the detriment of project effectiveness and stakeholder satisfaction with the project, once completed. Chapter 1 discusses some major drivers for stormwater or graywater capture and use projects, including water supply, water reliability, pollution prevention, energy savings, and environmental stewardship and education. Brief summaries of the potential for graywater or stormwater to address these drivers are provided in Tables 9-2 and 9-3. Related project objectives could include reduced use of potable water supplies, enhanced local control of water supplies, and delayed need for infrastructure investments. Environmental objectives could include reduced stormwater discharges during rain events, reduced discharges to combined sewer systems to reduce the number and magnitude of overflows, reduced energy use, and enhanced groundwater recharge to reduce damage to urban streams from high-volume flows.

Stormwater or graywater projects can often be designed to optimize particular objectives if they are clearly identified in advance. For example, if reduction of potable water use is the primary objective, stormwater capture and use systems can be designed with tanks that are sized to meet as much of the water demand as is feasible under local climate conditions. Using stormwater and/or graywater for indoor nonpotable applications and limiting irrigation to that needed to preserve native landscaping would maximize water conservation. However, in many regions—particularly those working to minimize combined sewer overflows—pollution prevention is the primary objective of stormwater projects, and water supply provides a secondary benefit. In fact, to reduce stormwater pollution most cost-effectively, projects might instead use distributed shallow groundwater infiltration rather than stormwater capture. However, systems can, in many cases, be designed to balance multiple, sometimes competing objectives. For example, in areas working to manage combined sewer overflows, real-time weather forecasting can be used to automatically drain a tank to the sewer system in advance of a storm so that sufficient tank storage is available to capture runoff from the predicted storm (see Chapter 6). Such a system sacrifices some water supply to enhance the use of the tank to minimize stormwater pollution effects. The level of desired benefits also needs to be considered. For example, a project to advance public education may not necessitate large benefits to be effective, but a project to meet regulated pollution reduction measures, such as those contained in combined sewer overflow consent decrees, would necessitate large quantifiable outcomes.

Opportunities and Constraints

Next, local opportunities and constraints should be identified. Understanding the legal and regulatory controls on stormwater and graywater use is a key first step. As discussed in Chapter 8, water rights may limit large- or small-scale stormwater or graywater use in the western United States. This issue is discussed in more detail later in the chapter in several project examples. If downstream water rights could be impacted by the project, a water right permit may need to be secured. Graywater and stormwater use is regulated only at the state or local level, and relevant regulations may be found in state plumbing codes, wastewater disposal regulations, environmental regulations, and state or local public health laws (see Tables B-1 and B-2 for major state regulations, although this list is not exhaustive). A few states provide water quality guidance to determine when treatment is necessary for nonpotable uses (see Chapter 8, Box 8-2). Most of the existing regulations govern household- or small-building-scale projects, and neighborhood/multi-residential projects or regional capture projects necessitate consultation with appropriate state and local agencies to determine project design requirements.

TABLE 9-2 Capacity to Address Drivers Through Graywater Reuse

	Household Scale	Neighborhood to Regional Scales
Water Supply (Quantity drivers)		
Water scarcity	Graywater can reduce indoor and outdoor household use (see Chapter 3), although indoor graywater use requires substantial treatment, dual plumbing, and rigorous maintenance. If water savings are the primary objective, then homeowners should first address outdoor irrigation demand by converting to native landscaping. Such efforts could reduce the size and complexity of the graywater irrigation systems needed while making more water available to downstream users.	Multi-residential buildings can achieve significant reductions in indoor water use through graywater reuse (approximately 24% when used for toilet flushing, and more if other nonpotable uses are included), with no impacts to the water available to downstream users. If water savings are the primary objective, then opportunities to reduce outdoor irrigation demand should be considered by converting to native landscaping.
Water supply reliability	During drought restrictions, graywater provides a modest but reliable water source for irrigation that can help maintain native landscaping.	
Water supply diversification	Not a major driver.	GW provides an additional drought-resistant supply to diversify a community's water portfolio.
Water Quality (Pollution drivers)		
Pollution prevention	Not a major driver.	Not a major driver.
Environmental Practices		
Energy savings and greenhouse gas reductions	Laundry-to-landscape systems should provide low-energy on-site reuse for irrigation. Larger systems with pumps and treatment may require more energy than conventional water sources, although life-cycle energy requirements of various systems remain unknown.	Large graywater systems with pumps and treatment likely require more energy than conventional drinking water sources, but graywater treatment, even at small scales, requires less energy than municipal wastewater treatment. Overall, the life-cycle energy requirements of various systems remain unknown.
Environmental Stewardship	May be a major driver at the household scale. However, ways to minimize outdoor water use should also be considered when optimizing environmental stewardship.	May be an important driver for developers, who sometimes benefit from higher rental or resale values for green buildings.
Hydromodification	Not a driver.	
Other		
Extend life of existing infrastructure	Graywater systems have been cited as ways to extend the life of septic systems, although the committee was unable to find data to support this claim.	In dense urban areas, graywater reuse can extend the life of existing wastewater infrastructure by allowing additional development without expanding conveyance capacity.
Financial benefits	Rebates may be offered in some locations. Cost savings may be feasible for simple laundry-to-landscape systems based on potable water savings.	Incentives may be available for large projects that extend the life of existing urban wastewater infrastructure. Cost savings may be feasible based on potable water savings, although capital and maintenance costs of large-scale systems are not well defined.

Once the legal and regulatory framework is understood, the opportunities for on-site use can be considered within existing constraints. Chapter 2 presents an array of applications for nonpotable water at the household, neighborhood, and regional scales. Opportunities should be considered that address multiple objectives, where possible, and that deliver a wide range of benefits, including some that are not easily monetized, such as aesthetic enhancements, public education, and aspirational value (see Chapter 7).

Site Characterization: Water Availability and Quality

Understanding water availability and quality to meet the intended uses is an essential next step in the planning process.

Quantity

The total annual quantity of graywater and/or stormwater from various sources should be assessed, along with its inter-annual variability. Key questions include the following:

- Is sufficient stormwater and/or graywater available on an average annual basis to meet water supply objectives considering the target end uses? If not, then is supplemental water use acceptable?
- What is the timing of the water availability relative to the water demands? What storage capacity is needed to provide consistent water availability?

TABLE 9-3 Capacity to Address Drivers Through the Beneficial Use of Stormwater

	Household Scale	Neighborhood to Regional Scales
Water Supply (Quantity drivers)		
Water scarcity	Stormwater can reduce indoor and outdoor household use, although indoor stormwater use requires treatment, dual plumbing, and rigorous maintenance. If water savings are the primary objective, then opportunities to reduce outdoor irrigation demand should be considered by converting to native landscaping	Substantial potential exists to enhance regional water supplies by capturing and recharging stormwater, if suitable aquifers and recharge conditions exist, although water rights may need to be acquired. Under suitable climatic conditions, multi-residential buildings can achieve significant reductions in indoor water use (up to 24% when used for toilet flushing, and more if other nonpotable uses are included), with no impacts to the water available to downstream users. If water savings are the primary objective, opportunities to reduce outdoor irrigation demand should be considered by converting to native landscaping.
Water supply reliability	Not a major driver; during drought, roof runoff could provide some irrigation supply, but household tanks are rarely large enough to provide reliability during an extended dry spell, and during drought conditions, the roof runoff amounts available will be less than normal because of the lack of rain.	Neighborhood or regional stormwater recharge during wet periods can significantly enhance groundwater availability during times of drought. Neighborhood-scale stormwater capture using large tanks can also enhance water reliability.
Water supply diversification	Not a major driver.	Large-scale stormwater recharge provides a means to diversify a community's water portfolio.
Water Quality (Pollution drivers)		
Pollution prevention	Stormwater capture at the household scales can reduce runoff from the site, particularly with larger tanks, but pollution prevention is not usually a major driver at this scale.	Reduction of stormwater pollution is often a major driver behind large stormwater capture and use projects, which aim to provide multiple benefits from large required investments to reduce stormwater runoff.
Environmental Practices		
Energy savings and greenhouse gas reductions	Household-scale irrigation systems without pumps or treatment require minimal energy. Larger systems with pumps and treatment may require more energy than conventional water sources, although life-cycle energy requirements of various systems remain unknown.	Large graywater systems with pumps and treatment likely require more energy than conventional water sources, although life-cycle energy requirements of various systems remain unknown.
Environmental stewardship	May be a major driver at the household scale. However, ways to minimize irrigation use should also be considered.	May be an important driver for developers, who sometimes benefit from higher rental or resale values for green buildings.
Hydromodification	Not typically a major driver at the household-scale because the benefits are small.	Large-scale stormwater capture or recharge systems can reduce stormwater runoff, improve the timing of surface water flows, and reduce erosion, which may be an important driver in urban areas with degraded streams.
Other		
Extend life of existing infrastructure	Not typically a major driver.	Neighborhood and regional stormwater infiltration or capture systems can be part of distributed strategies to address combined sewer overflows, in place of an expensive new separate storm sewer system.
Financial incentives	Rebates for rain barrels and tanks may be offered locally. Long-term cost savings may be feasible based on potable water savings.	Incentives may be available for large projects that contribute to regional water quality goals. Long-term cost savings may be feasible based on potable water savings.

Chapter 3 describes the quantities and timing of stormwater and graywater availability in six locations in the United States based on 1995-1999 precipitation data for a medium-density residential scenario. The chapter also broadly discusses the potential for graywater and stormwater to address water supply needs by examining graywater and stormwater use scenarios for irrigation and/or toilet flushing. The information in Chapter 3, although not intended for site-specific planning, illuminates how local climate, storage capacity, and on-site applications all affect how on-site water resources can reduce potable water use. For cities located in the central and eastern United States, where the timing of rainfall is better matched to irrigation demand, both graywater reuse and roof runoff capture with moderate

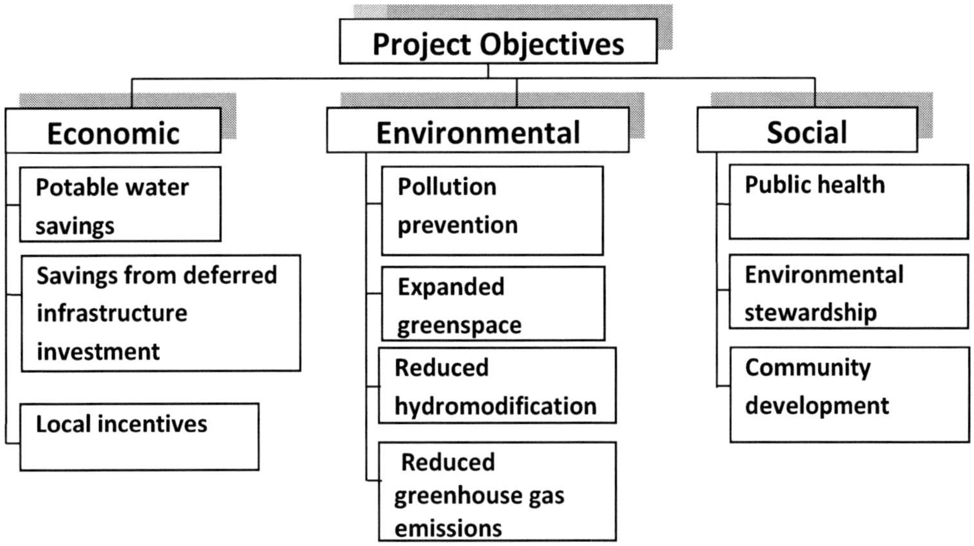

FIGURE 9-2 Example decision hierarchy, within which each objective can be scored according to the extent to which a project alternative addresses the objective.

BOX 9-1 Stakeholder Engagement in Decision Making: Example of Sonoma County

The Sonoma County Water Agency in California is undertaking scoping studies to identify stormwater management/groundwater recharge projects that would be located in the Sonoma Valley and Petaluma River watersheds. The process involves (1) articulation of the key project purpose; (2) screening of which project alternatives are not suitable for the project purpose; and (3) prioritization of the alternatives based on the ability to fulfill the objectives and the weight of the importance of the objective relative to other objectives. Based on the water agency's 2010 Water Supply Strategies Action Plan and the 2007 Groundwater Action Plan, the key project purpose is two-fold—to reduce flood hazards and to increase opportunities for groundwater recharge.

A screening process eliminates projects that do not meet the two-fold objective (e.g., levees and floodwalls do not address the groundwater recharge objective). The next steps involve community engagement to acquire a sense of the public's interest and preference as expressed by weighting objectives. This is achieved by county-wide meetings as well as local public workshops that reflect the interests of the region for which the project is intended. Supporting objectives were developed in consultation with stakeholders. Attendees were asked to prioritize elements of the two core project objectives and seven supporting objectives. Figure 9-1-1 shows the results for the Upper Petaluma Watershed, from which relative weights can be assigned and used to evaluate different project alternatives.

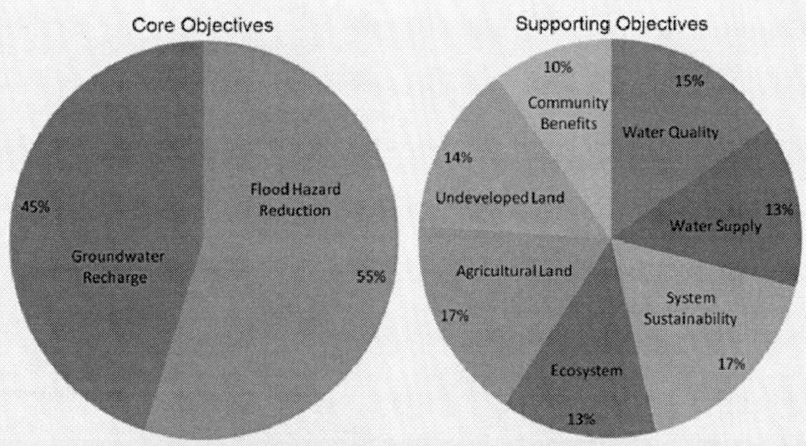

FIGURE 9-1-1 Example of stakeholder engagement to assess public interests and relative importance for project decision making. SOURCE: RMC (2011).

tank sizes can lead to substantial potential reductions in total water demand (up to 26 percent for whole-house graywater and 28 percent for stormwater for the medium-density residential scenarios analyzed; see Tables 5-1 and 5-5). In contrast, in the arid Southwest, where precipitation is limited and concentrated in winter months when irrigation demand is low, very large storage capacity is needed to significantly reduce potable water demand through stormwater capture. In these areas, large-scale groundwater recharge is an attractive water supply management alternative, if appropriate conditions for infiltration are available. In the arid Southwest, whole-house graywater can provide a substantial and consistent water source, although small relative to average outdoor irrigation demand. If reducing potable water demand is the primary objective, then conservation efforts to convert nonnative vegetation to xeriscaping should be encouraged, because reductions in outdoor water use provide the greatest opportunities for overall water savings, particularly in the arid Southwest.

Quality

An advantage of on-site graywater or stormwater use is the capacity to match treatment needs to the end use, with the potential for minimal or no treatment for some uses with little or no human exposures. Planning, therefore, requires an understanding of the quality of the local graywater or stormwater (Chapter 4) and the potential human exposures (Chapter 5) to calculate the potential risks associated with those exposures. These risks can then be used to assess the need for additional treatment. For graywater, multi-residential-scale systems have an averaging effect on graywater quality, and representative data are available on physical and chemical properties, although pathogen data are more limited. At the household scale, quality can vary widely based on whether best management practices for source control are implemented, although the additional risk of pathogenic illness from untreated household-scale graywater is lower considering the other potential pathways for disease spread within a household. For stormwater, a wide array of factors affect water quality (e.g., climate, rainfall intensity, land use, properties of surface materials), and there remains a significant shortage of information on human pathogens in stormwater (Chapter 4). For most nonpotable uses, human pathogens are the primary concern, although groundwater infiltration projects, particularly at a large scale, also necessitate a thorough characterization of organic and inorganic chemical constituents, including salts, to determine an appropriate system design (Chapter 4). When alternate water sources are considered for irrigation use, the salt content of the source water should be considered along with local soil conditions. Some applications may not be appropriate when salt content is high and local soil has a high clay content and/or elevated background salt content. In addition, opportunities for source control of pollutants can be considered.

Identify Candidate Strategies and Components

With a firm understanding of the project objectives (Chapter 1), opportunities for on-site use (Chapter 2), available water quantity (Chapter 3) and water quality (Chapter 4), potential human exposures (Chapter 5), and legal and regulatory constraints (Chapter 8), planners can appropriately narrow the suite of design and treatment options (see Chapter 6). For cities and water utilities, a key question to consider is project scale and whether to emphasize larger-scale projects (neighborhood or regional) or incentivize household- and building-scale projects. Household-scale projects are relatively easy to implement, could be partially subsidized by utilities, and would not require land purchases. However, household-scale, on-site, graywater or stormwater capture and use projects must be maintained by individual homeowners, and the use of these systems (and therefore the associated benefits) could be challenging to assess. For example, if stormwater tanks are not routinely used for irrigation or other nonpotable uses, then they provide neither the water supply nor pollution prevention benefits that are intended. Neighborhood-scale projects are typically paid for and managed by a water utility or a facility owner that can provide periodic maintenance and oversight, allowing the benefits to be documented if so desired. Neighborhood-scale systems also typically include more extensive treatment so that the water can be used safely for a wider range of beneficial uses. Neighborhood- and regional-scale stormwater and graywater projects may provide efficiencies of scale. Recent estimates from LADWP (2014) for stormwater capture suggest that neighborhood and regional stormwater infiltration systems can be comparable to other new water supply alternatives, while offering an array of additional benefits, such as pollution control and expanded greenspace in the urban environment (see Chapter 7). The availability of land and appropriate geology to support such projects should be determined in areas considering this option.

System Design Selection

Final design selection involves weighing how well the project objectives are achieved, overall costs (including capital and operations and maintenance costs minus any subsidies or incentives), and an assessment of financial, societal, and economic benefits. Projects at all scales should consider the acceptability to stakeholders.

For larger projects, where many stakeholders are involved, structured decision tools, such as multi-criteria decision analysis, can be used. These tools provide a structured and transparent mechanism by which various alternatives are evaluated to support final project design selection. The stakeholders and subject matter experts collaborate to create a decision hierarchy, as illustrated in Figure 9-2, that summarizes the key factors (or criteria) that affect the attractiveness of various options. This hierarchy can then be used to rate each alternative by establishing quantitative or semi-quantitative measures of each factor, and each factor is weighted by stakeholders based on its relative importance. The resulting computation blends the technical qualities of each option, as determined by the subject matter experts, and the relative importance of each factor, reflecting the values of the stakeholders (see Box 9-1).

The process is not complete when the relative value of each option is computed, for several reasons. First, not all stakeholders will have the same value set. This can be addressed by assigning different relative weights, thereby allowing the value of each option to be calculated and reflect the preference of individual stakeholders. Second, the results can lead to insights into which criteria are most important in distinguishing the relevant options. The most highly weighted criteria that are also the most different from a technical perspective will create the greatest difference in the assessment of alternatives. Both of these outcomes can be used to develop further options which, for example, can incorporate the most desirable elements of some of the original options as well as the key concerns of multiple stakeholders. Therefore, one can understand that it is not only the computation of scores but also the discussion that the process elicits that is important when allowing a diverse group of stakeholders to reach a decision that all can support.

Implementation

Implementation issues vary in complexity, depending on the project's scale. Project implementation issues include securing financing, working with regulators for necessary permits, developing mechanisms for system maintenance, and establishing appropriate system monitoring. Routine monitoring to assess treatment performance is a critical component of system implementation to minimize health risks. Depending on the extent of exposures, real-time monitoring may be appropriate to provide quality assurance, with an automatic shut-off when the water treatment system malfunctions (see Box 2-2). Monitoring on-site stormwater use may also be valuable, to ensure that the system is providing the intended benefits. Long-term maintenance and operations plans are also essential for effective operation.

San Francisco Public Utilities Commission has developed a guidebook for implementing alternative on-site water supplies that identifies relevant requirements for permitting, design and construction, inspection, maintenance, and monitoring (SFPUC, 2015). A few other localities have developed specific, consolidated, on-site, water reuse programs (see SFPUC, 2014, for a blueprint on developing such programs); in other locations, project permitting may require approval from numerous local agencies.

DECISION CONSIDERATIONS AT HOUSEHOLD AND NEIGHBORHOOD SCALES

The decision framework outlined in Figure 9-1 and described in the preceding section can be used at a range of scales to determine whether on-site graywater and/or stormwater capture and use is a sound alternative considering the risks, costs, and benefits, and if so, what designs are most appropriate to implement. In the following section, the committee uses this framework to examine decisions in a household scale example and two neighborhood-scale examples. This information is presented to help synthesize information described elsewhere in the report (Chapters 1-8) for specific decision-making contexts.

Household-scale Example

Define Objectives

Defining objectives is the first step of any on-site water supply decision process. Typical drivers behind graywater and stormwater use projects and the extent to which household-scale projects address these drivers are summarized in Tables 9-2 and 9-3. At a household scale, common objectives of on-site alternative water supply systems include reducing potable water demand, environmental stewardship, cost savings from reduced potable water and wastewater fees, pollution prevention (stormwater), and reliability of water supply during drought (graywater) (see Chapter 1). Some conflicts may exist among objectives, such as the objective to reduce stormwater pollution versus the objective to maximize water conservation and preserve supplies for downstream users. The relative importance of any one of these objectives over the others may determine the most applicable strategies. For example, if reliability of irrigation water during periods of extended droughts is a key objective, then only graywater systems offer a constant supply of water regardless of climatic conditions at a household scale.

Identify Opportunities and Constraints

The legal and regulatory framework should be understood because some states have specific permitting requirements or limit the potential uses of stormwater and graywater. Water rights permitting can be a key constraint to stormwater capture and use in western states (Chapter 7), but several states offer permitting exemptions for household-scale projects. For example, California, Utah, and Washington currently allow capture of rooftop runoff without a water rights permit for small projects (see Table B-1). In Arizona and Texas, onsite capture from rooftops, paved surfaces, and landscaped areas is exempt from water rights permitting. Although no specific laws address stormwater use in Idaho and New Mexico, state publications encourage the capture and use of rooftop runoff. The capture and use of roof runoff is generally not permitted in Colorado unless a water right permit is secured (see Table B-1 for exemptions).

State and local regulations may impact the potential end uses of graywater and stormwater and require specific best management practices or design standards (see Table B-3). Additionally, wastewater disposal regulations, environmental regulations, and state or local public health laws could impact implementation. Although Chapter 7 and Appendix B attempt to summarize the major legal and regulatory frameworks affecting the on-site beneficial use of graywater and stormwater, state and local laws are likely to continue to evolve as more people express interest in these practices and as courts continue to assess the legal implications. Therefore, interested homeowners should seek clarity from state or local government agencies on the latest legal and regulatory context for on-site graywater and stormwater use.

Along with constraints, opportunities for on-site use should be identified. At the household scale, the most common use of graywater and stormwater is landscape irrigation. Use for washing or toilet flushing may be feasible but will require more extensive treatment because of the potential human exposures.

Site Characterization—Water Availability and Quality

When considering stormwater or graywater capture and beneficial use at a household or building scale, it is critical to understand the amount of water available relative to the intended uses. At the household level, a water availability assessment for graywater is fairly straight-forward, given available water use data and the number of people living in the home. On average, 9.6 gpcd of graywater is provided from laundry water, with as low as 4 gpcd provided from high-efficiency washers with 14 gpcd or more from older, low-efficiency washers (Figure 3-4). On average, 26 gpcd graywater is available from all household water use, including bathroom faucets, showers, bathtubs, and laundry (see DeOreo et al., 2016). The use of water saving appliances and fixtures may reduce the amount of graywater available (see Figure 3-4). Use of graywater for toilet flushing requires dual plumbing and diligent maintenance. Therefore, most homeowners use graywater for irrigation and not toilet flushing. Chapter 3 outlines potential water savings for a medium-density, residential development in six cities considering conservation irrigation of turfgrass, but an individual's outdoor water use will vary with the local climate, type of vegetation, irrigation rates and frequency, and other behavioral factors. Homeowners should consider available graywater supply versus irrigation needs, recognizing that in arid climates, available graywater may only provide a small fraction of outdoor water demands for typical, non-native vegetation. However, graywater may be sufficient to provide a reliable supply of irrigation water for water-efficient landscaping.

Stormwater availability for beneficial use at the household level will vary widely based on local climate conditions, the source area available for stormwater capture (i.e., square footage of roof area), the storage volume, and the timing of rainfall relative to water demands. In the arid Southwest, the volume and timing of rainfall is poorly matched to the irrigation demand (see Figure 3-2), and extremely large storage tanks are needed to substantially reduce potable water use. In California, for example, beneficial use of large volumes of stormwater can be achieved by neighborhood- or regional-scale capture facilities and storage by groundwater recharge. In the committee's scenario analysis of potable water savings potential in Los Angeles (see Chapter 3), two rain barrels used for irrigation only reduced potable water use by 1 percent, and a moderate 2,200-gallon storage tank reduced potable water use by 4 percent. In contrast, in Lincoln, Nebraska, moderate-size storage tanks used only for outdoor irrigation resulted in a potential 21 percent reduction in overall water use, while two rain barrels resulted in 5 percent potential savings (see Tables 3-5 and 3-6).

Graywater and stormwater use present several water quality concerns at the household level. To minimize the risks of untreated graywater reuse at the household scale, residents should comply with best management practices and use subsurface (including landscape-covered drip) irrigation and only irrigate non-food crops. For stormwater use, roof runoff is primarily used because of its preferred water quality, but use of water from roofs with copper or galvanized steel materials should be avoided because of elevated metal content (see Chapter 4). Additionally, tree cover over roofs (as habitat for squirrels and birds) may result in high levels of indicator bacteria in the runoff, although the occurrence of human pathogens in roof runoff and other stormwater remains poorly understood (see Chapter 5). Human health risks can be reduced by minimizing uses that may cause inadvertent exposure (see Chapter 5).

Identify Candidate Strategies and System Components

With the objectives, constraints, and opportunities identified and the site characterized, candidate strategies can be considered in more detail. Graywater systems at the household scale include simple, low-cost, laundry-to-landscape systems to more complex, whole-house systems that require a storage tank and pump. If the system is used for applications with potential human exposures (e.g., spray irrigation, toilet flushing), then disinfection is also required. Whole-house systems that include treatment and disinfection require substantial maintenance that is usually beyond the skills of the typical homeowner (see Chapter 6).

For stormwater capture, household options include capturing roof runoff in cisterns or rain barrels or constructing rain gardens for shallow groundwater infiltration. Pitt et al. (2011) describes how to calculate benefits provided by various tank sizes for various roof areas and precipitation rates. Additionally, local or state regulations may influence system design requirements. In most cases, treatment is not necessary for irrigation, although human exposures should be minimized to reduce health risks. Disinfection may be desirable for spray irrigation at commercial buildings or other areas with substantial potential human contact (Chapter 5). Household roof runoff capture systems without treatment are relatively easy to implement and require minimal maintenance. If installed, then treatment systems would need to be relatively simple and not require significant expertise or attention (see Chapter 6).

Select System Design

Once identified, alternative designs and treatment options can be assessed for their capacity to deliver water supply benefits, water reliability, pollution control, and other project objectives relative to the cost. A full range of benefits, including social and environmental benefits (see Box 7-1), should be considered, although the data to assess these benefits may not always be available. For example, energy savings may also be possible for on-site graywater and stormwater systems, but the data are lacking to quantify life-cycle energy benefits (or costs) at the household scale.

Costs for household-scale, on-site, graywater and stormwater use can range widely depending on conveyance systems, tank size, whether treatment is included, and whether the system is self-installed or professionally installed (see Chapter 7). The committee calculated payback periods based on costs reported in the available literature and modeled potential water savings from the scenario analyses in Chapter 3. The payback periods vary depending on the uses and climate factors. For conservation irrigation use[1] only, calculated payback periods based on the scenario analysis and the many associated assumptions (see Chapter 3) range from 5 to 26 years for rain barrels, 14 years to more than 50 years for a self-installed 2,200-gallon (8,300 liter) tank, and 2.5 to 6.0 years for a laundry-to-landscape system, assuming water use is actually reduced by the amount of graywater or stormwater utilized (see Chapter 7). These payback periods address only equipment and do not include the value of homeowner labor or the costs of maintenance. Local costs and benefits are needed to inform decision making for on-site reuse, because cost and benefits can vary substantially by location.

Homeowners must weigh the various benefits against their own objectives and budgets. Consider, for example, a homeowner who lives in the arid Southwest and whose primary objectives are sustainability and water conservation. The largest water savings would be provided by approaches to reduce or eliminate potable water demand for irrigation, such as the use of xeriscaping and other types of climate-appropriate, low-water-use landscapes (Mayer et al., 2015). Graywater irrigation through a simple laundry-to-landscape system could help maintain those landscapes, particularly during extended droughts, and reduce the costs associated with irrigation. In arid climates, simple laundry-to-landscape graywater systems have much shorter payback periods than do rain barrels or cisterns. If that homeowner lives in a city that already reuses wastewater through a centralized water reclamation facility, then graywater reuse would not change regional water savings, although simple laundry-to-landscape systems could reduce total energy use (definitive data are not available). Based on the scenario analyses, in Lincoln, Nebraska, rain barrels and laundry-to-landscape graywater systems are fairly comparable in terms of potential payback periods, although the volume of water conserved is smaller than other system designs. If sustainability and water pollution control are important objectives, then moderate-sized cisterns can provide substantial water savings (although with longer payback periods) and can reduce the adverse environmental effects of stormwater runoff. There is no single "best" configuration to maximize on-site water supply at the household scale, because of site-specific factors and individual differences in overall project objectives.

[1] The committee recognizes that many residents irrigate at rates far above that required to meet the evapotranspiration deficit; thus, potential potable water savings could be higher than reported here. However, behavioral factors that would lower actual water savings were not considered.

Implementation

At the household scale, installation can be performed by skilled do-it-yourselfers or professional installers. Maintenance needs should be well understood, because even the simplest rain barrel systems require periodic maintenance to remove sediment and ensure proper functioning. Owners should be aware of how water use impacts the desired project benefits. For example, stormwater capture systems provide minimal pollution prevention benefits if the tanks are not regularly emptied so that they are available to capture runoff from the next storm. Likewise, homeowners who increase the extent of landscaping to take advantage of newly available graywater supplies can ultimately increase potable water use even with the installation of graywater systems.

Neighborhood-scale Examples

For the neighborhood scale, two examples are considered in the context of the decision framework presented in Figure 9-1—a multi-residential building development and an office-park development. These two examples provide different opportunities and considerations.

Define Objectives

The objectives of a neighborhood-scale project (either a multi-residential development or a business park with many distributed buildings) might include cost savings through reduced potable water and wastewater fees, financial incentives related to stormwater management or extending the capacity of existing water and wastewater infrastructure, enhanced water reliability, environmental stewardship, public education, pollution prevention, and projecting a "green" image that could attract residents or businesses (see Chapter 1 and Tables 9-2 and 9-3). The relative priority of these objectives may influence the on-site water supply strategy selected.

Identify Opportunities and Constraints

Stormwater capture for beneficial use at the neighborhood scale can be constrained by water rights laws and local and state regulations. As discussed in the household-scale example, several states (e.g., California, Utah, Washington) exempt small rooftop capture systems, but a large multi-residential development or business park would likely exceed the capacity limits of these exemptions. Arizona and Texas water law appears to allow for stormwater capture before it has entered a natural water course (see Chapter 8 for more details). Coastal cities with no downstream users (e.g., San Francisco, Los Angeles) may be exempt from water rights permitting requirements. In prior appropriation states without exemptions for stormwater capture, a water rights permit must be acquired. Local and state regulations may also constrain potential applications for captured stormwater. Some states and localities do not permit stormwater use for toilet flushing (or for any use other than irrigation) (see Chapter 8).

Graywater regulations vary significantly from state to state. Five states (Arizona, California, New Mexico, Oregon, and Washington) have tiered regulatory frameworks that prescribe increased requirements for large facilities (see Table 8-3). States without a tiered framework may require additional consultation so that state and local agencies are comfortable that the project is adequately protective of public health. Several states (e.g., Idaho, Nevada, Ohio, Utah) only allow graywater to be used for irrigation, and because multi-residential units (particularly high-rise buildings) are likely to generate much more graywater than needed for landscape irrigation, such restrictions would limit the usefulness of these projects. However, state and local laws on graywater and stormwater use are evolving quickly, so interested developers should consult with local and state government agencies to understand the implications of the current legal and regulatory framework.

In a multi-residential building or a business park development, possible applications include landscape irrigation, shallow groundwater recharge, toilet flushing, ornamental water features, heating, ventilation, and air conditions (HVAC) cooling water, and washing (see Chapter 2).

Site Characterization—Water Availability and Quality

Stormwater can be captured from rooftops, driveways, and parking areas, although on-site stormwater capture is typically limited to rooftop runoff, because it tends to have the highest quality (see Chapter 4). Potential on-site water supply benefits can be calculated based on the stormwater capture area, local climate conditions, tank size or infiltration basin design, and the timing of water demands (see Chapter 3 for regional specifics or Pitt et al., 2011 for water availability calculations). High-rise buildings, which have a small area of capture, provide limited water supply relative to the overall on-site water use, but office parks may have substantial roof area. Stormwater availability relative to tank size will be greatest where rainfall timing is well-matched to water demand for the intended applications. Among the cities analyzed, Lincoln, Madison, and Newark showed the best match between stormwater availability and irrigation demand. When considering the capacity of stormwater to address a continuous demand, such as toilet flushing, Madison, Newark, Lincoln, and Birmingham provided the largest reductions in potable water use because of the substantial, near-year-round precipitation (see Table 3-6 and Figure 3-2). In the arid Southwest the highly seasonal rainfall that occurs when irrigation is typically not needed makes stormwater capture for beneficial use more challenging. Careful consideration of roofing materials is advised to minimize metal contamination (Box 4-1).

If a new multi-residential building is constructed with dual plumbing to capture all graywater from bathroom faucets, showers, bathtubs, and laundry, then approximately 26 gpcd of graywater (or 45 percent of indoor water used) could be available for reuse for indoor or outdoor nonpotable uses. In cases of multi-residential buildings, irrigation demand is often small relative to the amount of graywater generated. In such cases, graywater use for toilet flushing may be a more viable option to achieve reduced demand for potable water. For a large multi-residential building, because source control becomes more difficult to manage and potential human exposures increase, graywater treatment is needed for most applications. For a business park, a separate analysis of on-site sources of graywater is advised prior to further planning because graywater represents a fairly small percentage of total wastewater generated in most businesses and institutional buildings (see Chapter 3) unless laundry or showers represent a significant part of average water use.

Identify Candidate Strategies and System Components

In a business park or institutional setting with large rooftop collection areas, stormwater could be captured, stored in large tanks, and treated for various on-site uses, or used to recharge groundwater (see Chapter 6). The design alternatives would need to be developed considering overall objectives, water availability, potential nonpotable uses, and water demand. For example, a system designed to capture all runoff from a 1-inch storm may be quite different from a system designed to optimize potable water savings. In the arid Southwest, where rainfall is concentrated during periods of low irrigation demand, very large stormwater storage tanks are typically needed to significantly reduce potable water use on an annual basis. Under appropriate hydrogeologic conditions, on-site or neighborhood-scale groundwater infiltration can instead be used to enhance regional water supply while reducing stormwater pollution. Infiltration projects tend to have significantly lower costs compared to large-scale stormwater capture and use projects, although the benefits of such projects are distributed regionally rather than to the building developer. Source areas for infiltration projects should be selected to minimize contamination (see Chapter 4), although infiltration basins can be designed to provide additional water quality treatment during infiltration (Chapter 6). If stormwater is captured and used on-site, then the level of service provided by the on-site, non-potable water system should be considered in the system design to optimize its use (relative to potable water). For example, at a fire station where stormwater is used for washing and tank filling, the use of nonpotable water could be encouraged by adjusting the flow rate and pressure to be greater than the potable water system.

New multi-residential buildings may be good candidates for dual-plumbed facilities that use treated graywater and/or stormwater for toilet flushing and other possible uses, such as laundry or HVAC. A building-wide treatment system can be managed and routinely maintained by trained operators, minimizing overall risk. Chapter 6 discusses the treatment necessary for specific applications of graywater (Figure 6-5) and stormwater (Figure 6-8), and an array of technologies are available at the multi-residential building scale to address the treatment objectives (Table 6-3) for potential end uses and exposures (see Chapter 5). However, few localities have specified treatment guidelines or requirements objectives, so developers may have to work closely with local public health agencies to develop treatment strategies that are protective of public health until such guidance is developed. Both graywater and stormwater can be used to meet nonpotable water demands, but combined systems typically involve separate treatment of the two sources before they are combined into a single collection (see Chapter 6). In some new developments, wastewater from toilets is also captured on site to maximize energy recovery (see Box 2-2). Different system designs can be developed to maximize various project objectives.

Select System Design

The potential benefits of on-site graywater and/or stormwater capture and use (including potable water savings, averted wastewater fees, other incentives, pollution prevention, energy recovery, environmental stewardship, public education, and improved public image) can be compared to the costs of various design alternatives and to a conventional water and wastewater system in the context of the overall project objectives. A full range of benefits, including social and environmental benefits (see Box 7-1), should be considered, although not all benefits have been thoroughly quantified. The financial costs and benefits of projects at this scale are site-specific and can be calculated by design engineers for the purpose of comparison among alternatives. General cost and benefit information for comparable projects would be helpful to inform decision making, but such information is currently not readily available.

Implementation

Project implementation will necessitate close coordination with several local agencies (e.g., water, public health, building) for appropriate permitting of an on-site water capture and use project. SFPUC (2014, 2015) outlined a streamlined permitting process for on-site use of alternative water supplies in the San Francisco region and produced a "blueprint" for other cities that want to develop an on-site water program to encourage large-scale implementation. System maintenance and monitoring to assess treatment performance is essential to minimize human health risks. Additionally, residents should be informed about source control strategies to help maintain good system operation and minimize public health risk (see Chapter 2).

CONCLUSIONS

There is no single best way to use graywater or stormwater to address local water needs, because project drivers and objectives, legal and regulatory constraints, potential applications, site conditions, source water availability, and project budgets, all vary widely. This chapter lays out a decision framework that can be used when considering the use of graywater or stormwater to meet various objectives and summarizes information from the report relevant to key decision steps at both the household and neighborhood scales.

Information is generally available to support water management decision making for on-site, nonpotable applications for simple, household-scale, graywater and/stormwater systems with minimal human exposures. However, additional research would enhance decision making for larger systems or those with treatment requirements. Adequate information is available (or could be obtained) on graywater and stormwater availability for small-scale systems such as basic water quality parameters, system design and treatment technology effectiveness, and the existing regulatory framework. This information can be used to assess the capacity for on-site or local alternative water supplies to meet water demands while providing other benefits. However, as projects grow in size and scope, detailed analysis is required to explore options, assess siting issues, and address concerns about water quality and availability. Therefore, key uncertainties affect the capacity to make fully informed decisions on appropriate and cost-effective designs for larger or more complex graywater or stormwater beneficial use systems. These uncertainties, which could be reduced by additional research, include:

- Water quality objectives for various uses that are protective of public health;
- The occurrence and fate of pathogens in stormwater and graywater;
- Costs and benefits for neighborhood- and regional-scale systems, including nonmonetized benefits, such as water pollution control and community amenities;
- Energy implications of on-site alternative water supplies; and
- Long-term system performance and maintenance needs.

Lack of clarity on water rights and legal and regulatory inconsistencies are also impediments to water management decision making in some states. More discussion on research needs is provided in Chapter 10.

Stakeholder engagement is crucial to the evaluation, selection, and implementation of any urban water management system and is particularly important when new options for distribution throughout the urban area, such as stormwater and graywater reuse, are being considered. The first step is understanding that stakeholders are those groups and individuals that can affect selection and implementation of the relevant system. Fortunately, effective and proven approaches exist to identify and engage appropriate stakeholders in the process, leading to the selection of implementable solutions. Experience shows that co-benefits, such as expanded greenspace or environmental stewardship, are often important to gaining stakeholder support. As noted in the discussion, experience shows that stakeholder engagement founded on trust and shared responsibility can be effective in planning and implementing projects, while stakeholder groups comprised of intractable adversaries are likely to have just the opposite effect.

10

Priorities for Research

There is substantial potential for graywater and stormwater to contribute to local water supply needs while providing other benefits such as stormwater pollution reduction, water supply diversification, and increased local control of water supplies. Graywater and stormwater use could be an important part of a broader effort to reimagine urban water infrastructure to efficiently use water, energy, and financial resources while enhancing water supply reliability, resiliency, and the livability of cities. However, as discussed in Chapter 9, major gaps exist in our understanding that make decision making more difficult, particularly with regard to neighborhood- and regional-scale stormwater and graywater capture initiatives.

This chapter highlights the major research needs identified by the committee that should be addressed to better support decision making. Additionally, this chapter presents research needs that look forward to ways to improve the water and energy efficiency of our nation's water infrastructure and maximize financial, environmental, and social benefits. These research needs, if addressed, have the potential to advance the use of graywater and stormwater to expand local water supplies and ensure its safe and reliable use. Twelve research needs are categorized according to five themes outlined in Box 10-1:

1. Risk and water quality
2. Treatment technology
3. Infrastructure
4. Social science and decision analysis
5. Policy and regulatory

RISK AND WATER QUALITY

Although no documented reports of adverse human health effects from the use of stormwater or graywater have been identified, additional examination of risk from microbial and chemical contaminants is necessary to support safe and appropriate design and implementation of stormwater and graywater use systems—particularly for large-scale systems. Such efforts can also facilitate adoption of water-saving practices in areas lacking a regulatory framework.

1. Assess the occurrence and fate of human pathogens in graywater and stormwater

Pathogens represent the most significant risks in nonpotable graywater and stormwater applications, considering the potential for adverse health effects from a single exposure to a small volume of water. Currently, most of the information on microorganisms in graywater and stormwater is limited to the occurrence of indicator microorganisms (e.g., total and fecal coliform bacteria, enterococci) rather than human pathogenic organisms. However, it is well documented that no consistent quantitative relationship exists between the occurrence or concentrations of indicator organisms and pathogenic organisms. This is particularly true for roof runoff, in which the sources of indicator bacteria may be primarily derived from animal waste that may or may not contain human pathogens. Depending on the project scale and contributing source areas, human waste from leaking sewers or faulty septic systems may contribute to the pathogen loads of stormwater. Therefore, additional work is needed to characterize the occurrence of pathogens in stormwater from various source areas (e.g., rooftops [tree covered and unshaded], open space, mixed use) and scales to inform guidance on appropriate treatment or exposure control to protect human health.

In graywater, for which the source of pathogens is typically human waste from laundry or showers, the usefulness of indicator data to predict risk will vary based on the scale of the project and other vehicles of disease spread (graywater would represent one of many potential vehicles of infectious disease within a single household). Pathogen data from graywater are extremely limited and insufficient for comprehensive risk analysis. Research is needed to assess the variabil-

BOX 10-1 Summary of Research Needs to Enhance the Safe and Reliable Use of Graywater and Stormwater and Conserve Water, Energy, Environmental, and Financial Resources

Risk and Water Quality

1. Assess the occurrence and fate of pathogens in graywater and stormwater
2. Assess the occurrence and fate of chemical contaminants in stormwater
3. Understand the implications of enhanced water conservation on graywater quality and use
4. Develop risk-based water quality guidance for various uses that could serve as a basis to develop standards of practice
5. Develop monitoring technology and strategies to assure compliance with water quality criteria

Treatment Technology

6. Develop treatment systems to meet tailored (fit-for-purpose) water quality objectives across a range of scales
7. Understand the long-term performance and reliability of graywater and stormwater treatment systems (from small to large scales)

Infrastructure

8. Envision opportunities for water- and energy-conserving infrastructure designs in new construction and demonstrate their performance
9. Identify strategies to retrofit existing infrastructure for enhanced beneficial use of stormwater

Social Science and Decision Analysis

10. Understand behavioral impacts on overall water use in the context of graywater and stormwater projects
11. Collect performance data (including cost, energy, water savings, water quality, and other benefits) in support of integrated water supply management, decision making, and refinement of decision tools

Policy and Regulatory Issues

12. Identify incentives and various regulatory strategies that have proven effective in the implementation of stormwater or graywater systems to conserve water supplies

ity in pathogen concentrations in graywater under different scales, storage conditions, and with and without source control practices, with the goal of bracketing likely and possible pathogen concentrations for the basis of broad risk calculations. These data are needed to identify appropriate practices to limit exposure when untreated graywater is used and to devise effective treatment systems where appropriate. Additionally, the implications of scale to the calculation of risk need to be examined.

Because hundreds of potential pathogens could be present in graywater or stormwater, it is necessary to choose representative pathogens for which the water is to be analyzed. The choice of pathogens could be based on a "worst-case" approach, or other approach based on the specific local situation.

2. Assess the occurrence and fate of chemical contaminants in stormwater

Stormwater quality is highly variable over space and time and is a direct function of land use, catchment size, and climatic and seasonal factors. Thus, stormwater water quality is difficult to predict. Organic contaminants, pathogens, and salts pose a particular concern for groundwater recharge. Some organic contaminants in urban stormwater are not common to municipal wastewater, and therefore their occurrence and fate remain poorly understood. Future research should investigate the occurrence and persistence of hazardous organic contaminants in urban stormwater. This research should identify those contaminants, such as urban pesticides and additives used in automotive and commercial applications, that are likely to be present in urban stormwater and may pose human health risks when stormwater is used for groundwater recharge. Information on the occurrence and fate of these compounds in urban settings can inform the development of appropriate source control and/or treatment strategies. In the case of large-scale projects, field studies and demonstration projects should proceed in tandem with planning for greater beneficial use of stormwater for water supply.

3. Understand the implications of enhanced water conservation on graywater quality and use

Indoor water conservation campaigns have had the most impact on use of water for flushing toilets and laundry

(Chapter 3). As discussed in Chapter 3, decreased water use, particularly in laundry machines, has the potential to impact the quality of graywater, making it more concentrated. This has the most impact on laundry-to-landscape systems, which are the most practical graywater systems for implementation in existing development. Research to date on impacts of graywater irrigation on soil quality has focused on graywater systems that include water from multiple sources (bathroom and laundry water; Sharvelle et al., 2012). New improvements in laundry machines continue to be made, reducing their water use. Such advances may render laundry-to-landscape programs obsolete. If homeowners are willing to adopt ultra-low water use washing machines, then this may be a more efficient way to reduce water demand than using laundry water for irrigation. The impact of high-efficiency laundry machines use for graywater irrigation needs to be better understood and evaluated in terms of water availability, water quality, and subsequent impact to soil quality.

4. Develop risk-based water quality guidance for various uses that could serve as a basis to develop standards of practice

Risk-based water quality guidance should be developed for various stormwater and graywater uses based on an understanding of anticipated human exposures from nonpotable uses of graywater and stormwater and possible contaminant concentrations (to be informed by research needs 1 and 2). Additional research is needed to better understand the impacts on risk and reliability associated with organic matter and turbidity levels recommended by the National Sanitation Foundation (NSF) International NSF 350 standard for graywater and stormwater for toilet flushing at a range of scales, because these factors significantly affect treatment costs. This information is needed to help localities that are struggling with the lack of existing standards or guidance and project developers (e.g., multi-residential buildings using graywater or stormwater for toilet flushing) who want to know what level of treatment is necessary and appropriate to protect public health. Such guidance, if rigorously developed based on comprehensive water quality and exposure data, would also reassure stakeholders that stormwater and graywater nonpotable use projects meet common acceptable-risk standards. Once risk-based water quality guidance is developed, standards of practice can be developed to meet these objectives, which can allow for more focused and cost-effective technology development and production, ultimately reducing costs for end users. Risk-based guidance also applies to large-scale groundwater recharge projects, although such guidance could vary with hydrogeologic conditions.

5. Develop monitoring technologies and strategies to assure compliance with regulatory criteria

The beneficial use of stormwater and graywater is achieved through distributed systems of varying scales. Because stormwater capture and recharge systems may be distributed over many locations, advanced monitoring and online control technologies could help reduce the costs associated with assessing system function and use. Research is needed to develop approaches for real-time monitoring of graywater and stormwater treatment systems for projects that have substantial human exposure. Technology advancement on rapid pathogen detection would be beneficial for large-scale graywater reuse as well as other sectors of water management and reuse, although surrogate detection systems may be feasible.

Stormwater capture and use systems are often installed as part of a broader effort to control stormwater discharges and combined sewer overflows. Advanced distributed monitoring systems could also assess the timing of water use relative to the demand for stormwater capture to assess the benefits provided and to identify strategies to optimize multiple benefits.

TREATMENT TECHNOLOGY

As graywater and stormwater use expands in scale and for uses other than subsurface or restricted access irrigation, treatment will likely be necessary to protect public health (Chapters 5 and 6). Additionally, managed recharge of stormwater may require treatment to prevent groundwater contamination (Chapter 5). Although treatment technology is relatively advanced, some focused research and development could improve the cost-effectiveness and reliability of graywater and stormwater use.

6. Develop treatment systems to meet tailored (fit-for-purpose) water quality across a range of scales

There is a need to assess and design treatment systems for various beneficial uses of graywater and stormwater. In the absence of fit-for-purpose water quality guidance, common standards of practice for graywater and stormwater treatment for various applications have yet to be developed. A wide array of technologies is applied, with varying levels of treatment. For managed recharge of stormwater, existing unit processes and the sequence in which they are applied must be tailored to achieve efficient treatment in terms of cost, energy, and maintenance. Until a standard of practice for stormwater treatment for water supply augmentation is developed, systems will continue to be developed as one-off designs that remain costly and may lack sufficient proof-of-concept for wide-scale adoption.

The current lack of a standard of practice results in a time-consuming design phase (particularly for neighborhood- or regional-scale systems), which is often followed by iterations of design modifications once the system is in operation. Achieving a standard of practice that includes treatment process trains that are appropriate for stormwater or graywater use projects requires extensive application of technologies through demonstration projects. Rigorous monitoring of pilot and demonstration projects with extensive data collection is needed to adequately synthesize findings and to develop a standard of practice. Treatment system development, where feasible, should examine the capacity of natural, passive, or low-energy processes (e.g., wetlands, soils, engineered media, aeration) to reliably meet contaminant removal objectives while reducing energy use.

7. Understand the long-term performance and reliability of graywater and stormwater systems (from small to large scales)

Data on long-term performance of graywater and stormwater systems are lacking. Many systems do not include water quality monitoring, and therefore performance of systems remains unreported and unknown. Concerns regarding long-term performance and reliability of graywater systems continue to be a limitation for acceptance of such systems. Public health departments remain wary of treatment system performance and the potential health risks that may result from lack of maintenance or system failures. An extensive effort to characterize long-term performance and reliability is needed, which can include monitoring and data collection from systems that are currently operational as well as installation and monitoring of new pilot or demonstration projects, especially at neighborhood or regional scales. Development of an online database such as the International Stormwater BMP Database[1] is recommended to serve as a portal for collecting and sharing information on the costs and performance of graywater and stormwater capture and beneficial use systems (see research need 11). Research on long-term performance should also assess the human behavioral dimensions of operation and maintenance of graywater and stormwater systems at various scales to assess long-term risks and develop strategies for better training or oversight, if needed.

With respect to the groundwater recharge of stormwater, there is a need to better understand water quality improvements during storage and infiltration, including long-term performance of natural treatment processes. Research is needed to assess the level of water quality improvement that can be expected in urban stormwater during infiltration and storage (including pathogen and organic contaminant removal) to determine additional treatment or system maintenance needs. Regulatory flexibility may be needed to allow demonstration tests at reasonable scales in the field. Demonstration tests will inform risk management and operational procedures for improved resilience and reliability.

[1] See http://www.bmpdatabase.org.

INFRASTRUCTURE

Today's urban water infrastructure is not designed for large-scale graywater or stormwater use. In many places infrastructure does not exist for urban stormwater or graywater capture, requiring new visions of water infrastructure to reach potential water and energy saving efficiencies. These issues suggest the following research needs with respect to infrastructure:

8. Envision new opportunities for water- and energy-conserving infrastructure and demonstrate their performance

Conventional design of stormwater and wastewater systems was developed many years ago when the beneficial uses were not envisioned, and retrofitting existing urban water infrastructure to accommodate beneficial uses of on-site water sources is expensive. However, new construction at building or neighborhood scales could include these features at a relatively small incremental cost. New urban water management models that incorporate graywater and stormwater use offer the potential to be much more water- and resource-efficient than traditional models. Research is needed to continue to develop new visions for urban water infrastructure that conserve water, reduce energy use (and generate energy where feasible), and reduce waste by recycling nutrients and other valuable resources. Important questions exist relative to key system components, including resource recovery and treatment technologies. Yet, this highly significant transformation can be achieved only if this potential is demonstrated. These systems could be incorporated into new urban development, and practical learning from the demonstration of these models could subsequently be incorporated into urban redevelopment.

9. Identify strategies to retrofit existing infrastructure for enhanced beneficial use of on-site water sources

Research is needed on cost-effective strategies for retrofitting these drainage and delivery components in existing developed areas to meet additional objectives, including improving urban hydrology, reducing pollutant discharges, and enhancing water availability. Methods should be evaluated for retrofitting existing stormwater systems for enhanced stormwater capture, such as exploring options for conversion of stormwater detention ponds to enable stormwater capture. Given investments in existing systems and the future build-out of water reuse facilities in some areas, the benefits and feasibility of joint reclaimed wastewater and stormwater infiltration systems should be assessed. Additionally, the implications of extensive graywater use on existing wastewater conveyance infrastructure need to be better understood and the costs of managing such impacts evaluated.

SOCIAL SCIENCE AND DECISION ANALYSIS

Understanding the human dimensions of graywater and stormwater use will inform better project design and implementation. In addition, compiling existing performance data in a centralized database could improve support for decision making. Key priorities for research include

10. Understand behavioral impacts on overall water use in the context of graywater and stormwater projects

A major unresolved issue is the extent to which behavioral factors affect the benefits provided by graywater and stormwater projects. As noted in Chapter 3, two pilot projects on laundry-to-landscape graywater systems showed no reduction in potable water use, on average, and one of these studies showed increased potable water use. Similar studies on Australian household-scale stormwater capture projects have shown that actual water savings are only 0.3 to 0.7 times the theoretical value (see Box 3-2). It is well known that many rain barrels and cisterns are installed but not used sufficiently, compromising both water savings and pollution prevention benefits. The availability of a new low- or zero-cost water supply could cause households to plant more water-intensive landscaping or simply maintain existing landscaping more intensively. Similarly, knowledge that laundry water is being reused for irrigation may lead residents to do more laundry because the net cost of each load has been reduced. It is also possible that the positive emotional feeling (or "warm glow") from making investments in green infrastructure may impact potable water consumption elsewhere, either positively or negatively. Therefore, research is needed to understand how such systems affect water use behavior.

Research to assess user knowledge and experience with on-site graywater and stormwater use systems is also important to assess the extent to which best management practices are understood, systems are properly installed and maintained, and appropriate source control practices are used. Additionally, research to understand homeowner-scale applications of stormwater and graywater would be useful to assess probable contaminant exposures as opposed to those anticipated if only best practices are followed. Such research would provide an improved understanding of household-scale graywater and stormwater risks and benefits and could be used to identify opportunities for public outreach and education to maximize potential benefits.

11. Collect performance data (including cost, energy, water savings, and water quality) in support of integrated water supply management, decision making, and refinement of decision tools

Because of the absence of ample documentation of costs, performance, risks, and co-benefits (see Chapter 7), many utilities are hesitant to integrate graywater or stormwater capture and use into their long-term water resource plans. Alternatively, some individuals and organizations may be over optimistic about the comparative benefits and costs of such projects for their communities. The U.S. Environmental Protection Agency (EPA) has a National Menu of Stormwater Best Management Practices, but this menu contains limited case studies and performance data on stormwater capture projects. Additional data are needed to quantify the multiple benefits and costs of small- to large-scale stormwater and graywater projects.

Improved financial cost data will reduce uncertainty for cost and benefits analyses and facilitate comparisons among alternatives. Well-documented case studies can better clarify capital costs and maintenance requirements, including documented energy costs and savings. Costs are easier to quantify than benefits, but benefits are equally important because graywater and stormwater use can provide multiple benefits beyond water savings or supply, including many that are not easily monetized. Systematic approaches to accounting for the full range of benefits (including multi-sectorial benefits and costs) should be developed to facilitate synthesis of project-specific information. In addition, water quality performance data, potential water supply capture benefits, and the broad array of potential co-benefits should be collected in a database of graywater and stormwater projects. In many cases, the benefits information may be missing or quite limited, in which case a systematic effort to develop such information will be of considerable value. Many innovative types of small-scale stormwater capture projects have minimal documentation of performance metrics.

To advance the state of the art, this information should be synthesized, including both smaller- and larger-scale graywater and stormwater projects, so that utilities, cities, building developers, and residents can understand the state of the practice and the relative costs and benefits (e.g., water supply, energy savings, water quality, aesthetics). Such information could also be incorporated into decision tools to improve the statistical basis of the many variables considered for on-site water systems.

POLICY AND REGULATORY ISSUES

Beneficial use of graywater and stormwater is a relatively new and growing practice in many regions of the United States, and, as a result, legal and regulatory policies have not evolved as quickly as the practices. Consequently, communities would benefit from additional research on effective practices in regulatory programs and policies.

12. Examine how incentives and various regulatory strategies have proven effective in the implementation of stormwater or graywater systems to conserve water supplies

Implementation of stormwater capture and graywater strategies may be constrained by institutional "silos" that include multi-jurisdictional inefficiencies that may hinder the maximal use of these two potential supply sources. In addition, regulatory challenges to optimizing the capture and storage of stormwater for water supply purposes can be a significant hurdle for implementing an integrated stormwater capture and recharge strategy. Local governmental entities that have responsibility for stormwater or wastewater management vary significantly throughout the United States, so the institutional structure and constraints will likely vary from state to state and also within states.

These issues point to the need to examine various regulatory strategies for effective practices that can enhance broader implementation of stormwater and graywater systems to conserve or enhance potable water supplies. Research should assess regulatory innovations to increase on-site water use including market incentives, pollution credit markets, and integrated watershed management planning that optimizes local sustainable water supplies. Research should include an assessment of regulatory barriers, including those that may arise if conveyances and delivery systems cross property lines (or water management boundaries). This research should also assess the legal challenges and water rights issues for coastal and inland states and identify specific low barriers that can be overcome.

References

Adamowicz, W.L., P. Boxall, J. Louviere, J. Swait, and M. Williams. 1998. Stated preference Methods for valuing environmental amenities. Pp. 460-479 in Valuing Environmental Preferences: Theory and Practice of the Contingent Valuation Method in the US, EC and Developing Countries, I. Bateman and K. Willis, eds. London, UK: Oxford University Press.

Ahmed, W., and S. Toze. 2015. Microbiological quality and associated health risks with the use of roof-captured rainwater. Pp. 229-252 in Rainwater Tank Systems for Urban Water Supply: Design, Yield, Energy, Health Risks, Economics and Social Perceptions, A. K. Sharma, D. Begbie, and T. Gardner, eds. London: IWA Publishing.

Ahmed, W., F. Huygens, A. Goonetilleke, and T. Gardner. 2008. Real-time PCR detection of pathogenic microorganisms in roof-harvested rainwater in Southeast Queensland, Australia. Applied and Environmental Microbiology 74(17):5490-5496.

Ahmed, W., A. Vieritz, A. Goonetilleke, and T. Gardner. 2010. Health risk from the use of roof-harvested rainwater in Southeast Queensland, Australia, as potable or non-potable water, determined using quantitative microbial risk assessment. Applied and Environmental Microbiology 76(22):7382-7391.

Ahmed, W., L. Hodgers, J.P.S. Sidhu, and S. Toze. 2012. Fecal indicators and zoonotic pathogens in household drinking water taps fed from rainwater tanks in Southeast Queensland, Australia. Applied and Environmental Microbiology 78(1):219–226.

Ahmed, W., H. Brandes, P. Gyawali, J.P.S. Sidhu, and S. Toze. 2014. Opportunistic pathogens in roof-captured rainwater samples, determined using quantitative PCR. Water Research 53:361-369.

Alan Plummer Associates, Inc. 2010. Stormwater Harvesting Guidance Document for the Texas Water Development Board. Fort Worth, TX: Alan Plummer Associates, Inc.

Albrechtsen, H.J. 2002. Microbiological investigations of rainwater and graywater collected for toilet flushing. Water Science and Technology 46(6–7):311-316.

Alcubilla, R.G., and J.R. Lund. 2006. Derived willingness-to-pay for household water use with price and probabilistic supply. Journal of Water Resources Planning and Management 132(6):424-433.

Allen, L., and C. Woelfle-Erskine. 2011. From Landry to Landscape: Tap into Greywater. Mother Earth News, August/September 2011. Available at http://www.motherearthnews.com/green-homes/home-design/greywater-zm0z11zphe.aspx.

Allen, L., J. Christian-Smith, and M. Palaniappan. 2010. Overview of Greywater Reuse: The Potential of Greywater Systems to Aid Sustainable Water Management. Oakland, CA: Pacific Institute.

Allen, R.G., I.A. Walter, R.L. Elliot, T.A. Howell, D. Itenfisu, M.E. Jensen, and R. Snyder. 2005. The ASCE Standardized Reference Evapotranspiration Equation. ASCE and American Society of Civil Engineers. Available at http://www.kimberly.uidaho.edu/water/asceewri/ascestzdetmain2005.pdf.

Appleyard, S.J. 1993. Impact of stormwater infiltration basins on groundwater quality, Perth metropolitan region, Western Australia. Environmental Geology 21:227-236.

Aronson, L.J., A.J. Gold, R.J. Hull, and J.L. Cisar. 1987. Evapotranspiration of cool season turfgrasses in the humid Northeast. Agronomy Journal 79:901-905.

ASCE (American Society of Civil Engineers). 2009. Report Card for America's Infrastructure. Reston, VA: ASCE.

Ashbolt, N.J., W.O.K. Grabow, and M. Snozzi. 2001. Indicators of microbial water quality. Chapter 13 in Water Quality: Guidelines, Standards, and Health, World Health Organization. London: IWA Publishing.

Auerbach, E.A., E.E. Seyfried, and K.D. McMahon. 2007. Tetracycline resistance genes in activated sludge wastewater treatment plants. Water Research 41:1143-1151.

Australian and New Zealand Environment and Conservation Council and the Agriculture and Resource Management Council of Australia and New Zealand. 2000. Australian and New Zealand Guidelines for Fresh and Marine Water Quality, Volume 1. Commonwealth of Australia, Canberra. Available at https://www.environment.gov.au/system/files/resources/53cda9ea-7ec2-49d4-af29-d1dde09e96ef/files/nwqms-guidelines-4-vol1.pdf.

Australian SCEW (Standing Council on Environment and Water). 2009. Australian Guidelines for Water Recycling: Stormwater Harvesting and Reuse, National Water Quality Management Strategy Document No 23. Natural Resource Management Ministerial Council Environment Protection and Heritage Council, and National Health and Medical Research Council. July. Available at http://www.scew.gov.au/system/files/resources/5fe5174a-bdec-a194-79ad-86586fd19601/files/wq-agwr-gl-stormwater-harvesting-and-reuse-final-200907.pdf.

AWWA (American Water Works Association). 2009. Planning for the Distribution of Reclaimed Water, 3rd edition: Manual M24. Denver, CO: AWWA.

Backhaus, T. 2014. Medicines, shaken and stirred: A critical review on the ecotoxicology of pharmaceutical mixtures. Philosophical Transactions of the Royal Society B-Biological Sciences 369(1656):20130585.

Bambic, D., G. McBride, W. Miller, R. Stott, and S. Wuertz. 2011. Quantification of Pathogens and Sources of Microbial Indicators for QMRA in Recreational Waters. Alexandria, VA: Water Environmental Research Foundation.

Bannerman, R., K. Baun, M. Bohn, P.E. Hughes, and D.A. Graczyk. 1983. Evaluation of Urban Nonpoint Source Pollution Management in Milwaukee County, Wisconsin, Vol. I. PB 84-114164. Prepared for U.S. Environmental Protection Agency, Water Planning Division, by Wisconsin Department of Natural Resources. November.

Bannerman, R., S. Corsi, and J. Lorman. 2014. Road Salt Impacts-A Sleeping Giant. Iowa State Stormwater Conference. April.

Barnes, T.G., and L. Adams. 1998. Creating Urban Stormwater Control Ponds for Water Quality and Wildlife Habitat. College of Agriculture, University of Kentucky Extension Service. November.

Barnett, T.P., and W. Pennell, eds. 2004. Impact of Global Warming on Western U.S. Water Supplies. Climate Change 62 (Spec. Vol.).

Barraud, S., A. Gautier, J.P. Bardin, and V. Riou. 1999. The impact of intentional stormwater infiltration on soil and groundwater. Water Science and Technology 39(2):185-192.

Beal, C.C., M.N. Chong, J. Fyfe, A. Turner, and T. Gardner. 2105. Quantifying mains water savings from residential rainwater tanks. Pp. 47-72 in Rainwater Tank Systems for Urban Water Supply: Design, Yield, Energy, Health Risks, Economics, and Social Perceptions, A. K. Sharma, D. Begbie, and T. Gardner, eds. London: IWA Publishing.

Beecher, J., and T. Chesnutt. 2012. Declining Water Sales and Utility Revenues: A Framework for Understanding and Adapting. Alliance for Water Efficiency. Available at http://www.ipu.msu.edu/research/pdfs/Summit-Summary-and-Declining-Water-Sales-and-Utility-Revenues-2012-12-16.pdf.

Bender & Company. 2014. Little NEPAs: State Environmental Policy Laws Using NEPA Approach. In Treatise on Environmental Law. LexisNexis database. Accessed April 6, 2014.

Bergdolt, J., S. Sharvelle, and L.A. Roesner. 2011. Guidance Manual for Separation of Graywater from Blackwater for Graywater Reuse. WERF Project INFR4SG09a. Alexandria, VA: Water Environment Research Foundation.

Bijoor, N.S., D.E. Pataki, D. Haver, and J.S. Famiglietti. 2014. A comparative study of the water budgets of lawns under three management scenarios. Urban Ecosystems 17(4):1095-1117.

Birks, R., and S. Hills. 2007. Characterization of indicator organisms and pathogens in domestic greywater for recycling. Environmental Monitoring and Assessment 129(1-3):61-69.

Bishop, C.A., J. Struger, D.R. Barton, L.J. Shirose, L. Dunn, A. Lang, and D. Shepherd. 2000a. Contamination and wildlife communities in stormwater detention ponds in Guelph and the Greater Toronto area Ontario, 1997 and 1998, Part I. Wildlife communities. Water Quality Research Journal of Canada 35:399-435.

Bishop, C.A., J. Struger, L.J. Shirose, L. Dunn, and G.D. Campbell. 2000b. Contamination and wildlife communities in stormwater detention ponds in Guelph and the Greater Toronto Area Ontario, 1997 and 1998, Part II. Contamination and biological effects of contamination. Water Quality Research Journal of Canada 35:437-474.

Blackhurst, M., C. Hendrickson, and H.S. Matthews. 2010. Cost-effectiveness of green roofs. Journal of Architectural Engineering 16(4):136-143.

Blanchoud, H., F. Farrugia, and J.M. Mouchel. 2004. Pesticide uses and transfers in urbanised catchments. Chemosphere 55(6):905-913.

Bonham, C.H. 2006. Perspectives from the field: A review of western instream flow issues and recommendations for a new water future. Environmental Law 36:1205-1235.

Boston Water and Sewer Commission. 2013. Stormwater Best Management Practices: Guidance Documents. Available at http://www.bwsc.org/ABOUT_BWSC/systems/stormwater_mgt/Stormwater%20BMP%20Guidance_2013.pdf.

Brand, A.B., and J.W. Snodgrass. 2010. Value of artificial habitats for amphibian reproduction in altered landscapes. Conservation Biology 24:295-301.

Brosseau, G. 2007. Summarizing the Report Fresno-Cloves Storm Water Quality Management Program Fresno Metropolitan Flood Control District, November 2006- Evaluation of Basin EK Effectives. Presentation to the NRC Committee on Reducing Stormwater Pollution, December 17, 2007, Irvine, CA.

Brown, P.W., C.F. Mancino, M.H. Young, T.L. Thompson, P.J. Wierenga, and D. Kopec. 2001. Penman-Monteith crop coefficients for use with desert turf systems. Crop Science 41:1197-1206.

Brownell, M., V. Harwood, R. Kurz, S. McQuaig, J. Lukasik, and T. Scott. 2007. Confirmation of putative stormwater impact on water quality at a Florida beach by microbial source tracking methods and structure of indicator organism populations. Water Research 41(16):3747-3757.

Buchheister, A., C.F. Bonzek, J. Gartland, and R.J. Latour. 2013. Patterns and drivers of the demersal fish community of Chesapeake Bay. Marine Ecology Progress Series 481:161-180.

Burton, G.A., Jr., and R.E. Pitt. 2002. Stormwater Effects Handbook: A Toolbox for Watershed Managers, Scientists, and Engineers. Boca Raton, FL: CRC Press.

Byappanahalli, M.N., R.L. Whitman, D.A. Shively, M.J. Sadowsky, and S. Ishii. 2006. Population structure, persistence, and seasonality of autochthonous Escherichia coli in temperate, coastal forest soil from a Great Lakes watershed. Environmental Microbiology 8(3):504-513.

Cadavid, C., and A.W. Ando. 2013. Valuing preferences over stormwater management outcomes given state-dependent preferences and heterogeneous status quo. Water Resources Research 49(7):4114-4125.

CADWR (California Department of Water Resources). 2014. DWR Drops State Water Project Allocation to Zero, Seeks to Preserve Remaining Supplies. Press Release: January 31, 2014. Available at http://www.water.ca.gov/news/newsreleases/2014/013114pressrelease.pdf. Accessed July 23, 2014.

Carpenter, F., M. Rietschier, B. Morris, E.U. Earn, D. Freedman, et al. 2009. Georgia Rainwater Harvesting Guidelines. Available at http://www.dca.state.ga.us/development/constructioncodes/programs/downloads/GeorgiaRainWaterHarvestingGuidelines_2009.pdf.

Casanova, L.M., C.P. Gerba, and M. Karpiscak. 2001. Chemical and microbial characterization of household graywater. Journal of Environmental Science and Health, Part A 36:395-401.

CEC (California Energy Commission). 2015. Energy Commission Approves Water Appliance Standards to Save More than 100 Billion Gallons per Year. News Release: April 8, 2015. Available at http://www.energy.ca.gov/releases/2015_releases/2015-04-08_water_appliance_standards_nr.html. Accessed November 13, 2015.

CH2M-Hill. 2006. Operations and Maintenance Manual, Sun Valley Watershed—Sun Valley Park Drain and Infiltration System. Available at https://dpw.lacounty.gov/wmd/svw/SVP.aspx.

Chambers Group. 2014. Initial Study Checklist: Pacoima Spreading Grounds Improvement Project. Prepared for Los Angeles Flood Control District, Alhambra, CA, December. Available at https://dpw.lacounty.gov/wrd/Projects/PacoimaSG/PacoimaSG-Final-IS.pdf.

Characklis, G.W., M.J. Dilts, O.D. Simmons, C.A. Likirdopulos, L.A.H. Krometis, and M.D. Sobsey. 2005. Microbial partitioning to settleable particles in stormwater. Water Research 39(9):1773-1782.

Chino Basin Watermaster. 2001. Chino Basin Recharge Master Plan. Chino Basin Watermaster, Rancho Cucamonga, CA. August.

Christova-Boal, D., R.E. Eden, and S. McFarlane. 1996. An investigation into greywater reuse for urban residential properties. Desalination 106(1-3):391-397.

CIMIS (California Irrigation Management Information System). 2011. California Irrigation Management Information System. Available at http://wwwcimis.water.ca.gov/cimis. Accessed July 21, 2011.

City and County of San Francisco. 2014. San Francisco's Non-potable Water Program: Guidebook for the Collection, Treatment, and Reuse of Alternate Water Sources in San Francisco. Available at http://www.sfwater.org/modules/showdocument.aspx?documentid=4962.

City of Bellingham. 2012. Rainwater Harvesting: Guidance toward a Sustainable Water Future. Available at https://www.cob.org/documents/pw/environment/water-conservation/rainwater-harvesting-booklet.pdf.

City of Guelph. 2012. Guelph Residential Greywater Field Test Final Report. Available at http://guelph.ca/wp-content/uploads/GreywaterFieldTestReport.pdf.

City of Long Beach. 2011. Laundry to Landscape' Graywater Pilot Program. Available at http://www.longbeach.gov/news/display news.asp?NewsID=5414&targetid=55.

City of Long Beach. 2013. "Laundry to Landscape" Graywater Pilot Program Report. July 2013. Pp. 56-58 in Water Reuse and Conservation in the City of Rosemount. Available at http://rcp.umn.edu/wp-content/uploads/2015/07/PUBH-6132-Water-Reuse-Report.pdf.

City of Los Angeles. 1992. Graywater Pilot Project: Final Project Report. Office of Reclamation. November.

City of Los Angeles. 2015. Sustainability City Plan. Los Angeles, CA: Office of the Mayor.

City of Tucson. 2005. Water Harvesting Guidance Manual. Available at http://www.tucsonaz.gov/files/transportation/2006WaterHarv esting.pdf.

Clark, S. 2000. Urban Stormwater Filtration: Optimization of Design Parameters and a Pilot-Scale Evaluation. Ph.D. Dissertation, University of Alabama at Birmingham. 430 pp.

Clark, S.E., and R. Pitt. 2007. Influencing factors and a proposed evaluation methodology for predicting groundwater contamination potential from stormwater infiltration activities. Water Environment Research 79(1):29-36.

Clark, S.E., and R. Pitt. 2009. Solids removal in storm-water filters modeled using a power equation. Journal of Environmental Engineering 135(9):896-899.

Clark. S.E., and R. Pitt. 2012. Targeting treatment technologies to address specific stormwater pollutants and numeric discharge limits. Water Research 46(20):6715-6730.

Clark, S.E., K.A. Franklin, J.M. Hafera, and J.T. Kirby. 2007. Roofing as a Contributor to Urban Runoff Pollution. World Environmental and Water Resources Congress 2007: Restoring Our Natural Habitat, May 15-19, 2007, Tampa, FL. CD-ROM.

Clark, S.E., B.V. Long, C.Y.S. Siu, J. Spicher, and K.A. Steele. 2008a. Early-life roof runoff quality: Green vs traditional roofs. Pp.1-10 in The Low Impact Development for Urban Ecosystem and Habitat Protection: Conference Proceedings, November 16-19, 2008, Seattle, WA, N. She, and M. Char, eds. Reston, VA: American Society of Civil Engineers.

Clark, S.E., K.A. Steele, J. Spicher, C.Y.S. Siu, M.M. Lalor, R. Pitt, and J.T. Kirby. 2008b. Roofing materials' contributions to storm-water runoff pollution. Journal of Irrigation and Drainage Engineering 134(5):638-645.

Clark, S.E., B. Pitt, and R. Field. 2010. Groundwater contamination potential from infiltration of urban stormwater runoff. Pp. 119-164 in Effects of Urbanization on Groundwater: An Engineering Case-based Approach for Sustainable Development, N.-B. Chang, ed. Reston, VA: American Society of Civil Engineers.

Clary, J., R. Pitt, and B. Steets, eds. 2014. Pathogens in Urban Stormwater Systems. Reston, VA: ASCE. 289 pp.

Colby, S.L., and J.M. Ortman. 2014. Projections of the Size and Composition of the U.S. Population: 2014 to 2060, Current Population Reports. P25-1143. U.S. Census Bureau, Washington, DC.

Conn, K., K. Lowe, J.E. Drewes, C. Hoppe-Jones, and M.B. Tucholke. 2010. Occurrence of pharmaceuticals and consumer product chemicals in raw wastewater and septic tank effluent from single-family houses. Environmental Engineering Science 27(4): 347-356.

Coomes, P., T. Rockaway, J. Rivard, and B. Kornstein. 2010. North America Residential Water Usage Trends Since 1992. Denver, CO: WRF. Available at http://www.waterrf.org/PublicReportLibrary/4031.pdf. Accessed October 17, 2014.

Copeland, C. 2014. Energy-Water Nexus: The Water Sector's Energy Use. Congressional Research Service Report R43200. Available at https://www.fas.org/sgp/crs/misc/R43200.pdf. Accessed October 29, 2015.

Cordery-Cotter, S., and S. Sharvelle. 2014. Workshop presentation: Graywater Reuse, A Case Study at Colorado State University. Available at https://www.watereuse.org/wp-content/uploads/2015/09/Presentation-Colorado-2014-Workshop-Graywater-Reuse-A-Case-Study-at-Colorado-State-University.pdf.

Corsi, S.R., S.R. Greb, R.T. Bannerman, and R.E. Pitt. 1999. Evaluation of the Multi-chambered Treatment Train, A Retrofit Water-Quality Management Device. U.S. Geological Survey Open-File Report 99-270. 24 pp.

Crabtree, K.D., R.H. Ruskin, S.B. Shaw, and J.B. Rose. 1996. The detection of Cryptosporidium oocysts and Giardia cysts in cistern water in the U.S. Virgin Islands. Water Research 30(1):208-216.

Craig, R.K. 2008. Climate change, regulatory fragmentation, and water triage. University of Colorado Law Review 79:825-876.

CREST (Cleaner Rivers through Effective Stakeholder-led TMDLs). 2009. Los Angeles River Bacteria Source Identification Study: Final Report.

CSBE (Center for the Study of the Built Environment). 2003. Graywater Reuse in Other Countries and its Applicability to Jordan. Available at http://www.csbe.org/assets/Greywater-Publications/Gray water-in-other-countries-E.pdf. Accessed November 20, 2015.

Cunliffe, D.A. 2004. Guidance on the Use of Rainwater Tanks. Enhealth. Australian Government. Available at https://www.health.gov.au/internet/main/publishing.nsf/Content/D3E8C00F3041304BCA257BF0001C685A/$File/env_rainwater.pdf.

Daigger, G.T. 2009. State-of-the-art review: Evolving urban water and residuals management paradigms: Water reclamation and reuse, decentralization, resource recovery. Water Environment Research 81(8):809-823.

Daigger, G.T. 2011. Interview. International Water Association, August 27, 2011. Available at http://www.research-europe.com/index.php/2011/08/dr-glen-daigger-president-of-the-international-water-association.

Dandy, G., A. Ganji, J. Kandulu, D. Hatton MacDonald, A. Marchi, H. Maier, A. Mankad, and C.E. Schmidt. 2014. Managed Aquifer Recharge and Stormwater Use Options: Net Benefits Report. Technical Report Series No. 14/1. Goyder Institute for Water Research, Adelaide, Australia. 181 pp. Available at http://goyderinstitute.org/index.php?id=20.

Davis, A., and M. Burns, 1999. Evaluation of lead concentrations in runoff from painted structures. Water Research 33(13):2949-2958.

Davis, A.P., M. Shokouhian, and S. Ni. 2001. Loading estimates of lead, copper, cadmium, and zinc in urban runoff from specific sources. Chemosphere 44(5):997-1009.

DDOE (District Department of Environment). 2013. Stormwater Management Guidebook. Available at http://doee.dc.gov/sw-guidebook.

Debroux, J.F., J.A. Soller, M.H. Plumlee, and L.J. Kennedy. 2012. Human health risk assessment of non-regulated xenobiotics in recycled water: A review. Human and Ecological Risk Assessment 18(3):517-546.

DeOreo, B., P. Mayer, B. Dziegielewski, and J. Kiefer. 2016. Residential End Uses of Water, Version 2. Water Research Foundation.

DeOreo, W.B., P.W. Maye, L. Martien, M. Hayden, A. Funk, M. Kramer-Duffield, and R. Davis. 2011. California Single-Family Water Use Efficiency Study. Aquacraft Water Engineering and Management, Boulder, CO. Available at http://water.cityofdavis.org/Media/PublicWorks/Documents/PDF/PW/Water/Documents/California-Single-Family-Home-Water-Use-Efficiency-Study-20110420.pdf.

DeSemple, D. 2006. Rio Salado Environmental Restoration Project. Pp. 2839-2852 in Proceedings WEFTEC 79th Annual Conference, October 21-25, 2006, Dallas, TX. Alexandria, VA: Water Environment Foundation. Available at http://acwi.gov/swrr/Rpt_Pubs/wef06_session36/Rio-Salado-224.pdf.

Despins, C. 2010. Ontario Guidelines for Residential Rainwater Harvesting Systems. Available at http://www.arcsa-edu.org/Files/ONTARIO_RWH_HANDBOOK_2010.pdf. Accessed June 19, 2015.

Despins, C., K. Farahbakhsh, and C. Leidl. 2009. Assessment of rainwater quality from rainwater harvesting systems in Ontario, Canada. Journal of Water Supply: Research and Technology-AQUA 58(2):117-134.

Dierkes, C., and W.F. Geiger. 1999. Pollution retention capabilities of roadside soils. Water Scienceand Technology 39(2):201-208.

DOE (Department of Energy). 2014. The Water-Energy Nexus: Challenges and Opportunities. June 2014. Available at http://www.energy.gov/sites/prod/files/2014/07/f17/WaterEnergy Nexus Full Report July 2014.pdf.

Donnelly, K., and J. Christian-Smith. 2013. California Water Rates and the "New Normal." The Pacific Institute. Available at http://www.pacinst.org/publication/water-rates-series.

Donner, E., E. Eriksson, D.M. Revitt, L. Scholes, H.C. Holten-Lutzhof, and A. Ledin. 2010. Presence and fate of priority substances in domestic greywater treatment and reuse systems. The Science of the Total Environment 408(12):2444-2451.

Dorfman, M., and A. Haren. 2013. Testing the Waters: 23rd Annual Report. Washington, DC: Natural Resources Defense Council.

Dreiseitl, H. 1998. The role of water in cities. Pp. 25-29 in Sustaining Urban Water Resources in the 21st Century, September 7-12, 1997, Malmo, Sweden, A.C. Rowney, P. Stahre, and L.A. Roesner, eds. New York; ASCE.

Duffield, J.M. 1986. Waterbird use of urban stormwater wetland system in central California, USA. Colonial Waterbirds 9(2):227-235.

Dupont, H., C. Chappell, C. Sterling, P. Okhuysen, J. Rose, and W. Jakubowski. 1995. Infectivity of Cryptosporidium parvum in healthy volunteers. New England Journal of Medicine 332(13):855-859.

Dziegielewski, B., J. Kiefer, E.M. Opitz, G.A. Porter, G.L. Lantz, W.B. DeOreo, and P.W. Mayer. 2000. Commercial and Institutional End-uses of Water. AWWA Research Foundation and American Water Works Association, Denver, CO. Available at http://ufdc.ufl.edu/WC13511002/00001/5j.

Dzombak, D.A., and F.M. Morel. 1990. Surface Complexation Modeling: Hydrous Ferric Oxide. Hoboken, NJ: John Wiley & Sons.

EBMUD (East Bay Municipal Utility District). 2009. Summary of Newly Adopted Graywater Code Fact Sheet. East Bay Municipal Utility District.

Echeverria, J.D. 2010. Is regulation of water a constitutional taking? Vermont Journal of Environmental Law 11(3):581-583.

Eisenberg, J.N.S., J.A. Soller, J. Scott, D.M. Eisenberg, and J.M. Colford. 2004. A dynamic model to assess microbial health risks associated with beneficial uses of biosolids. Risk Analysis 24(1):221-236.

EPA (U.S. Environmental Protection Agency). 1983. Results of the Nationwide Urban Runoff Program. Washington, DC: EPA, Water Planning Division. December. Available at http://www3.epa.gov/npdes/pubs/sw_nurp_vol_1_finalreport.pdf. Accessed June 18, 2015.

EPA. 1992. Use of Microbial Risk Assessment in Setting U.S. Drinking Water Standards. EPA 814/S-92-001. Available at http://nepis.epa.gov/Exe/ZyPDF.cgi?Dockey=20001W58.PDF.

EPA. 1994. National Water Quality Inventory: 1994 Report to Congress. Washington, DC: EPA.

EPA. 1998. Guidelines for Ecological Risk Assessment. EPA/630/R-95/002-F. Available at http://www.epa.gov/sites/production/files/2014-11/documents/eco_risk_assessment1998.pdf.

EPA. 2003. Cross-Connection Control Manual. EPA/816/R-03/002. Available at http://www.epa.gov/ogwdw/pdfs/crossconnection/crossconnection.pdf.

EPA. 2004a. Report to Congress: Impacts and Control of CSOs and SSOs. EPA 833-R-04-001. Available at cfpub.epa.gov/npdes/cso/cpolicy_report2004.cfm.

EPA. 2004b. Guidelines for Water Reuse. EPA/625/R-04/108. Available at http://www.epa.gov/ord/NRMRL/pubs/625r04108/625r04108.htm.

EPA. 2008a. Managing Wet Weather with Green Infrastructure: Action Strategy. Available at http://water.epa.gov/infrastructure/greeninfrastructure/upload/gi_action_strategy.pdf.

EPA. 2008b. Clarification on which stormwater infiltration practices/technologies have the potential to be regulated as "Class V" wells by the Underground Injection Control Program. Memorandum, June 13, 2008. Available at http://www.ecy.wa.gov/programs/wq/stormwater/municipal/resources/EPAmemoinfiltrationclassvwells.pdf. Accessed January 22, 2015.

EPA. 2009. Technical Guidance on Implementing the Stormwater Runoff Requirements for Federal Projects under Section 438 of the Energy Independence and Security Act. EPA 841-B-09-001. Washington, DC: EPA. Available at http://www.epa.gov/sites/production/files/2015-09/documents/eisa-438.pdf.

EPA. 2010. Green Infrastructure Case Studies: Municipal Policies for Managing Stormwater with Green Infrastructure. Available at http://www.epa.gov/nscep/index.html.

EPA. 2011. eGRID2010 Version 1.1 Year 2007 Summary Tables. Washington, DC: EPA.

EPA. 2012a. Guidelines for Water Reuse. EPA/600/R-12/618. Available at http://nepis.epa.gov/Adobe/PDF/P100FS7K.pdf.

EPA. 2012b. Administrative Order for Compliance on Consent. Docket No. CWA-03-2012-0264. EPA Region 3, Philadelphia, PA.

EPA. 2012c. Stormwater Program. In National Pollutant Discharge System (NPDES). Available at http://cfpub.epa.gov/npdes/home.cfm?program_id=6.

EPA. 2012d. National Menu of BMPs. In National Pollutant Discharge Elimination System (NPDES). Available at http://cfpub.epa.gov/npdes/stormwater/menuofbmps/index.cfm.

EPA. 2013a. Green Infrastructure Strategic Agenda. Available at http://water.epa.gov/infrastructure/greeninfrastructure/upload/2013_GI_FINAL_Agenda_101713.pdf.

EPA. 2013b. Rainwater Harvesting: Conservation, Credit, Codes, and Cost Literature Review and Case Studies. EPA-841-R-13-002. Washington, DC: EPA. Available at http://www.epa.gov/sites/production/files/2015-11/documents/rainharvesting.pdf.

EPA. 2013c. Storm Water Drainage Wells. Available at http://water.epa.gov/type/groundwater/uic/class5/types_stormwater.cfm. Accessed January 22, 2015.

EPA. 2014a. Low Impact Development (LID) and Other Green Design Strategies. In National Pollutant Discharge Elimination System (NPDES). Available at http://cfpub.epa.gov/npdes/stormwater/menuofbmps/index.cfm?action=factsheet_results&view=specific&bmp=124.

EPA. 2015. Water Sense: Types of Facilities. Available at http://www3.epa.gov/watersense/commercial/types.html.

Eriksson, E., K. Auffarth, M. Henze, and A. Ledin. 2002. Characteristics of grey wastewater. Urban Water 4(1):85-104.

Eriksson, E., A. Baun, L. Scholes, A. Ledin, S. Ahlmann, M. Revitt, C. Noutsopoulus, and P. S. Mikkelsen. 2007. Selected stormwater priority pollutants—a European perspective. Science of the Total Environment 383(1-3):41-51.

Etchepare, R., and J.-P. van der Hoek. 2015. Health risk assessment of organic micropollutants in graywater for potable reuse. Water Research 72:186-198.

Faller, M., and D. Reiss. 2005. Runoff behavior of metallic materials used for roofs and facades—a 5 year exposure study in Switzerland. Materials and Corrosion 56(4):244-249.

Ferguson, B.C., N. Frantzeskaki, and R.R. Brown. 2013. A strategic program for transitioning to a water sensitive city. Landscape and Urban Planning 117:32-45.

Fischer, D., E.G. Charles, and A.L. Baehr. 2003. Effects of stormwater infiltration on quality of groundwater beneath retention and detention basins. Journal of Environmental Engineering 129(5):464-471.

Förster, J. 1999.Variability of roof runoff quality PB. Water Science and Technology 39(5):137-144.

Friedler, E., and M. Hadari. 2006. Economic feasibility of on-site greywater reuse in multi-storey buildings. Desalination 190(1): 221-234.

Friedler, E., R. Kovalio, and A. Ben-Zvi. 2006. Comparative study of the microbial quality of greywater treated by three on-site treatment systems. Environmental Technology 27(6):653-663.

Gan, J., S. Bondarenko, L. Oki, D. Haver, and J.X. Li. 2012. Occurrence of fipronil and its biologically active derivatives in urban residential runoff. Environmental Science and Technology 46(3):1489-1495.

Gan, K., and M. Redhead. 2013. Melbourne Residential Water Use Studies, 10TR5-01. Smart Water Fund. Available at https://www.yvw.com.au/yvw/groups/public/documents/document/yvw1004065.pdf.

Garland, J.L., L.H. Levine, N.C. Yorio, J.L. Adams, and K.L. Cook. 2000. Graywater processing in recirculating hydroponic systems: Phytotoxicity, surfactant degradation, and bacterial dynamics. Water Research 34(12):3075-3086.

Getches, D.H. 2001. The metamorphosis of western water policy: Have federal laws and local decisions eclipsed the states' role? Stanford Environmental Law Journal 20(3):3-72.

Getches, D.H. 2009. Water Law in a Nutshell. St. Paul, MN: West Publishing Co.

Ghimire, S.R., J.M. Johnston, W.W. Ingwersen, and T.R. Hawkins. 2014. Life cycle assessment of domestic and agricultural rainwater harvesting systems. Environmental Science and Technology 48(7):4069-4077.

Gilliom, R.J., J.E. Barbash, C.G. Crawford, P.A. Hamilton, J.D. Martin, N. Nakagaki, et al. 2007. The Quality of Our Nation's Waters Pesticides in the Nation's Streams and Ground Water, 1992–2001. Circular 1291. Reston, VA: U.S. Geological Survey.

GLACIRWM (Greater Los Angeles County Integrated Regional Management). 2013. Greater Los Angeles County Integrated Regional Water Management Plan: 2013 Update. Prepared by the Leadership Committee of the Greater Los Angeles County Integrated Regional Management Region.

Gleick, P. H., D. Haasz, C. Henges-Jeck, V. Srinivasan, G.H. Wolff, K.K. Cushing, and A. Mann. 2003. Waste Not, Want Not: The Potential for Urban Water Conservation in California. Oakland, CA: Pacific Institute for Studies in Development, Environment, and Security.

Good, J.C. 1993. Roof runoff as a diffuse source of metals and aquatic toxicity in storm water. Water Science and Technology 28(3-5):317-321.

Goodman, J.A., and L. Austin. 2011. Predicting hydromodification impacts using a four factor approach. Pp. 499-508 in Proceedings of World Environment and Water Resources Congress: Bearing Knowledge for Sustainability, R.E. Beighley, II, and M.W. Killgore, eds. Reston, VA: ASCE.

Governor of the State of California. 2014. Governor Edmund G. Brown, Jr., Declares Drought State of Emergency. Available at http://gov.ca.gov/news.php?id=18368. Accessed July 23, 2014.

Grebel, J.E., S.K. Mohanty, A.A. Torkelson, A.B. Boehm, C.H. Higgins, R.M. Maxwell, K.L. Nelson, and D.L. Sedlak. 2013. Engineered infiltration systems for urban stormwater reclamation. Environmental Engineering Science 30(8):437-454.

Grohmann, G.S, N.J. Ashbolt, M.S. Genova, G. Logan, P. Cox, and C.S.W. Kueh. 1993. Detection of viruses in coastal and river water-systems in Sydney. Australian Water Science Technology 27:457-461.

Gromaire-Mertz, M.C., S. Garnaud, A. Gonzalez, and G. Chebbo. 1999. Characterization of urban runoff pollution in Paris. Water Science and Technology 39(2):1-8.

Gross, A., N. Azulai, G. Oron, Z. Ronen, M. Arnold, and A. Nejidat. 2005. Environmental impact and health risks associated with greywater irrigation: A case study. Water Science and Technology 52(8):161-169.

Gross, A., D. Kaplan, and K. Baker. 2007. Removal of chemical and microbiological contaminants from domestic greywater using a recycled vertical flow bioreactor (RVFB). Ecological Engineering 31:107-114.

Gual, M., A. Moia, and J. March. 2008. Monitoring of an indoor pilot plant for osmosis rejection and greywater reuse to flush toilets in a hotel. Desalination 219:81-88.

Guest, J. S., S.J. Skerlos, J.L. Barnard, M.B. Beck, G.T. Daigger, H. Hilger, S.J. Jackson, K. Karvazy, L. Kelly, L. Macpherson, J.R. Mihelcic, A. Pramanik, L. Raskin, M.C.M. van Loosdrecht, D. Yeh, and N.G. Love. 2009. A new planning and design paradigm to achieve sustainable resource recovery from wastewater. Environmental Scienceand Technology 43(16):6126-6130.

Gumbs, A.F., and F.E. Dierberg. 1985. Heavy metals in the drinking water from cisterns supplying single-family dwellings. Water International 10:22-28

Haas, C.N., J.B. Rose, and C.P. Gerba. 2014. Quantitative Microbial Risk Assessment, 2nd Edition. Hoboken, NJ: Wiley & Sons.

Hagekhalil, A., S. Kharaghani, W. Tam, R. Haimann, and K. Susilo. 2014. Becoming the green-blue city. Civil Engineering Magazine 84(12):68-79, 90-91.

Hambly, A.C., R.K. Henderson, A. Baker, R.M. Stuetz, and S.J. Khan. 2012. Cross-connection detection in Australian dual reticulation systems by monitoring inherent fluorescent organic matter. Environmental Technology Reviews 1(1):67-80.

Hampson, P.S. 1986. Effects of Detention on Water Quality of Two Stormwater Detention Ponds Receiving Highway Surface Runoff in Jacksonville, Florida. U.S. Geological Survey Water-Resources Investigations Report 86-4151. Prepared in Cooperation with the Florida Department of Transportation. Denver, CO: USGS.

Hart, C., and D. White. 2006. Water quality and construction materials in rainwater catchments across Alaska. Journal of Environmental Engineering Science 5:S19-S25.

Hatt, B.E., A. Deletic, and T.D. Fletcher. 2006. Integrated treatment and recycling of stormwater: A review of Australian practice. Journal of Environmental Management 79:102-113.

Heijerick, D.G., C.R. Janssen, C. Karlen, I.O.Wallinder, and C. Leygraf. 2002. Bioavailability of zinc in runoff water from roofing materials. Chemosphere 47(10):1073-1080.

Hendrickson, T.P., M.T. Nguyen, M. Sukardi, A. Miot, A. Horvath, and K.L. Nelson. 2015. Life-cycle energy use and greenhouse gas emissions of a building-scale wastewater treatment and nonpotable reuse system. Environmental Science and Technology 49(17):10303-10311.

Herald Sun. 2007. Staff ill from recycled water. Available at http://www.herald.sun.com.au/news/Victoria/staff-ill-from-recycled-water/story-e6frf7kx-1111113190490. Accessed October 15, 2014.

Hering, J.G., T.D. Waite, R.G. Luthy, J.E. Drewes, and D.L. Sedlak. 2013. A changing framework for urban water systems. Environmental Science and Technology 47(19):10,721–10,726.

Higgins, C.P., and R.G. Luthy. 2006. Sorption of perfluoroinated surfactants on sediments. Environmental Science and Technology 40(23):7251-7256.

Hodgson, B. 2012. Development of a Cost Effective and Energy Efficient Treatment Systems for Graywater Reuse for Toilet Flushing at the Multi-Residential Scale. M.S. Thesis, Colorado State University.

Holmstrom, B.K., D.L. Olson, and B.R. Ellefson. 1995. Water Resources Data, Wisconsin Water Year 1995: U.S. Geological Survey Water – Data Report WI-95-1. Available at http://pubs.usgs.gov/wdr/1995/wi-95-1/report.pdf.

Holmstrom, B.K., D.L. Olson, and B.R. Ellefson. 1996. Water Resources Data, Wisconsin, Water Year 1996: U.S. Geological Survey Water – Data Report WI-96-1.

ICC (International Code Council). 2012. International Fire Code. Available at http://publicecodes.cyberregs.com/icod/ifc/2012/index.htm.

ICC. 2012. International Plumbing Code 2012. Available at http://publicecodes.cyberregs.com/icod/ipc/2012.

IEUA (Inland Empire Utilities Agency). 2014. Groundwater Recharge Annual Report for Calendar Year 2013. May 2014.

Jartun, M., R.T. Ottesen, E. Steinnes, and T. Volden. 2008. Runoff of particle bound pollutants from urban impervious surfaces studied by analysis of sediments from stormwater traps. Science of the Total Environment 396(2-3):147-163.

Jarvis, I.W.H., K. Dreij, A. Mattson, B. Jemstrom, and U. Stenius. 2014. Interactions between polycyclic aromatic hydrocarbons in complex mixtures and implications for cancer risk assessment. Toxicology 321(3):27-39.

Jensen, M.E., R.D. Burman, and R.G. Allen, eds. 1990. Evapotranspiration and Irrigation Water Requirements. New York: ASCE.

Jiang, S.C., W. Chu, B.H. Olson, J.W. He, S. Choi, J. Zhang, J.Y. Le, and P.B. Gedalanga. 2007. Microbial source tracking in a small southern California urban watershed indicates wild animals and growth as the source of fecal bacteria. Applied Microbiology and Biotechnology 76(4):927-934.

John, D.E., and J.B. Rose. 2005. Review of factors affecting microbial survival in groundwater. Environmental Science and Technology 39(19):7345-7356.

Johnson, T. 2011. The Increasing Role of Storm Water for Groundwater Recharge in the Central Basin, Los Angeles Coastal Plain, Southern California. Managed Aquifer Recharge Symposium, January 25-26, Irvine, CA.

Jones, M.P., and W.F. Hunt. 2014. Urban Waterways. Rainwater Harvesting: Guidance for Homeowners. North Carolina Cooperative Extension Service. Available at: http://content.ces.ncsu.edu/rainwater-harvesting-guidance-for-homeowners.pdf Accessed June 19, 2015.

Kakuturu, S.P., and S.E. Clark. 2015. Clogging mechanism of stormwater filter media by NaCl as a deicing salt. Environmental Engineering Science 32(2):141-152.

Karlen, C., I.O.,Wallinder, D. Heijerick, and C. Leygraf. 2002. Runoff rates, chemical speciation and bioavailability of copper released from naturally patinated copper. Environmental Pollution 120(3):691-700.

Kazner, C., T. Wintgens, and P. Dillon. 2012. Water Reclamation Technologies for Safe Managed Aquifer Recharge. London: IWA Publishing.

Kenny, J.F., and K.E. Juracek. 2012. Description of 2005–10 Domestic Water Use for Selected U.S. Cities and Guidance for Estimating Domestic Water Use. U.S. Geological Survey Scientific Investigations Report 2012–5163. 31 pp.

Kinde, H., M. Adelson, A. Ardans, E.H. Little, D. Willoughby, D. Berchtold, D.H. Read, R. Breitmeyer, D. Kerr, R. Tarbell, and E. Hughes. 1997. Prevalence of Salmonella in municipal sewage treatment plant effluents in southern California. Avian Diseases 41:392-398.

King County DNRP (Department of Natural Resources and Parks). 2009. King County Washington's Stormwater Design Manual. Available at http://www.kingcounty.gov/environment/waterandland/stormwater/documents/surface-water-design-manual.aspx. Accessed June 19, 2015.

Ku, H.F., and D.L. Simmons. 1986. Effect of Urban Storm Water Runoff on Groundwater Beneath Recharge Basins on Long Island, New York. Water-Resources Investigations Report 85-4088. U.S. Geological Survey. Available at http://pubs.usgs.gov/wri/1985/4088/report.pdf.

Kuru, B., and M. Luettengen. 2012. Investigation of residential water reuse technologies. Pp. 1-14 in Water Smart Innovations Conference, October 3-5, 2012, Las Vegas, NV.

LACFCD (Los Angeles County Flood Control District). 2012. Sun Valley Watershed, Rory M. Shaw Wetlands Park. Available at http://dpw.lacounty.gov/wmd/svw/docs/RoryMShawWetlands_Factsheet.pdf. Accessed May 14, 2015.

LADWP (Los Angeles Department of Water and Power). 2010. Urban Water Management Plan. Available at https://www.ladwp.com/cs/idcplg?IdcService=GET_FILE&dDocName=QOELLADWP005416&RevisionSelectionMethod=LatestReleased. Accessed March 11, 2013.

LADWP. 2014. Stormwater Capture Master Plan Interim Report, Draft. Prepared by Geosyntec Consultants. December.

LADWP. 2015. Stormwater Capture Master Plan: The Master Planning Process—Final Report. Available at: https://www.ladwp.com/cs/idcplg?IdcService=GET_FILE&dDocName=OPLADWPCCB421767&RevisionSelectionMethod=LatestReleased. Accessed November 11, 2015.

Lafont, M., L. Grapentine, Q. Rochfort, J. Marsalek, G. Tixier, and P. Breil. 2007. Bioassessment of wet-weather pollution on fine sediments in urban waters by benthic indices and the sediment quality triad. Water Science and Technology 56:13-20.

LAO (Legislative Analyst's Office). 2006. Reforming the CALFED Bay-Delta Program. February. Available at http://www.lao.ca.gov/analysis_2006/resources/res_02_anl06.html.

Lawson, S., A. LaBranche-Tucker, H. Otto-Wack, R. Hall, B. Sojka, E. Crawford, D. Crawford, and C. Brand. 2009. Virginia Rainwater Harvesting Manual. Available at https://www.radford.edu/content/dam/departments/administrative/Sustainability/Documents/Rainwater-Manual.pdf. Accessed June 19, 2015.

Le Viol, I., J. Mocq, R. Julliard, and C. Kerbiriou. 2009. The contribution of motorway stormwater retention ponds to the biodiversity of aquatic macroinvertebrates. Biological Conservation 142:3163-3171.

Little, V.L. 2000. Residential Graywater Reuse: The Good, The Bad, The Healthy in Pima Arizona: A Survey of Current Graywater Reuse. Water Resources Research Center, University of Arizona.

Liu, H., R. Ramnarayanan, and B.E. Logan. 2004. Production of electricity during wastewater treatment using a single chamber microbial fuel cell. Environmental Science and Technology 38(7): 2281-2285.

Lock, M.J. 2000. Braving the waters of Supreme Court takings jurisprudence: Will the Fifth Amendment protect western water rights from federal environmental regulation? University of Denver Water Law Review 4:76-78.

Lockie, S., and S. Rockloff. 2005. Decision Frameworks: Assessment of the Social Aspects of Decision Frameworks and Development of a Conceptual Model. Coastal CRC Discussion Paper. Norman Gardens, Australia: Central Queensland University.

Los Angeles Department of Public Health. 2011. Guidelines for Harvesting Rainwater, Stormwater, and Urban Runoff for Outdoor Non-Potable Uses, September 2011. Available at http://www.smgov.net/uploadedFiles/Departments/OSE/Categories/Urban_Runoff/Guidelines_Harvesting_Rainwater_Stormwater_Urban_Runoff_Outdoor_Non-Potable_LA_County.pdf. Accessed June 22, 2015.

Luthy, R.G., and D.L. Sedlak. 2015. Urban water supply reinvention. Daedalus 144(3):72-82.

MacKenzie, W.R., N.J. Hoxie, M.E. Proctor, M.S. Gradus, K.A. Blair, D.E. Peterson, J.J. Kazmierczak, D.G. Addiss, K.R. Fox, J.B. Rose, and J.P. Davis. 1994. A massive outbreak in Milwaukee of Cryptosporidium infection transmitted through the public water supply. New England Journal of Medicine 331:161-167.

Macomber, P.S.H. 2010. Guidelines on Rainwater Catchment Systems for Hawaii. University of Hawaii at Manoa. Available at http://www.ctahr.hawaii.edu/hawaiirain/Library/Guides&Manuals/HI_Guidelines_2010.pdf.

Maestre, A. 2005. Stormwater Characteristics as Described in the National Stormwater Quality Database. Ph.D. Dissertation, Department of Civil, Construction, and Environmental Engineering, University of Alabama.

Maestre, A., R. Pitt, and J. Clary. 2015. The National Stormwater Quality Database (NSQD), Version 4. American Water Resources Conference, November 16-19, 2015, Denver, CO.

Maimon, A., A. Tal, E. Friedler, and A. Gross. 2010. Safe on-site reuse of greywater for irrigation—a critical review of current guidelines. Environmental Science and Technology 44:3213-3220.

Maimone, M., D.E. O'Rourke, J.O. Knighton, and C.P. Thomas. 2011. Potential impacts of extensive stormwater infiltration in Philadelphia. Environmental Engineer: Applied Research and Practice 14:1-12.

Marsden, J. 2013. Economic Viability of Water Recycling Systems. Australian Water Recycling Centre of Excellence. 62 pp. Available at http://www.australianwaterrecycling.com.au/research-publications.html.

Martinez, C.J. 2013. Graywater Reuse in Florida. Doc. No. AE453. Gainesville, FL: Institute of Food and Agricultural Services, University of Florida. Available at http://edis.ifas.ufl.edu/pdf-files/AE/AE45300.pdf.

Maupin, M.A., J.F. Kenny, S.S. Hutson, J.K. Lovelace, N.L. Barber, and K.S. Linsey. 2014. Estimated Use of Water in the United States in 2010. U.S. Geological Survey Circular 1405. 56 pp. Available at http://dx.doi.org/10.3133/cir1405.

Mayer, P.W., W.B. DeOreo, E. M. Opitz, J.C. Kiefer, W.Y. Davis, B. Dziegielewski, and J.O. Nelson 1999. Residential End Uses of Water. Denver, CO: AWWA Research Foundation and American Water Works Association Available at http://www.waterrf.org/publicreportlibrary/rfr90781_1999_241a.pdf.

Mayer, P., P. Lander, and D. Glenn. 2015. Outdoor Water Savings Research Initiative, Phase 1. Analysis of Published Research. Chicago, IL: Alliance for Water Efficiency.

McCarty, P.L., J. Bae, and J. Kim. 2011. Domestic wastewater treatment as a net energy producer—can this be achieved? Environmental Science and Technology 45(17):7100-7106.

McNabola, A., and K. Shields. 2013. Efficient drain water heat recovery in horizontal domestic shower drains. Energy and Buildings 59:44-49.

MDE (Maryland Department of the Environment). 2012. Stormwater Design Guidance-Rainwater Harvesting. Available at http://www.mde.state.md.us/programs/Water/StormwaterManagementProgram/SedimentandStormwaterHome/Documents/ESDMEP%20Guidance%20RWH.pdf.

Mehlhart, G. 2005. Greywater Recycling – Planning Fundamentals and Operational Information. fbr – Information Sheet H 201. Fachvereinigung Betriebs and Regenwassernutzung fbr e.V.; Darmstadt, Germany. Available at http://www.fbr.de/fileadmin/user_upload/files/Englische_Seite/H201_fbr-Information_Sheet_Greywater-Recycling_neu.pdf.

Melbourne Water. 2013/2014. Water Use Data. Available at http://www.melbournewater.com.au/waterdata/wateruse/Pages/default.aspx.

Melillo, J.M., T.C. Richmond, and G.W. Yohe, eds. 2014. Climate Change Impacts in the United States: The Third National Climate Assessment. U.S. Global Change Research Program, Washington DC.

Memon, F.A., Z. Zheng, D. Butler, C. Shirley-Smith, S. Lui, C. Makropoulos, and L. Avery. 2007. Life cycle impact assessment of greywater recycling technologies for new developments. Environment Monitoring and Assessment 129:27-35.

Mendez, C.B., S. Bae, B. Chambers, S. Fakhreddine, T. Gloyna S. Keithley, L. Untung, M.E. Barrett, K. Kinney, and M.J. Kirisits. 2011. Effect of Roof Material on Water Quality for Rainwater Harvesting Systems–Additional Physical, Chemical, and Microbiological Data Report. Texas Water Development Board, Austin, TX. Available at https://www.twdb.texas.gov/innovativewater/rainwater/projects/rainquality/2011_02_rainquality_final_rpt.pdf.

Metropolitan Council. 2011. Stormwater Reuse Guide: A Guide for City Planners, Engineers, Homeowners and Green Thinkers. Fall 2011. Available at http://rfcd.pima.gov/pdd/lid/pdfs/31-mn-metropolitan-council-stormwater-reuse-guide-2011.pdf. Accessed December 11, 2015.

Metropolitan Council. 2013. Municipalities Reusing Stormwater See Multiple Benefits. August 14. 2013.

Mikkelsen, P.S., M. Hafliger, M. Ochs, J.C. Tjell, P. Jacobsen, and M. Boller. 1996. Experimental assessment of soil and groundwater contamination from two old infiltration systems for road run-off in Switzerland. The Science of the Total Environment 189/190:341-347.

Montana Department of Agriculture. 2011. Summary Report: Urban Storm Water Runoff Monitoring for Pesticides and Nitrate in Helena and Billings, 2011. Groundwater Protection Program. Available at http://agr.mt.gov/agr/Programs/NaturalResources/Groundwater/Reports/PDF/Stormwater_report.pdf.

Motavalli, J. 2012. Movement to banish copper from brake pads gains momentum. New York Times: February 3, 2012. Available at http://wheels.blogs.nytimes.com/2012/02/03/movement-to-banish-copper-from-brake-pads-gains-momentum/?_r=0.

MPCA (Minnesota Pollution Control Agency). 2012. Beneficial Use of Wastewater and Stormwater: Capital Grants for Demonstration Projects, Minnesota Pollution Control Agency. February 2012. Available at http://www.pca.state.mn.us/index.php/view-document.html?gid=17179. Accessed December 11, 2015.

MPCA. 2015. Minnesota Stormwater Manuel: Stormwater Re-use and Rainwater Harvesting. Available at http://stormwater.pca.state.mn.us/index.php/Stormwater_re-use_and_rainwater_harvesting.

MPWMD (Monterey Peninsula Water Management District). 2014. Update: Water Supply for the Monterey Peninsula, Monterey County. IRWM Group, Monterey Peninsula Water Management District. June 2014. Available at http://www.water.ca.gov/irwm/grants/docs/PlanReviewProcess/MontereyPeninsula_IRWMP/IRWMplan-Appendices_1to2c_w_linksJuly21.pdf. Accessed December 11, 2015.

Mukheibir, P., T. Boyle, C. Moy, and S. White. 2013. Estimating the reliable residential water substitution from household rainwater tanks. In Proceedings of the 7th IWA International Conference on Efficient Use and Management of Water, October 22-25, 2013, Paris.

Mullaney, J.R., D.L. Lorenz, and A.D. Arntson. 2009. Chlorides in Groundwater and Surface Water in Areas Underlain by the Glacial Aquifer System, Northern United States. USGS Scientific Investigation Report 2009-5086. Reston, VA: USGS. 41pp. Available at http://pubs.usgs.gov/sir/2009/5086/pdf/sir2009-5086.pdf.

Murakami, M., F. Nakajima,, and H. Furumai. 2005. Size- and density-distributions and sources of polycyclic aromatic hydrocarbons in urban road dust. Chemosphere 61(6):783-791.

MWD IRP Technical Workgroup. 2009. Stormwater/Urban Runoff Issue Paper–Final, October 2009. Metropolitan Water District of Southern California, Los Angeles, CA.

MWDSC (Metropolitan Water District of Southern California). 2007. Groundwater Assessment Study. Chapter 4—Groundwater Basin Reports, Central Basin. Report Number 1308.

MWDSC. 2010. Integrated Water Resources Plan. Report No. 1373. Metropolitan Water District of Southern California, Los Angeles, CA. Available at: http://www.mwdh2o.com/Reports/2.4.1_Integrated_Resources_Plan.pdf. Accessed December 11, 2015.

MWH. 2003. Percolation Optimization Investigation. Los Angeles County Department of Public Works. December.

Nara, Y., R. Pitt, S.R. Durrans, and J. Kirby. 2006. Sediment transport in grass swales. Pp. 379-402 in Stormwater and Urban Water Systems Modeling, Monograph 14, W. James, K.N. Irvine, E.A. McBean, and R.E. Pitt, eds. Guelph, Ontario: CHI.

National Assessment Synthesis Team. 2001. Climate Change Impacts on the United States: The Potential Consequences of Climate Variability and Change. Report for the U.S. Global Change Research Program. Cambridge, UK: Cambridge University Press.

NCSU (North Carolina State University). 2011. Urban Waterways: Stormwater Wetlands and Ecosystem Services. Available at http://www.bae.ncsu.edu/stormwater/PublicationFiles/WetlandEcosystemServices2011.pdf. Accessed May 14, 2015.

Negahban-Azar, M., S. Sharvelle, M. Stromberger, C. Olson, and L. Roesner. 2012. Fate of graywater constituents after long-term application for landscape irrigation. Water, Air, and Soil Pollution 223(8):4733-4749.

Negahban-Azar, M., S. Sharvelle, Y. Qian, and A. Shogbon. 2013. Leachability of chemical constituents in soil-plant systems irrigated with synthetic graywater. Environmental Science: Processes and Impacts 14(4):760-762.

Nieber, J.L., A.J. Erickson, P.T. Weiss, J.S. Gulliver, and R.M. Hozalski. 2010. Infiltration practices. In Stormwater Treatment: Assessment and Maintenance. University of Minnesota, St. Anthony Falls Laboratory, Minneapolis, MN. Available at http://stormwaterbook.safl.umn.edu/content/infiltration-practices.

Nieber, J.L., C.N. Arika, L. Lahti, J.S. Gulliver, and P.T. Weiss. 2014. The Impact of Stormwater Infiltration Practices on Groundwater Quality, Department of Civil Engineering, University of Minnesota, Metropolitan Council. July. Available at http://www.metrocouncil.org/METC/files/9f/9f5e1d8d-30d0-44f2-ba83-aeca8f372c61.pdf.

Nightingale, H.I. 1987. Water quality beneath urban runoff water management basins. Journal of American Water Resources Association 23(2):197-205.

Nolde, E. 1999. Greywater reuse systems for toilet flushing in multistory buildings—over ten years of experience in Berlin. Urban Water 1(4):275-284.

NPD Group. 1999. Graywater Awareness and Usage Study. Port Washington, NY: Soap and Detergent Association. 52pp. Available at http://www.aciscience.org/docs/graywater_habits_&_practices_survey.pdf.

NRC (National Research Council). 1983. Risk Assessment in the Federal Government: Managing the Process. Washington, DC: The National Academies Press.

NRC. 1997. Valuing Groundwater: Economic Concepts and Approaches. Washington, DC: The National Academy Press.

NRC. 2005. Decision Making for the Environment: Social and Behavioral Science Research Priorities. Washington, DC: The National Academies Press.

NRC. 2008a. Desalination: A National Perspective. Washington, DC: The National Academies Press.

NRC. 2008b. Public Participation in Environmental Decision Making. Washington, DC: The National Academies Press.

NRC. 2009a. Urban Stormwater Management in the United States. Washington, DC: The National Academies Press.

NRC. 2009b. Summary of a Workshop on Water Issues in the Apalachicola-Chattahoochee-Flint and Alabama-Coosa-Tallapoosa (ACF-ACT) River Basins. Washington, DC: The National Academies Press.

NRC. 2009c. Science and Decisions: Advancing Risk Assessment. Washington, DC: The National Academies Press.

NRC. 2009d. Prospects for Managed Underground Storage of Recoverable Water. Washington, DC: The National Academies Press.

NRC. 2012a. Water Reuse: The Potential for Expanding the Nation's Water Supply through Reuse of Municipal Wastewater. Washington, DC: The National Academies Press.

NRC. 2012b. Science for Environmental Protection: The Road Ahead. Washington, DC: The National Academies Press.

NRMCC/EPHC (Natural Resource Management Ministerial Council, and Environment Protection and Heritage Council) and Australian Health Ministers' Conference. 2006. National Guidelines for Water Recycling: Managing Health and Environmental Risks. 415pp. Available at http://www.environment.gov.au/system/files/resources/044e7a7e-558a-4abf-b985-2e831d8f36d1/files/water-recycling-guidelines-health-environmental-21.pdf.

NRMMC/EPHC/NHMRC (Natural Resource Management Ministerial Council, Environment Protection and Heritage Council, and National Health and Medical Research Council). 2009a. Australian Guidelines for Water Recycling, Managing Health and Environmental Risks, Volume 2B. Stormwater Harvesting and Reuse. National Water Quality Management Strategy Document 23. Available at http://www.environment.gov.au/water/quality/publications/nwqms-australian-guidelines-water-recycling-managing-health-phase2-stormwater.

NRMMC/EPHC/NHMRC. 2009b. Australian Guidelines for Water Recycling, Managing Health and Environmental Risks, Volume 2C. Managed Aquifer Recharge. NWQMS Document 24. Available at http://www.environment.gov.au/water/quality/publications/nwqms-australian-guidelines-water-recycling-managing-health-phase2-aquifer-recharge.

NWRI (National Water Resources Institute). 2004. Report of Scientific Advisory Panel, Orange County Water District's Santa Ana River Water Quality and Health Study. August.

NYSDEC (New York State Department of Environmental Conservation). 2013. NYSDEC and NYCDEP Announce Ground Breaking Agreement to Reduce Combined Sewer Overflows Using Green Infrastructure in New York City. Press Release: March 13, 2013. Available at http://www.dec.ny.gov/press/80919.html.

NYSDEC. 2015. New York State Stormwater Design Manual. Available at http://www.dec.ny.gov/chemical/29072.html.

OCWD. 2014. Tenth Annual Santa Ana Water Quality Monitoring Report, October 2014.

OEHHA and SWRCB (Office of Environmental Health Hazard Assessment and the State Water Resources Control Board). 2009. Hydromodification: Principles, Problems, and Solutions. Available at http://oehha.ca.gov/ecotox/pdf/hydromodfacts102109.pdf. Accessed December 11, 2015.

Ogburn, O. 2013. Urban Stormwater Contamination Associated with Gutter and Pipe Material Degradation. Ph.D. dissertation. Department of Civil, Construction, and Environmental Engineering, University of Alabama, Tuscaloosa, AL.

Ogoshi, M., Y. Suzuki, and T. Asano. 2001. Water reuse in Japan. Water Science and Technology 43(10):17-23.

Olivieri, V.P., C.W. Kruse, K. Kawata, and J.E. Smith. 1977. Microorganisms in Urban Stormwater. EPA-600/2-77-087. NTIS NO. PB-272245.

Olson, P.J. 2014. AWWA: Water Reuse Standards and Codes, Presentation to American Society of Plumbing Engineers. Available at https://www.aspe.org/sites/default/files/webfm/pdfs/AWWA.pdf.

O'Shea, M.L., and R. Field. 1992a. Detection and disinfection of pathogens in storm-generated flows. Canadian Journal of Microbiology 38(4):267-276.

O'Shea, M.L., and R. Field. 1992b. An evaluation of bacterial standards and disinfection practices used for the assessment and treatment of stormwater. Advances in Applied Microbiology 37:21-40.

O'Toole, J., M. Sinclair, S.F. Barker, and K. Leder. 2014. Advice to risk assessors modeling viral health risk associated with household graywater. Risk Analysis 34(5):797-780.

Ottoson, J., and T.A. Stenstrom. 2003. Faecal contamination of greywater and associated microbial risks. Water Research 37:645-655.

Pacific Institute and NRDC. 2014. Stormwater Capture Potential in Urban and Suburban California. Issue Brief IB:14-05-G. June 2014. Available at http://pacinst.org/wp-content/uploads/sites/21/2014/06/ca-water-stormwater.pdf.

Page, D., and K. Levett. 2010. Stormwater Harvesting and Reuse Risk Assessment for Various Uses. CSIRO: National Research Flagships Report. Available at http://www.clw.csiro.au/publications/waterforahealthycountry/2010/wfhc-stormwater-reuse-risk-assessment.pdf.

Page, D., D. Gonzalez, P. Dillon, J. Vanderzalm, G. Vadakattu, S. Toze, J. Sidhu, K. Miotlinski, S. Torkzaban, and K. Barry. 2013. Managed Aquifer Recharge and Urban Stormwater Use Options: Public Health and Environmental Risk Assessment Final Report. Available at http://www.goyderinstitute.org/uploads/Public%20Health%20Risk%20Assessment%20Final%20Report%20June%202014%20%282%29.pdf. Accessed November 11, 2015.

Parkin, R.T., J.A. Soller, and A.W. Olivieri. 2003. Incorporating susceptible subpopulations in microbial risk assessment: Pediatric exposures to enteroviruses in river water. Journal of Exposure Analysis and Environmental Epidemiology 13(2):161-168.

Paul, M.J., and J.L. Meyer. 2001. Streams in the urban landscape. Annual Review of Ecology and Systematics 32:333-365.

Payne, E.G.I., B.E. Hatt, A. Deletic, M.F. Dobbie, D.T. McCarthy, and G. I. Chandrasena. 2015. Adoption Guidelines for Stormwater Biofiltration Systems. Melbourne, Australia: Cooperative Research Centre for Water Sensitive Cities.

Pidou, M., L. Avery, T. Stephenson, P. Jeffrey, S.A. Parsons, S. Liu, F. A. Memon, and B. Jefferson. 2008. Chemical solutions for greywater recycling. Chemosphere 71(1):147-155. Cincinnati, OH. 467 pp.

Pitt, R. 1987. Small Storm Urban Flow and Particulate Washoff Contributions to Outfall Discharges, Ph.D. Dissertation, Civil and Environmental Engineering Department, University of Wisconsin, Madison, WI. November.

Pitt, R. 1997. Unique features of the Source Loading and Management Model (SLAMM). Pp. 13-37 in Advances in Modeling the Management of Stormwater Impacts, Volume 6, W. James, ed., Guelph, Ontario: CHI.

Pitt, R. 2011a. Regional WinSLAMM Calibrations using Standard Land Use Characteristics and Pollutant Sources. Prepared for Geosyntec, Cadmus, and EPA in Support of New Stormwater Regulations. April 18, 2011. Available at http://unix.eng.ua.edu/~rpitt/Publications/8_Stormwater_Management_and_Modeling/WinSLAMM_modeling_examples/Standard_Land_Use_file_descriptions_final_April_18_2011.pdf.

Pitt, R. 2011b. Lincoln, Nebraska, Retrofit Stormwater Management Options Performance and Relative Costs. Prepared for Wright Waters, Denver, CO. July 29, 2011. Available at http://rpitt.eng.ua.edu/Publications/8_Stormwater_Management_and_Modeling/Stormwater_management_and_decision_analyses/Stormwater%20Management%20Options%20Lincoln%20Jly%2029%202011.pdf.

Pitt, R. 2011c. Lincoln, Nebraska, Standard Land Use Characteristics and Pollutant Sources. Prepared for Wright Waters, Denver, CO. April 22, 2011. Available at http://rpitt.eng.ua.edu/Publications/4_Stormwater_Characteristics_Pollutant_Sources_and_Land_Development_Characteristics/Land_development_characteristics/Standard%20Land%20Use%20files%20and%20sources%20Lincoln%20April%2022%202011.pdf.

Pitt, R., and A. Maestre. 2014. Version 4 of the National Stormwater Quality Database.Available at http://www.bmpdatabase.org/nsqd.html.

Pitt, R., and J. McLean. 1986. Humber River Pilot Watershed Project. Ontario Ministry of the Environment, Toronto, Canada. June 1986. 483 pp.

Pitt, R., and L. Talebi. 2012a. Evaluation of Dry Wells and Cisterns for Stormwater Control: Milburn Township, New Jersey. EPA/600/R-12/600. Edison, NJ: Urban Watershed Management Branch, U.S. Environmental Protection Agency. October 2012. Available at http://nepis.epa.gov/Adobe/PDF/P100GLN2.pdf.

Pitt, R., and L. Talebi. 2012b. Evaluation and Demonstration of Stormwater Dry Wells and Cisterns in Milburn Township, New Jersey. Edison, NJ: EPA. 302 pp.

Pitt, R., S. Clark, R. Field, and K. Parmer. 1996. Groundwater Contamination from Stormwater. Chelsea, MI: Ann Arbor Press. 219 pp.

Pitt, R., S. Clark, and R. Field. 1999. Groundwater contamination potential from stormwater infiltration practices. Urban Water 1(3):217-236.

Pitt, R., R. Bannerman, S. Clark, and D. Williamson. 2005a. Sources of pollutants in urban areas Part 1. Older monitoring projects. Pp. 465-484, 507-530 in Effective Modeling of Urban Water Systems, Monograph 13, W. James, K.N. Irvine, E.A. McBean, and R.E. Pitt, eds. Guelph, Ontario: CHI. Pitt, R., R. Bannerman, S. Clark, and D. Williamson. 2005b. Sources of pollutants in urban areas Part 2—Recent sheetflow monitoring results. Pp. 485-530 in Effective Modeling of Urban Water Systems, Monograph 13. W. James, K.N. Irvine, E.A. McBean, and R.E. Pitt, eds. Guelph, Ontario: CHI.

Pitt, R., D. Williamson, and J. Voorhees. 2005c. Review of historical street dust and dirt accumulation and washoff data. Pp. 203-246 in Effective Modeling of Urban Water Systems, Monograph 13., W. James, K.N. Irvine, E.A. McBean, and R.E. Pitt, eds. Guelph, Ontario: CHI.

Pitt, R., J. Voorhees, and S. Clark. 2008. Evapotranspiration and related calculations for stormwater biofiltration devices: Proposed calculation scenario and data. Pp. 309-340 in Stormwater and Urban Water Systems Modeling, Monograph 16, W. James, E.A. McBean, R.E. Pitt, and S.J. Wright, eds. Guelph, Ontario: CHI.

Pitt, R., L. Talebi. R. Bean, and S. Clark. 2011. Stormwater Non-Potable Beneficial Uses and Effects on Urban Infrastructure. Report No. INFR3SG09. Alexandria, VA: Water Environment Research Foundation. 224 pp.

Pitt, R., L. Talebi, D. O'Bannon, D. Bambic, and J. Wright. 2014. Modeling of Green Infrastructure Components and Large Scale Test and Control Watersheds at Kansas City, MO. Prepared for EPA and Tetra Tech. 357pp.

Poor, P.J., K.L. Pessagno, and R.W. Paul. 2007. Exploring the hedonic value of ambient water quality: A local watershed-based study. Ecological Economics 60(4):797-806.

Prince George's County, Maryland. 1999. Low-Impact Development Design Strategies: An Integrated Design Approach. Department of Environmental Resources. June. Available at http://www.lowimpactdevelopment.org/pubs/LID_National_Manual.pdf. Accessed December 11, 2015.

Proskiw, G. 1998. Technology Profile: Residential Greywater Heat Recovery Systems. Available at http://www.gfxtechnology.com/Proskiw-NRCan.pdf.

PUB (Public Utilities Board). 2015. The Singapore Water Story. Available at http://www.pub.gov.sg/water/Pages/singaporewaterstory.aspx. Accessed August 5, 2015.

Quanrud, D., L. Abrell, R. Arnold, and E. Saez. 2010. Perfluoronated Compounds in Arizona Groundwater: Sources of Contamination. Progress Report 2010. State Water Resources Research Institute Program. Project ID 2010AZ380B. U.S. Geological Survey.

Qureshi, A.A., and B.J. Dutka. 1979. Microbiological studies on the quality of urban stormwater runoff in southern Ontario, Canada. Water Research 13(10):977-985.

Rajal, V.B., B.S. McSwain, D.E. Thompson, C.M. Leutenegger, and S. Wuertz. 2007. Molecular quantitative analysis of human viruses in California stormwater. Water Research 41(10):4287-4298.

Ramon, G., M. Green, R. Semiat, and C. Dosoretz. 2004. Low strength graywater characterization and treatment by direct membrane filtration. Desalination 170:241-250.

Rassier, P.J. 2008. Response of the Deputy Attorney General to Sen. Gary J. Schroeder regarding the capture and collection of rainwater under Idaho law. Available at http://www.harvesth2o.com/adobe_files/Idaho_Ltr%20Sen%20Schroeder%20081108.pdf. Accessed July 21, 2014.

Raucher, B., and G. Tchobanoglous. 2014. The Opportunities and Economics of Direct Potable Reuse. WateReuse.

Raucher, R., J. Clements, J. Cromwell, D. Mills, and L. Deck. 2009. A Triple Bottom Line Assessment of Traditional and Green Infrastructure Options for Controlling CSO Events in Philadelphia's Watersheds. Prepared by Stratus Consulting, Boulder, CO, for City of Philadelphia Water Department. August. Available at https://www.michigan.gov/documents/dnr/TBL.AssessmentGreenVsTraditionalStormwaterMgt_293337_7.pdf.

Rector, P., T. Duckworth, and D. Fonseca. 2014. Rain Barrels and Mosquitoes. Fact Sheet FS1240. Rutgers New Jersey Agricultural Experiment Station. Available at https://njaes.rutgers.edu/pubs/fs1240. Accessed November 3, 2015.

Regnery, J., J. Lee, P. Kitanidis, T. Illangasekare, J.O. Sharp, and J.E. Drewes. 2013. Integration of managed aquifer recharge for impaired water sources in urban settings—overcoming current limitations and engineering challenges. Environmental Engineering Science 30(8):409-420.

Reichel, B., S. Sharvelle, and L. Roesner. 2011. Development of the Integrated Urban Water Model. WERF Project No. INFR4SGc.

Riccardi, N. 2011. Firms turning to environmental law to combat rivals. L.A. Times, November 14, 2011. Available at http://articles.latimes.com/2011/nov/14/local/la-me-development-ceqa-20111114. Accessed April 6, 2014.

RMC. 2011. Upper Petaluma River Watershed Flood Control Project Scoping Study—Conceptual Alternatives Screening Evaluation, Draft Technical Memorandum, Prepared for Kent Gylfe, SCWA. December 6, 2011.

RMC Water and Environment. 2012. SFPUC Onsite Nonpotable Program: Final Nontechnical Summary. June.

RMC Water and Environment. 2014. Los Angeles Basin Stormwater Conservation Study, Task 2. Water Supply and Water Demand Projections. Prepared for County of Los Angeles Department of Public Works, Los Angeles County Flood Control District, U.S. Department of the Interior-Bureau of Reclamation. December.

Roa-Espinosa, A., and R.T. Bannerman. 1995. Monitoring BMP effectiveness at industrial sites. Pp. 467-486 in Stormwater NPDES Related Monitoring Needs, Proceedings of an Engineering Foundation Conference August 7-12, 1994, Mount Crested Butte, CO. New York: American Society of Civil Engineers.

Romero, C.C., and M.D. Dukes. 2008. Turfgrass Crop Coefficients in the U.S. Available at http://irrigationtoolbox.com/ReferenceDocuments/TechnicalPapers/IA/2008/2127translated.pdf.

Rose, J.B., G.S. Sun, C.P. Gerba, and N.A. Sinclair. 1991. Microbial quality and persistence of enteric pathogens in greywater from various household sources. Water Research 25(1):37-42.

Sabin, L.D., J.H. Lim, K.D. Stolzenbach, and K.C. Schiff. 2005. Contribution of trace metals from atmospheric deposition to stormwater runoff in a small impervious urban catchment. Water Research 39(16):3929-3937.

Salukazana, L., S. Jackson, N. Rodda, M. Smith, T. Gounden, N. Macleod, and C. Buckley. 2005. Plant Growth and Microbiological Safety of Plants Irrigated with Greywater. Paper presented at the Third International Conference on Ecological Sanitation, May 23-26, 2005, Durban, South Africa. Available at http://conference2005.ecosan.org/papers/salukazana_et_al.pdf.

Santa Ana River Watermaster. 2014. Forty-third Annual report of the Santa Ana River Watermaster for Water Year October 1, 2012-September 30, 2013. April 30, 2014. Available at http://www.wmwd.com/DocumentCenter/View/2117.

Santa Clara Valley. 2005. Hydromodification Management Plan, Chapter 6. Guidance for Selection and Design of Hydromodification Control Measures. Santa Clara Valley Urban Runoff Pollution Prevention Program. April 2005.

Schijven, J.F., and S.M. Hassanizadeh. 2000. Removal of viruses by soil passage: Overview of modeling, processes and parameters. Critical Reviews in Environmental Science and Technology 30(1):49125.

Schriewer, A., H. Horn, and B. Helmreich. 2008. Time focused measurements of roof runoff quality. Corrosion Science 50(2):384-391.

Schroeder, E., W. Stallard, D. Thompson, F. Loge, M. Deshusses, and H. Cox. 2002. Management of Pathogens Associated with Storm Drain Discharge. Center for Environmental and Water Resources Engineering, Department of Civil and Environmental Engineering, University of California, Davis. Prepared for Division of Environmental Analysis, California Department of Transportation.

SCVWD (Santa Clara Valley Water District). 2011. From Watts to Water. May 2011.

SCWC. 2012. Stormwater task force. In Stormwater Opportunities Paper on March 26, 2012.

Selbig, W.R., and N. Balster. 2010. Evaluation of Turf-Grass and Prairie-Vegetated Rain Gardens in a Clay and Sand Soil, Madison, Wisconsin, Water Years 2004-08. Reston, VA: U.S. Department of the Interior, U.S. Geological Survey.

Semizentral. 2015. Semizentral-Supply and Treatment Systems: Sub-project 3. Implementation of a RRC in Qingdao. Available at http://semizentral.de/en/projects/projects-china/implementation-of-semizentral-qingdao.

SFPUC (San Francisco Public Utilities Commission). 2009. San Francisco Stormwater Design Guidelines. Available at http://www.sfwater.org/Modules/ShowDocument.aspx?documentID=2779. Accessed December 11, 2015.

SFPUC. 2012. San Francisco Graywater Design Manual for Outdoor Irrigation. June. Available at http://sfwater.org/modules/showdocument.aspx?documentid=55.

SFPUC. 2013. District-scale Water Reuse Case Studies. October. Available at http://sfwater.org/modules/showdocument.aspx?documentid=6133.

SFPUC. 2014a. San Francisco's Non-potable Water System Projects. Available at http://www.sfwater.org/modules/showdocument.aspx?documentid=5499.

SFPUC. 2014b. Blueprint for Onsite Water Systems. September. Available at http://sfwater.org/modules/showdocument.aspx?documentid=6057.

SFPUC. 2015. San Francisco's Non-potable Water Program. San Francisco Water Power Sewer Services. Available at http://sfwater.org/index.aspx?page=686.

Sharma, A.K., D. Begbie, and T. Gardner, eds. 2015. Rainwater Tank Systems for Urban Water Supply: Design, Yield, Energy, Health Risks, Economics and Social Perceptions. London: IWA Publishing.

Sharvelle, S., L.A. Roesner, Y. Qian, M. Strobmerberger, and M. Negahban-Azar. 2012. Long-Term Study on Landscape Irrigation Using Household Graywater—Experimental Study. WERF Project 06-CTS-1CO. Available at http://greywateraction.org/wp-content/uploads/2014/12/Long-term-Study-on-Landscape-Irrigation-Using-Household-Graywater.pdf.

Sharvelle, S., L. Roesner, and R. Glenn. 2013. Treatment, Public Health, and Regulatory Issues Associated with Graywater Reuse. Alexandria, VA: WateReuse Foundation.

Shehabi, A., J.R. Stokes, and A. Horvath. 2012. Energy and air emission implications of a decentralized wastewater system. Environmental Research Letters 7(2).

Sheikh, B. 2010. White Paper on Graywater. WateReuse Association. Available at http://www.graywater.org.il/Documents/GraywaterFinal%20Report2010%20for%20the%20US%20-%20AWWA.pdf.

Shergill, S.S., and R. Pitt. 2004. Quantification of Escherichia coli and enterococci levels in wet weather and dry weather flows. 77th Annual Water Environment Federation Technical Exposition and Conference, October 4-7, 2004, New Orleans, LA. Available at http://water.epa.gov/polwaste/npdes/stormwater/upload/sw_idde_bacteria.pdf.

Showley, R. 2015. Recycling System Cuts Home Water Use. The San Diego Union-Tribune, June 22, 2015. Available at http://www.sandiegouniontribune.com/news/2015/jun/22/water-recycle-kb-homes. Accessed November 11, 2015.

Sidhu, J.S., S. Toze, L. Hodgers, M. Shackelton, K. Barry, D. Page, and P. Dillon. 2010. Pathogen inactivation during passage of stormwater through a constructed reedbed and aquifer storage, transfer and recovery. Water Science and Technology 62(5):1190-1197.

Simmons, G., V. Hope, G. Lewis, J. Whitmore, and W. Gao. 2001. Contamination of potable roof-collected rainwater in Auckland, New Zealand. Water Research 35(6):1518-1524.

Smith, I.E. 1975. Recovery and utilisation of heat from domestic waste water. Journal of Applied Energy 1(3):205-214.

Snodgrass, J.W., R.E. Casey, D. Joseph, and J.A. Simon. 2008. Microcosm investigations of stormwater pond sediment toxicity to embryonic and larval amphibians: Variation in sensitivity among species. Environment and Pollution 154(2):291-297.

Snodgrass, R.F.M. 2010. Greywater—the reuse of household water: A small step toward sustainable living and adaptation to climate change. Georgetown International Environmental Law Review 22:591-596.

Soller, J.A. 2009. The potential implications of person-to-person transmission of viral infection for U.S. EPA's Groundwater Rule. Journal of Water Health 7(2):208-223.

Sostar-Turk, S., I. Petrinic, and M. Simonic. 2005. Laundry wastewater treatment using coagulation and membrane filtration. Resources, Conservation and Recycling 44(2):185-196.

Sparling, D.W., J.D. Eisemann, and W. Kuenzel. 2004. Contaminant exposure and effects in red-winged inhabiting stormwater retention ponds. Environmental Management 33(5):719-729.

Spencer, J.L. 2013. The Carbon Impacts of Stormwater Capture, A Study Conducted for the California State Water Quality Control Board, Goldman School of Public Policy, University of California Santa Barbara. May 2013.

Stachel, B., J.-U. Holthuis, W. Schulz, W. Seitz, W.H. Weber, K.T. Tegge, and I. Dobner. 2010. Treatment techniques and analysis of stormwater run-off from roads in Hamburg, Germany. Pp. 445-461 in Xenobiotics in the Urban Water Cycle. D. Fatta-Kassinos, K. Bester, and K. Kümmerer, eds. Dordrecht, the Netherlands: Springer.

Stephenson, J.B., W.F. Zhou, B.F. Beck, and T.S. Green. 1999. Highway stormwater runoff in karst areas—preliminary results of baseline monitoring and design of a treatment system for a sinkhole in Knoxville, Tennessee. Engineering Geology 52(1-2):51-59.

Steuer, J., W. Selbig, N. Hornewer, and J. Prey. 1997. Sources of Contamination in an Urban Basin in Marquette, Michigan and an Analysis of Concentrations, Loads, and Data Quality, U.S. Geological Survey Water-Resources Investigations Report 97-4242. Middleton, WI: USGS.

Stevens, D., P. Dillon, D. Page, M. Warne, and G-G Ying. 2011. Assessing environmental risks of laundry detergents in greywater used for irrigation. International Water Association Journal of Water Reuse and Desalination 1(2):61-77.

Stokes, J.R., and A. Horvath. 2009. Energy and air emission effects of water supply. Environmental Science and Technology 43(8):2680-2687.

Stokes, J.R., and A. Horvath. 2010. Supply-chain environmental effects of wastewater utilities. Environmental Research Letters 5(1):014015.

Stokes, J.R., T.P. Hendrickson, and A. Horvath. 2014. Save water to save carbon and money: Developing abatement costs for expanded greenhouse gas reduction portfolios. Environmental Science and Technology 48(23):13583-13591.

Stoner, N., and C. Giles. 2011. Memorandum: Protecting Water Quality with Green Infrastructure in EPA Water Permitting and Enforcement Programs. Washington, DC: EPA. Available at http://water.epa.gov/infrastructure/greeninfrastructure/gi_regulatory.cfm.

Strecker, E.W., and A.L. Poresky. 2010. The feasibility and desirability of stormwater retention on site in California and on the West Coast. Pp. 1022-1035 in Proceedings of the 2010 International Low Impact Development Conference, April 11-14, 2010, San Francisco, CA, S. Struck, and K. Lichten, eds. Reston, VA: ASCE.

Strecker, E.W., W.C. Huber, J. P. Heaney, D. Bodine, J.J. Sansalone, M.M. Quigley, D. Pankani, M. Leisenring, and P. Thayumanavan. 2005. Critical Assessment of Stormwater Treatment and Control Selection Issues. Report No. 02-SW-1. Alexandria, VA: Water Environment Research Federation. 290 pp.

Sumandeep, S., and R. Pitt. 2004. Quantification of Escherichia coli and Enterococci Levels in Wet Weather and Dry Weather Flows. Available at http://www3.epa.gov/npdes/pubs/sw_idde_bacteria.pdf.

Sunding, D. 2013. Economic Analyses Conducted on the Bay Delta Conservation Plan. Presentation July 23, 2013. Available at http://www.mwdh2o.com/DocSvcsPubs/mwd_newsletter/Oct2013/Bay-Delta.pdf.

SWRCB. 2013. Recycled Water Policy, as modified by California State Water Board Resolution 2013-0003, January 22, 2013. Available at: http://www.swrcb.ca.gov/board_decisions/adopted_orders/resolutions/2013/rs2013_0003_a.pdf. Accessed December 11, 2015.

Talebi, L., and R. Pitt. 2013. Evaluation of Retrofitted Green Infrastructure Stormwater Controls at Cincinnati State College, the Cincinnati Zoo, and the Clark Montessori High School. Prepared for Metropolitan Sewer District of Greater Cincinnati and EPA. Available at http://unix.eng.ua.edu/~rpitt/Publications/5_Stormwater_Treatment/GI_controls/Cincy_modeling_report_Final.pdf.

Tampa Bay Water. 2008. Drought-Proof Water Supply Delivering Drinking Water, News Release: January 25, 2008. Available at http://www.tampabaywater.org/newsroom/agency-news/ArtMID/551/ArticleID/119/Drought-Proof-Water-Supply-Delivering-Drinking-Water.aspx.

Taylor, S. 1984. Making Bureaucracies Think: The Environmental Impact Statement Strategy of Administrative Reform. Stanford, CA: Stanford University Press.

Teerlink, J., A. Hering, C. Higgins, and J.E. Drewes. 2012. Variability of trace organic chemical concentrations in raw wastewater at three distinct sewershed scales. Water Research 46(10):3261-3271.

Thomas, R.B., M.J. Kirisits, D.J. Lye, and K.A. Kinney. 2014. Rainwater harvesting in the United States: A survey of common system practices. Journal of Cleaner Production 75:166-173.

Tixier, G., M. Lafont, L. Grapentine, Q. Rochfort, and J. Marsalek. 2011. Ecological risk assessment of urban stormwater ponds: Literature review and proposal of a new conceptual approach providing ecological quality goals and the associated bioassessment tools. Ecological Indicators 11(6):1497-1506.

Tobiason, S. 2004. Stormwater metals removal by media filtration: Field assessment case study. Pp. 1431-1448 in Watershed 2004; Proceedings of the Water Environment Federation.Alexandria, VA: Water Environment Federation.

Tobiason, S.A., and L. Logan., 2000. Stormwater Whole Effluent Toxicity (WET) testing and source tracing at SEA-TAC International Airport. Pp. 617-632 in WEFTEC 2000: Proceedings of the Water Environment Federation. Alexandria, VA: Water Environment Federation.

TRB (Transportation Research Board). 2006. Evaluation of Best Management Practices for Highway Runoff Control. National Cooperative Highway Research Program Report 565.

TWDB (Texas Water Development Board). 2005. The Texas Manual on Rainwater Harvesting, 3rd Edition. Available at http://www.twdb.texas.gov/publications/brochures/conservation/doc/RainwaterHarvestingManual_3rdedition.pdf.

TWDB. 2006. Rainwater Harvesting Potential and Guidelines for Texas. Texas Rainwater Harvesting Evaluation Committee Report to the 80th Legislature. Austin, TX. November.

Uba, B. N., and O. Aghogho. 2000. Rainwater quality from different roof catchments in the Port Harcourt district, Rivers State, Nigeria.Journal of Water Supply, Research and Technology-Aqua 49:281-288.

UDFCD (Urban Drainage and Flood Control District). 2013. Denver Urban Drainage and Flood Control's Criteria Manual. Available at http://www.udfcd.org/downloads/down_critmanual_volIII.htm. Accessed June 19, 2015.

Van Houtven, G., C. Mansfield, D. J. Phaneuf, R. von Haefen, B. Milstead, M.A. Kenney, and K. H. Reckhow. 2014. Combining expert elicitation and stated preference methods to value ecosystem services from improved lake water quality. Ecological Economics 99:40-52.

Vandegrift, J. 2014. Implementation of Graywater Reuse in the State of Colorado. M.S. Thesis, Colorado State University.

Vanderzalm, J., D. Page, D. Gonzalez, K. Barry, S. Toze, R. Bartak, Q. Shisong, W. Weiping, P. Dillon, and M.H. Lim. 2014. Managed Aquifer Recharge and Stormwater Use Options: Satellite Sites Stormwater Quality Monitoring and Treatment Requirements for Potable Supplies. Technical Report Series No. 14/10, Goyder Institute for Water Research, Adelaide, South Australia.

VanLoon, G., B.C. Anderson, W.E. Watt, and J. Marsalek. 2000. Characterizing stormwater sediments for ecotoxic risk. Water Quality Research Journal of Canada 35:341-364.

VDH (Virginia Department of Health). 2011. Virginia Rainwater Harvesting and Use Guidelines. March 31, 2011. Available at http://www.vdh.state.va.us/EnvironmentalHealth/ONSITE/gmp/documents/2011/pdf/GMP_154.pdf.

Vieira, A.S., C.D. Beal, E. Ghisi, and R.A. Stewart. 2014. Energy intensity of rainwater harvesting systems: A review. Renewable and Sustainable Energy Reviews 34:225-242.

Wallinder, O.I., Y. Hedberg, and P. Dromberg. 2009. Storm water runoff measurements of copper from a naturally patinated roof and from a parking space. Aspects on environmental fate and chemical speciation. Water Research 43(20):5031-5038.

Ward, R.L., D.L. Bernstein, E.C. Young, J.R. Sherwood, D.R. Knowlton, and G.M. Schiff. 1986. Human rotavirus studies in volunteers: Determination of infectious dose and serological response to infection. Journal of Infectious Diseases 154(5):871-880.

Warren, M. 2013. Rain and Stormwater Management at the Cincinnati Zoo and Botanical Gardens. Presentation to the EPA Conference on Technologies and Innovative Solutions for Harvesting and Nonpotable Use of Rain and Stormwater in Urban Settings, April 24-25, 2013, Cincinnati, OH. Available at http://www.arcsa-edu.org/epa_pdf%27s/Warren.pdf.

Waschbusch, R.J., R.T. Bannerman, and S.R. Greb. 1998. Yields of Selected Constituents in Street Runoff in Madison, Wisconsin, 1994-95. U.S. Geological Survey Water-Resources Investigation Report.

The Water Corporation. 2010. Perth Residential Water Use Study, 2008-2009. Available at https://www.water.wa.gov.au/__data/assets/pdf_file/0016/5272/98576.pdf. Accessed November 13, 2015.

Waterscan. 2014. Gray Water Recycling: Going Green to Save Your Business. Available at http://water-recycling-harvesting.co.uk/category/water-recycling.

Wathern, P. 1988. Environmental Impact Assessment: Theory and Practice. New York: Routledge.

WEF (Water Environment Federation). 2014. Green Infrastructure Implementation: A Special Report. Alexandria, VA: Water Environment Federation.

Weingaertner, D. 2013. Greywater. Characteristics, Biodegradability and Reuse of Some Greywaters. Ph. D. Dissertation. Institute of Technology, Karlsruhe, Germany.

Weiss, P.T., G. LeFevre, and J.S. Gulliver. 2008. Contamination of Soil and Groundwater Due to Stormwater Infiltration Practices: A Literature Review. Prepared for Minnesota Pollution Control Agency, by University of Minnesota. June 23, 2008.Available at http://www.pca.state.mn.us/index.php/view-document.html?gid=7732.

WERF (Water Environment Research Foundation). 2005. Critical Assessment of Stormwater Treatment and Control Selection Issues. Project No. 02-SW-1. Available at https://www.werf.org/a/ka/Search/ResearchProfile.aspx?ReportId=02-SW-1.

WERF. 2009. 21st Century Sustainable Water Infrastructure: Smart, Clean and Green. Water Environment Research Foundation, February 2009.

WERF. 2014. Transforming Our Cities: High-Performance Green Infrastructure. INFR1R11. Available at http://www.werf.org/a/ka/Search/ResearchProfile.aspx?ReportId=INFR1R11.

West, T., Y. Sim, and K. Kuyunz. 2014. Integrated Stormwater Management Demonstration Project Database. Presentation at the 3rd Annual Stormwater Task Force Workshop, June 26, 2014, Los Angeles, CA.

Weston, D.P., R.W. Holmes, and M.J. Lydy 2009. Residential runoff as a source of pyrethroid pesticides to urban creeks. Environmental Pollution 157(1):287-294.

Whelan, G., K. Kim, M.A. Pelton, J.A. Soller, K.J. Castleton, M. Molina, Y. Pachepsky, and R. Zepp. 2014. An integrated environmental modeling framework for performing quantitative microbial risk assessments. Environment Modelling and Software 55:77-91.

Wik, A., J. Lycken, and G. Dave. 2008. Sediment quality assessment of road runoff detention systems in Sweden and the potential contribution of tire wear. Water Air and Soil Pollution 194:301-314.

Wiles, K. 2013. Investigation of Pathogen Disinfection and Regrowth for a Lost Cost Graywater Reuse Treatment System for Toilet Flushing. M.S. Thesis, Colorado State University.

Winward, G., L.M. Avery, R. Frazer-Williams, M. Pidou, P. Jeffrey, T. Stephenson, and B. Jefferson. 2008. A study of the microbial quality of grey water and an evaluation of treatment technologies for reuse. Ecological Engineering 32(2):187-197.

Wohlsen, T., J. Bates, B. Gray, P. Aldridge, S. Stewart, M. Williams, and M. Katouli. 2006. The occurrence of Cryptosporidium and Giardia in the Lake Baroon catchment, Queensland, Australia. Journal of Water Supply Research and Technology-Aqua 55:357-366.

World Bank. 2012. Designing a Results Framework for Achieving Results: A How-to Guide. Available at http://siteresources.worldbank.org/EXTEVACAPDEV/Resources/designing_results_framework.pdf. Accessed February 28, 2013.

WQRA (Water Quality Research Australia). 2010. Coomera Cross-Connection Incident. Public Health Newsletter of Water Quality Research Australia 57:1-4.

WV DEP (West Virginia Department of Environmental Protection). 2012. West Virginia Stormwater Management and Design Guidance Manual. Available at http://www.dep.wv.gov/WWE/Programs/stormwater/MS4/Pages/StormwaterManagement-DesignandGuidanceManual.aspx.

Yamahara, K.M., B.A. Layton, A.E. Santoro, and A.B. Boehm, 2007. Beach sands along the California coast are diffuse sources of fecal bacteria to coastal waters. Environmental Science and Technology 41(13):4515-4521.

Yang, K., J. LeJeune, D. Alsdorf, B. Lu, C.K. Shum, and S. Liang. 2012. Global distribution of outbreaks of water-associated infectious diseases. PLoS Neglected Tropical Diseases 6(2):e1483.

Yates, M.V., S.R. Yates, and C.P. Gerba. 1988. Modeling microbial fate in the subsurface environment. CRC Critical Reviews in Environmental Control 17(4):307-344.

Yu, Z.L., A. Rahardianto, J.R. DeShazo, M.K. Stenstrom, and Y. Cohen. 2013. Critical review: Regulatory incentives and impediments for onsite graywater reuse in the United States. Water Environment Research. 85(7):650-662.

Zobrist, J., S.R. Muller, A. Ammann, T.D. Bucheli, V. Mottier, M. Ochs, R. Schoenenberger, J. Eugster, and M. Boller. 2000. Quality of roof runoff for groundwater infiltration. Water Research 34(5):1455-1462.

Zuma, B.M., R. Tandlich, K.J. Whittington-Jones, and J.E. Burgess. 2009. Mulch tower treatment system, Part I: Overall performance in greywater treatment. Desalination 242(1-3):38-56.

A

Calculating the Benefits of Rooftop Runoff Capture Systems

This appendix presents the methods used (with examples) to evaluate the beneficial uses of roof runoff harvesting for irrigation of landscaped areas and for toilet flushing. The Source Loading and Management Model, WinSLAMM (Pitt, 1997), was used to calculate the benefits of harvesting stormwater for storage and later beneficial uses. The methods were previously used and described by Pitt et al. (2011, 2014). WinSLAMM is a continuous model that evaluates a long series of rains for an area. WinSLAMM[1] is licensed for sale but is available free of charge to academic institutions. An evaluation license is also available to interested readers who wish to examine the model for a limited time. Input files used in this scenario analysis are available in the Academies' Public Access Records Office.

For this report, WinSLAMM focused on the capture of rooftop runoff and use for turfgrass irrigation and toilet flushing, as described in Box 3-1. Different storage tank volumes were also evaluated. Average monthly (and daily) irrigation requirements were calculated by subtracting average monthly rainfall from 1995 to 1999 (1996-1999 for Lincoln, because of missing data) from average monthly evapotranspiration (ET) values. Then using WinSLAMM and the 5-year precipitation time series, if rainfall was insufficient to meet the irrigation demand, then supplemental irrigation was required. If available, then the supplemental irrigation was supplied by previously stored roof runoff water in storage tanks. Toilet flushing requirements were based on typical national indoor water uses (11 gpcd; see Box 3-1). The following is a summary of the main calculations and data used for these analyses.

WINSLAMM

WinSLAMM evaluates stormwater runoff volumes and pollutant loads under an array of stormwater management practices including rain barrels and water tanks, although the committee did not assess pollutants in this analysis (Pitt, 1987). Using local rain records, the model calculates runoff volumes and pollutant loadings for each rain from individual source areas within various land use categories and sums the results over a given area or land use. Examples of runoff source areas considered by the model include roofs, streets, sidewalks, parking areas, and landscaped areas, which each have different runoff coefficients based on the type of surface, slope, and soil properties (Pitt, 1987). Example land use categories include commercial, industrial, institutional, open space, residential, and freeway/highway. The committee's scenario modeling exercise mainly focuses on roof runoff for small-scale stormwater harvesting and on land use runoff for larger scale stormwater harvesting.

Any length of rainfall record can be analyzed with WinSLAMM, from a single event to many decades. The rainfall files used in the committee's calculations were developed from hourly rainfall data obtained from the National Oceanic and Atmospheric Administration (NOAA) rainfall stations as published on EarthInfo CD-ROMs.

DATA REQUIREMENTS AND SOURCES OF INFORMATION

WinSLAMM uses various sets of information in its calculations. The main data required for the analyses in this report included rain data for the six locations examined (from NOAA weather stations), runoff coefficients for the source areas for different land uses, and land development characteristics for the land uses in each area examined.

Rainfall Data

As noted in the report, six areas of the country were examined to represent a range of climatic conditions:

- Los Angeles, California, having a median rainfall of about 12 inches per year over the long-term record (17 inches average during the 5-year calculation period)

[1] See http://winslamm.com.

- Seattle, Washington, having a median rainfall of about 37 inches of rainfall per year (42 inches average during the 5-year calculation period)
- Lincoln, Nebraska, having a median rainfall of about 26 inches of rainfall per year (28 inches average during the 4-year calculation period)
- Madison, Wisconsin, having a median rainfall of about 32 inches of rainfall per year (30 inches average during the 5-year calculation period)
- Birmingham, Alabama, having a median rainfall of about 54 inches of rainfall per year (50 inches average during the 5-year calculation period)
- Newark, New Jersey, having a median rainfall of about 43 inches of rainfall per year (44 inches average during the 5-year calculation period)

Most of the modeling calculations focused on recent 5 years of rainfall records (1995 through 1999 for all areas, except for Lincoln, where 1996 through 1999 rains were used due to many missing rains in the 1995 record).

The goal was to use a continuous period of actual rains that were similar to the long-term average conditions, because continuous simulations were needed to calculate the inter-event water demands based on the average ET values. The committee based its scenario analysis on 5-year rain periods to reduce data pre-processing demands and because long records are rarely available without data gaps. Moderate rain record lengths reduce these gap problems (although Lincoln was missing 1995) and have been used to reduce large year-to-year variabilities while attempting to match the average monthly ET values. In Table A-1 and Figure A-1, the committee compares the rainfall data from the 4- to 5-year calculation periods with the long-term precipitation record. Some variations are apparent even though the differences are not statistically significant. Some of these differences are discussed in the context of analysis uncertainties in Box 3-2.

The committee judges that the calculation methods and data used for these analyses represent reasonable conditions and present results that are useful for the comparative analysis presented in Chapter 3. However, the data are not intended as definitive predictions or as a basis for design guidance.

TABLE A-1 Comparison of Precipitation Annual Rain Totals and Rain Counts Between the Scenario Analysis Calculation Period and the Long-term Rainfall Record

	Los Angeles, CA	Seattle, WA	Lincoln, NE	Madison, WI	Birmingham, AL	Newark, NJ
Long-term rain record	1948-1999	1965-2012	1973-1999 (1995 gap)	1948-1999	1948-1999 (1978-1987 gap)	1948-1999
Scenario analysis calculation period	1995-1999	1995-1999	1996-1999	1995-1999	1995-1999	1995-1999
Long-term annual median rain total (in)	11.70	36.69	26.45	31.85	53.68	42.51
Scenario analysis calc. period median annual rain total (in)	15.82	42.10	28.62	31.19	52.40	41.28
p values (<0.05 indicates[a] significant difference)[a]	0.16	0.078	0.68	0.56	0.78	0.99
Comment of rain depth box and whisker plot comparisons	The calculation period has greater rains and a wider variation than the long term conditions	The calculation period has greater rains but similar variations as the long term conditions	The calculation period has similar rain depths per year and the variations are similar	The calculation period has smaller rain depths per year and the variations are similar	The calculation period has similar rain depths per year and the variations are similar	The calculation period has similar rain depths per year and the variations are similar
Long-term annual median rain counts	29	138	97	109	106	103
Scenario analysis calc. period median annual rain counts	32	140	97	103	97	97
p values (<0.05 indicates[a] significant difference)[a]	0.27	0.62	0.82	0.08	0.20	0.17

[a]Mann-Whitney rank sum p values (not independent data sets because the calculation period was included in the total period). Deemed acceptable as the hypothesis was to compare the full set with the subset. None of the rain depth or rain count comparisons indicated significant differences for the number of data available.

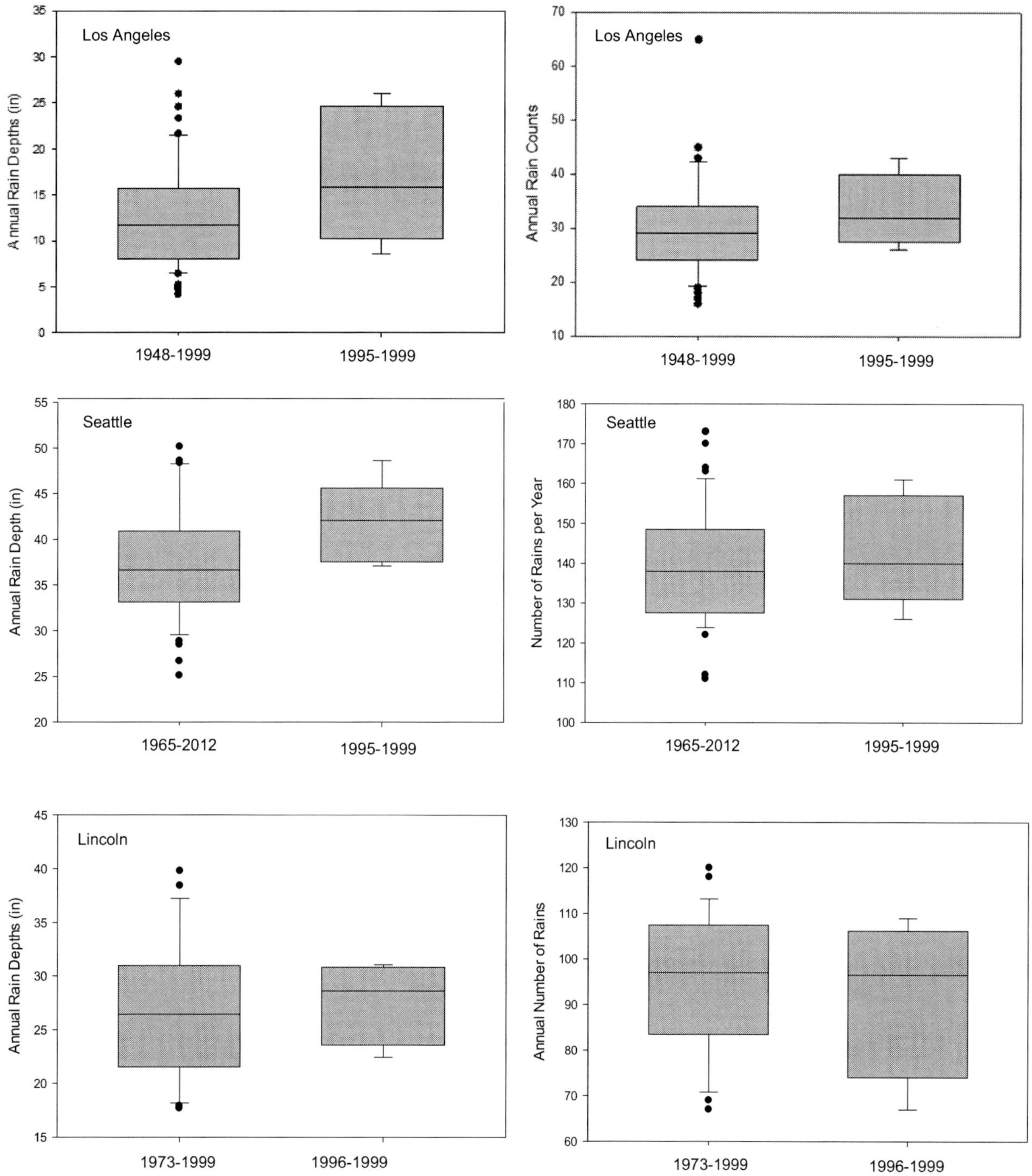

FIGURE A-1 Comparisons of the period of record with the scenario analysis period in terms of annual rain depth and number of rainfall events per year.

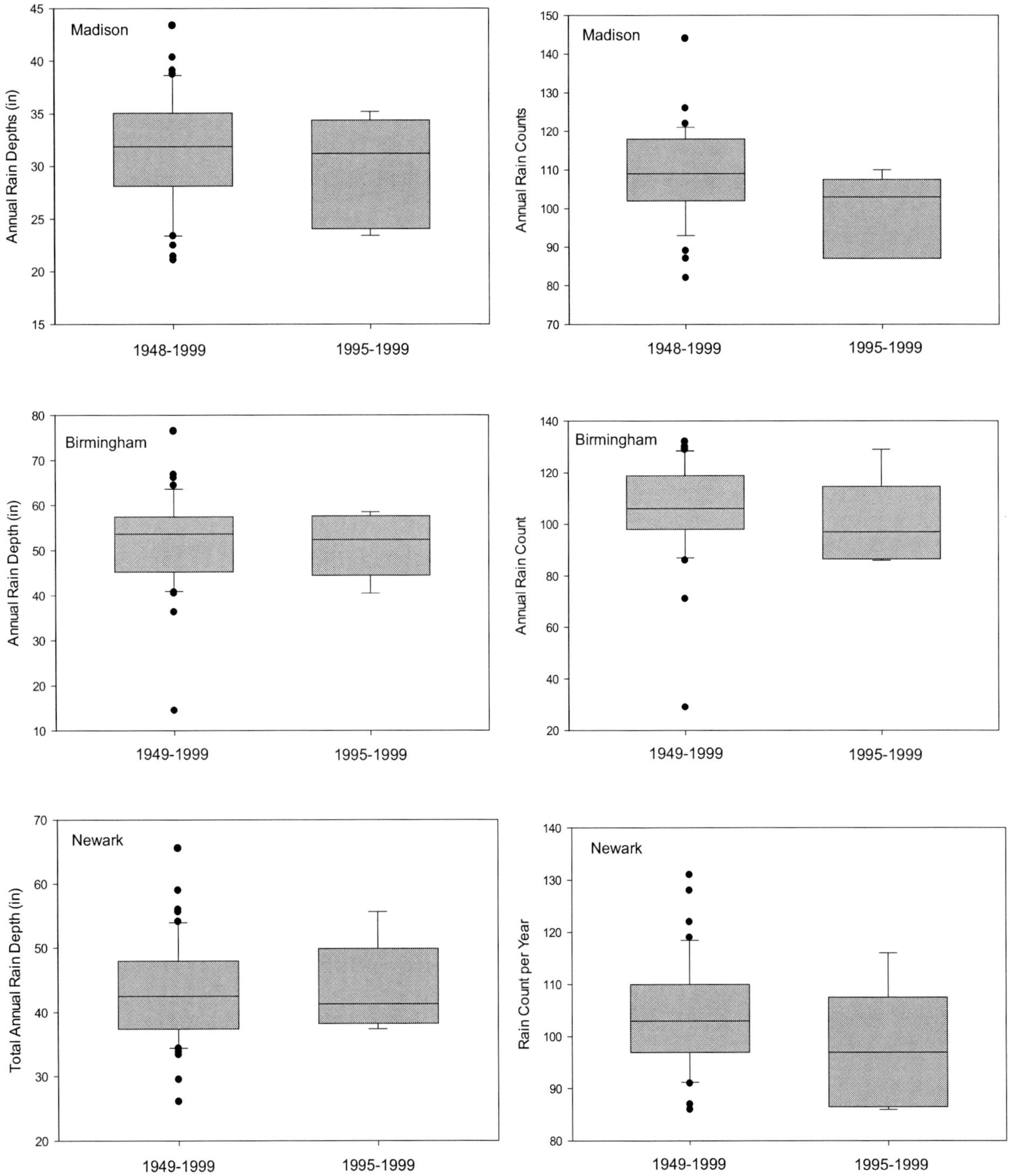

FIGURE A-1 Continued

Land Development Characteristics

An important element in calculating stormwater beneficial use opportunities using harvested roof runoff for landscaping irrigation is to know the typical areas of the roofs and the landscaped areas that are present in the different land uses and study locations. For larger-scale beneficial-use calculations, the areas of the other source areas in the land uses also need to be known. These areas were obtained from prior summaries conducted to support the U.S. Environmental Protection Agency's (EPA's) development of potential future stormwater regulations. These typical land development characteristics throughout the country, described in Pitt (2011a,b,c), are summarized in Table 3-3. Pitt (2011a) contains the citations and sources for the original data sources. More than 100 monitored locations were reviewed using site mapping and aerial photographs, along with concurrent monitoring data.

For irrigation beneficial uses of stormwater, the most suitable source for the collected water is from the building roofs because of its generally better water quality, high unit area runoff yield, and elevation above storage tanks and irrigated land. The landscaped areas represent the amount of area that can be irrigated with the harvested roof runoff water. Therefore, areas having relatively large roofs and small landscaped areas are most likely to have most of the irrigation demand in the area satisfied (but may not reduce the overall stormwater discharges as much as areas having small roofs and large irrigable land). Table A-2 shows the roof and landscaped areas for these six land uses for the Los Angeles area. Commercial areas generally have the smallest ratios of landscaped to roof areas and therefore are more likely to be able to meet irrigation requirements with the abundance of roof runoff. In contrast, it would be much more challenging to replace much of the irrigation water currently supplied by potable water supplies using roof runoff in low-density areas because the amount of roof runoff water is a much smaller portion of the total irrigation requirements. There are some differences in these development characteristics by region, and the rainfall patterns and evapotranspiration requirements vary greatly by area. Table A-3 shows the percentage of landscaped and roof areas and typical housing densities for medium-density, residential land uses (the focus of the committee's analysis) in each of the six locations of the country examined.

ROOF RUNOFF CALCULATIONS

The following sections describe an example set of calculations used to develop the analyses used in this report. These examples focus on medium-density, residential land use in Los Angeles.

TABLE A-2 Roof and Landscaped Areas for Los Angeles Land Uses

	Roof Area (%)	Landscaped Area (%)	Ratio of Landscaped Area to Roof Area
Commercial	28.1	14.9	0.53
High-density residential	20.7	46.4	2.24
Medium-density residential	18.0	52.5	2.92
Low-density residential	8.0	79.6	9.95
Industrial	20.2	24.3	1.20
Institutional	19.4	41.2	2.12

TABLE A-3 Landscaped and Roof Area, and Number of Households for Medium-Density, Residential Land Use in Six U.S. Locations

	Landscaped Areas (%)	Roof Areas (%)	# Roofs/100 Acres at 1,500 ft^2 Each
Los Angeles, CA	52.5	18.0	523
Seattle, WA	63.5	17.1	497
Lincoln, NE	62.8	18.1	526
Madison, WI	63.3	15.0	436
Birmingham, AL	81.3	8.8	256
Newark, NJ	56.2	15.9	462

Runoff Quantity

Table A-4 is a small portion of the WinSLAMM modeled scenario output showing runoff volume contributions for a 100-acre medium-density residential area in Los Angeles. These analyses were repeated for six major land use areas (commercial, high-density residential, medium-density residential, low-density residential, industrial, and institutional) and six U.S. locations. During this 5-year period examined (1995-1999), a total of about 84 inches fell, with rains as large as 3.5 inches (Table A-5). About 47 percent of the rainfall occurred as direct runoff for this area (or a the volumetric runoff coefficient [Rv] of 0.47), higher than for most residential areas, because these analyses assumed directly connected roof drainage, as would be the case for roof runoff harvesting. Most of the runoff volumes in this medium-density residential land use analysis originated from the street and roof areas, with smaller (and about equal amounts) from driveways, sidewalks, and landscaped areas. These relationships vary for different land uses and different geographical areas based on the local development characteristics, soils, and rain patterns.

Based on the 1995-1999 period, 100 acres of medium-density, residential area in Los Angeles produces about 14 million ft^3 of runoff, while the roofs in the area contribute about 5.2 million ft^3 of that runoff. These can be converted to inches of runoff over the drainage area for the 5-year period, for example:

$$\frac{13{,}969{,}610\,ft^3}{100\,ac} * \frac{acre}{43{,}560\,ft^2} * \frac{12\,in}{ft} = 38.48\,inches, or\ 7.70\,inches\ per\ year\ on\ average$$

For the roof area alone (which comprises 18 percent of the land use, or 18 acres):

$$\frac{5{,}170{,}000\,ft^3}{18\,ac} * \frac{acre}{43{,}560\,ft^2} * \frac{12\,in}{ft} = 79.13\,inches, or\ 15.83\,inches\ per\ year$$

The total rain depth for the 5 years is 83.67 inches, or 16.73 inches per year. The volumetric runoff coefficient (Rv) is the ratio of the runoff total to the rain total. Therefore, for the whole area, the total flow-weighted annual Rv is:

$$\frac{7.70\ inches}{16.73\ inches} = 0.45$$

while the Rv for the roof area alone is:

$$\frac{15.83\ inches}{16.73\ inches} = 0.95$$

TABLE A-4 Portion of WinSLAMM Model Output for Southwest, Medium-Density, Residential Areas (100-acre area) Showing Runoff Amounts (ft^3) from Different Sources Areas for Each Event and for All Areas Combined (5 years rain series)

			Runoff Amounts (ft^3)							
Month	Start Date	Rain Total (in.)	Land Use Totals	Roofs	Driveways	Sidewalks/ Walks	Street Area	Small Landscaped Area	Volumetric Runoff Coeff. (Rv)	Total Losses (in.)
1	1/3/1995	0.75	119,655	46,952	13,767	8,850	44,408	5,677	0.44	0.42
1	1/4/1995	3.5	716,277	226,403	83,501	53,679	219,291	133,402	0.56	1.53
1	1/7/1995	1.29	217,432	82,603	27,004	17,360	77,546	12,920	0.46	0.69
1	1/8/1995	0.4	56,379	24,323	6,521	4,192	19,719	1,623	0.39	0.24
1	1/10/1995	2.93	595,083	189,532	68,824	44,244	180,806	111,677	0.56	1.29
1	1/11/1995	0.14	16,023	7,115	1,812	1,165	5,931	0	0.32	0.1
1	1/11/1995	0.4	56,379	24,323	6,521	4,192	19,719	1,623	0.39	0.24
1	1/14/1995	0.12	13,386	5,899	1,499	964	5,024	0	0.31	0.08
1	1/20/1995	0.16	18,776	8,397	2,144	1,378	6,858	0	0.32	0.11
About 150 events between these two dates are not shown on this summary table										
4	4/11/1999	1.36	229,828	87,085	28,777	18,499	81,753	13,713	0.47	0.73
6	6/1/1999	0.52	77,811	32,034	8,880	5,709	28,535	2,653	0.41	0.31
6	6/2/1999	0.05	3,334	1,152	476	306	1,401	0	0.18	0.04
6	6/3/1999	0.02	314.4	166	90	58	0	0	0.04	0.02
11	11/8/1999	0.27	34,913	15,520	4,046	2,601	12,344	402	0.36	0.17
11	11/17/1999	0.01	78.6	41	23	15	0	0	0.02	0.01

TABLE A-5 Summary of All Events in 5-Year Rain Series in WinSLAMM Model Output for Southwest, Medium-Density, Residential Areas (100 acre area)

		Runoff Amounts (ft^3)							
	Rain Total (in.)	Land Use Totals	Roofs	Driveways	Sidewalks/ Walks	Street Area	Small Landscaped Area	Volumetric Runoff Coeff. (Rv)	Total Losses (in.)
Minimum	0.01	79	41	23	15	0	0	0.02	0.01
Maximum	3.5	716,277	226,403	83,501	53,679	219,291	133,402	0.56	1.53
Average	0.51	85,703	31,728	10,152	6,527	29,112	8,184	0.47	0.75
Total	83.67	13,969,610	5,170,000	1,655,000	1,064,000	4,745,000	1,334,000	n/a	45.18

TABLE A-6 Overall Summary of Runoff Volume Contributions by Source Area and Month for Los Angeles Medium Density Residential Areas

Five-Year Average Flows by Month	Rain Total (in.)	Land Use Totals	Roofs	Driveways	Sidewalks/ Walks	Street Area	Small Landscaped Area
Area (% of total land use)	n/a	100.00	18.00	7.00	4.50	18.00	52.50
Avg Jan runoff volume (in/mo)	4.89	2.26	4.65	3.85	3.85	4.26	0.41
Avg Feb runoff volume (in/mo)	3.76	1.88	3.63	3.13	3.13	3.38	0.49
Avg March runoff volume (in/mo)	2.48	1.13	2.33	1.88	1.88	2.12	0.22
Avg April runoff volume (in/mo)	0.86	0.35	0.78	0.60	0.60	0.70	0.03
Avg May runoff volume (in/mo)	0.59	0.24	0.54	0.42	0.42	0.49	0.02
Avg June runoff volume (in/mo)	0.25	0.09	0.21	0.15	0.15	0.18	0.00
Avg July runoff volume (in/mo)	0.01	0.00	0.01	0.00	0.00	0.01	0.00
Avg Aug runoff volume (in/mo)	0.00	0.00	0.00	0.00	0.00	0.00	0.00
Avg Sept runoff volume (in/mo)	0.05	0.02	0.05	0.03	0.03	0.04	0.00
Avg Oct runoff volume (in/mo)	0.29	0.13	0.28	0.24	0.24	0.26	0.02
Avg Nov runoff volume (in/mo)	1.30	0.56	1.22	0.95	0.95	1.11	0.05
Avg Dec runoff volume (in/mo)	2.24	1.03	2.14	1.77	1.77	1.97	0.16

All of the event data were sorted by month and then averaged to develop 5-year averaged monthly summaries of runoff volumes (average inches of runoff per month). Table A-6 is an overall summary showing these runoff volume contributions from each of the Los Angeles, medium-density, residential, source areas and the total annual flow conditions, expressed in average watershed-inches per month.

Evapotranspiration and Irrigation Demands

Evapotranspiration (ET) is defined as the rate at which readily available water is removed from the soil and plant surfaces, expressed as the rate of latent heat transfer per unit area or as a depth of water evaporated and transpired from a reference crop (Jensen et al., 1990). In the United States, ET monitoring is primarily focused in agricultural and wild land environments. With educational advancements stressing water conservation in urban areas, there is a new desire to apply ET data as a part of stormwater harvesting options for supplemental irrigation and to fine-tune actual irrigation requirements based on soil moisture and plant needs. Climate-based equations are the most common method used to determine ET. ET potential, ET_0, is only relevant for a standard condition that reflects normalized agricultural conditions. The ET_0 value is therefore adjusted according to the microclimate, soils, plants, and growing season conditions. Most of these adjustment factors were developed for agricultural situations, and their use in highly disturbed urban environments has not been well documented. However, it is becoming more common to directly measure urban area ET as part of stormwater management projects. As an example, Selbig and Balster (2010) directly measured ET in an urban setting in Madison, Wisconsin, as part of a stormwater management project for a variety of soil and plant conditions, including when the plants were mostly covered with snow.

The California Irrigation Management Information System (CIMIS) is a comprehensive example for determining ET rates within a state. Its web services are capable of producing an array of useful information about most locations and regions in California. The stations monitored are not limited to traditional agricultural areas, with some monitoring data also available in urban areas.

The ASCE Standardized Reference Equation (Allen et al., 2005) is an example of an ET equation that has been adopted for reference ET_0 calculations. Both the ASCE and Food and Agriculture Organization (FAO-56) have approved versions of the equation with only minor differences (standard crop height being the major difference). ASCE reference ET_0 can be calculated for only two specific crop heights—short (grasses) and tall (alfalfa). The data used in this report were calculated for a short reference crop, most relevant to typical home lawns.

The monthly rainfalls (or soil moisture additions due to the rainfall) for each geographical area, expressed in inches/month, were compared to the evapotranspiration rate requirements for landscaped area plants to determine the irrigation requirements to meet the plant's minimum moisture needs. The reference evapotranspiration rates (ET_0) were obtained from CIMIS for the southwest near Los Angeles and from the ASCE standardized reference equations for the other locations, as shown on Table A-7. The ET_0 values are given in inches/day and were therefore converted to inches/month for direct comparison to the monthly rainfall (or soil moisture addition) values. The Los Angeles and Seattle rainfall monitoring locations were represented by two ET_0 stations that were averaged for these analyses. The other areas only had single ET_0 stations representing their rainfall monitoring locations. Table A-8 shows the monthly values, while Figure A-2 is a plot comparing the seasonal evapotranspiration values for these six rainfall monitoring locations. The ET_0 patterns are similar for all locations with the greatest values (maximums of about 5 to 6.5 inches/month) occurring in the summer months, while the minimum winter ET_0 values are less than 2 inches/month. Seattle has the lowest ET_0 values for most months (annual total of about 28 inches), while Los Angeles has the highest values for most months (annual total of about 49 inches). Specific details on modeling evapotranspiration are also given by Pitt, et al. (2008).

Tables A-9 through A-14, along with Figure 3-2, show the calculations and resulting plots indicating the average monthly irrigation requirements (based on 1995-1999 rainfall; 1995-1999 for Lincoln) to meet the long-term average monthly ET values. A plant's actual ET is calculated by multiplying ET_0 rates by coefficients for each plant type providing a daily moisture estimate for the crop under well-watered conditions. Romero and Dukes (2008) prepared a summary of crop coefficients for the Southwest Florida Water Management District and the Florida Agricultural Experiment Station, which lists turfgrass coefficients for warm and humid areas that ranged from about 0.55 to 0.79 for warm

TABLE A-7 Evapotranspiration Reference Rate (ET_0) Stations Used for Beneficial Use Calculations

Rain Gage Location	ETo Data Source	Station Name	Latitude	Longitude	Elev. (ft)
Los Angeles Airport Weather Service Office, CA	CIMIS Average Monthly Rates, 1989-2011	Glendale, CA	34.197	-118.230	1,111
		Long Beach, CA	33.799	-118.095	17
Seattle Tacoma Airport, WA	ASCE Std. Ref. Eq., 2005-2010	Quilcene, WA	47.82	-122.88	62
		Enumclaw, WA	47.2	-121.96	771
Lincoln Airport, NE	ASCE Std. Ref. Eq., 2008-2011	Rainwater Basin NE	40.57	-98.17	1,790
Madison Dane Co Airport, WI	ASCE Std. Ref. Eq., 2005-2011	Wautoma, WI	43.1	-89.333	857
Birmingham Airport, AL	ASCE Std. Ref. Eq., 2003-2011	Talladega, AL	33.44	-86.081	600
Newark International Airport, NJ	ASCE Std. Ref. Eq., 2005-2011	New Middlesex County NJ	40.41	-74.494	116

TABLE A-8 Monthly ET_0 Values for Study Locations (inches/month)

	Jan	Feb	March	April	May	June	July	Aug	Sept	Oct	Nov	Dec	Annual total (inches/yr)
Los Angeles, CA	1.86	2.40	3.60	4.80	5.27	5.85	6.36	6.20	4.80	3.57	2.40	1.86	48.96
Seattle, WA	0.78	0.99	1.80	2.85	3.26	4.05	4.81	3.88	2.25	1.71	1.20	0.78	28.33
Lincoln, NE	0.93	1.41	3.00	4.50	5.58	6.30	6.20	5.27	4.50	3.72	2.10	0.93	44.44
Madison, WI	0.31	0.57	1.50	3.60	4.96	5.10	5.58	4.34	3.00	2.17	1.20	0.31	32.64
Birmingham, AL	1.24	2.26	3.30	4.50	4.96	4.80	4.96	4.65	4.20	3.72	2.10	1.55	42.24
Newark, NJ	0.62	0.85	2.70	4.20	5.27	5.10	5.58	4.96	4.20	3.10	2.70	1.24	40.52

season grasses. Aronson et al. (1987) listed coefficients for cool season grasses in the humid Northeast that ranged from about 0.6 to 1.04. Brown et al. (2001) presented a summary for arid areas with turfgrass coefficients ranging from about 0.8 to 0.9. For the calculations in this report, a turfgrass coefficient of 0.8 was used for all conditions.

Tables A-9 through A-14 show the monthly ET_0 reference values, the 0.8 turf grass coefficient that reduces the reference ET_0 values to obtain the actual expected evapotranspiration for typical turf grass, along with the average monthly rainfall amounts (based on 1995-1999 precipitation data for five locations and 1996-1999 data for Lincoln). The irrigation requirements shown here are simply the average amounts of water needed monthly in addition to rainfall to meet the ET requirements. Other calculations also considered the moisture added to the soil for each rain instead of the total rainfall, because not all of the rain infiltrates and is available for the plants. These tables show the actual differences between the average ET and rainfall values, and some (especially in the wetter months, or months having low ET requirements) have negative values (the rainfall is greater than the ET requirements). The actual average irrigation requirements per month ignore these negative values, as months with excessive rainfall cannot benefit months requiring irrigation, unless the excess runoff is stored for later beneficial uses (as indicated below in the storage tank modeling descriptions). Figure 3-2 graphically illustrates the average monthly irrigation requirements for the landscaped areas for each of these locations, which were then used in the model to calculate the effects of storage and roof runoff volumes for the different land uses on the resulting domestic water savings.

Table A-15 shows the amount of landscaped area as a percentage of the total land use for different areas in the Los Angeles. The monthly irrigation needs in ft^3 of water per acre of land use per month was calculated by unit conversions using the landscaped area percentage of the land use and the irrigation requirements in inches/month. Small rounding effects may be reflected in the summary tables and example calculations because the model and spreadsheet calculations are high precision, while the summaries and example calculations used truncated significant digits. Also shown on Table A-15 are the total runoff amounts from the roofs and for the whole area for these land uses in Los Angeles. Except for the commercial and industrial areas, land use runoff is not sufficient to completely satisfy the irrigation requirements, and roof runoff alone is close to meeting the irrigation needs only for the commercial areas (simply on a total volume comparison, assuming sufficient storage is provided). The effects of storage tanks also need to be considered, as described below. Other geographical areas with differing rain and ET patterns, plus different land development characteristics, result in dif-

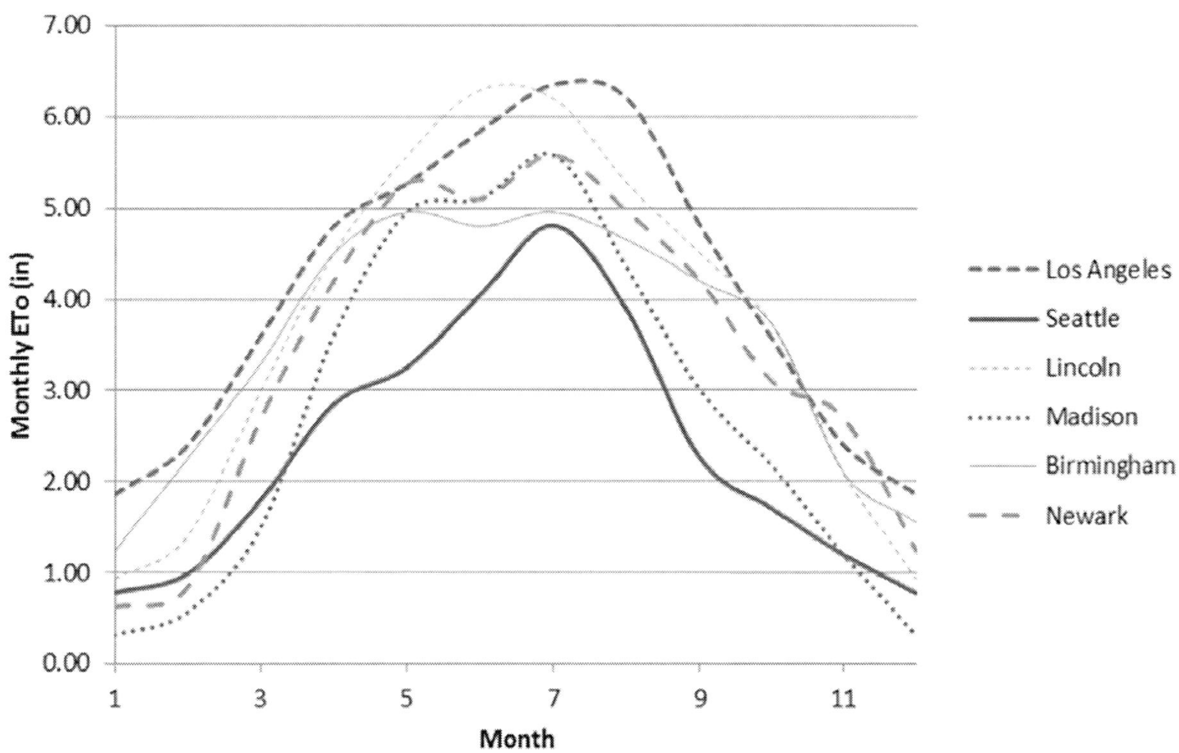

FIGURE A-2 Monthly reference evapotranspiration rates for six study areas.

ferent conclusions. Table A-16 shows the irrigation requirement, expressed in gallons per day per 100 acres of the land use, as used by the model as the water demand for three of the six locations analyzed.

Domestic Water Savings Due to Roof Runoff Harvesting

Two volumes corresponding to typical water storage scenarios (two water barrels per household and one large water storage tank per household) were examined with WinSLAMM corresponding to typical runoff harvesting scenarios. Table A-17 shows the storage volume calculations for the two water storage tank options examined, shown for the Los Angeles example. The model calculates the stormwater runoff volume reductions using continuous simulations for the study period. The water storage tanks are continuously modeled based on additions of roof runoff for each rain and withdrawals to meet monthly average irrigation demand to meet the ET deficits, considering rainfall-induced changes in soil moisture. Overall indoor and outdoor water use behavior was assumed to be the unchanged with the addition of low-cost onsite sources of water. If the tank is full while runoff is still occurring, then the excess runoff is discharged to the drainage system and is not available for beneficial use. If the tank empties due to water withdrawals, then supplemental potable water would be needed to meet additional water demands. Small tanks overflow and are empty more frequently than larger tanks and therefore supply less water for beneficial uses.

The Los Angeles water savings are calculated based on the runoff reductions (with 153,731 ft^3 of water storage volume per 100 acres, corresponding to a single 2,200-gallon water tank at each home). The model calculated 11.6 percent stormwater runoff reductions using this size tank for irrigation in this medium-density, residential land use area. The total average annual runoff for the medium-density, residential area was also calculated to be 27,940 ft^3 per acre. The average domestic water savings by using harvested roof runoff for this scenario analysis is therefore:

$$\frac{11.6\%\ MDR\ runoff\ reduction}{100} * \frac{27{,}940\ ft^3\ runoff}{acre - year} = \frac{3{,}241\ ft^3\ water\ savings}{acre - year}$$

or 2.42 millions of gallons (Mgal) per year for 100 acres.

Indoor Use of Roof Runoff for Toilet Flushing for Medium Density Residential Areas

Toilet flushing water use is based on a per capita water use of 11 gallons per capita per day. With 12 persons/acre and 100 acres of area, this is therefore

$$100\ acres * \frac{12\ persons}{acre} * \frac{11\ gallons}{person - day} = 13{,}200\ gallons/day$$

TABLE A-9 Los Angeles Irrigation Requirements to Meet ET Deficit

	Jan	Feb	Mar	Apr	May	Jun	Jul	Aug	Sept	Oct	Nov	Dec	Annual
LA ETo, in/mo (reference)	1.86	2.405	3.6	4.8	5.27	5.85	6.355	6.2	4.8	3.565	2.4	1.86	
Turf grass coefficient	0.8	0.8	0.8	0.8	0.8	0.8	0.8	0.8	0.8	0.8	0.8	0.8	
LA ET, in/mo (corrected for turf grass)	1.488	1.921	2.88	3.84	4.216	4.68	5.084	4.96	3.84	2.852	1.92	1.488	39.169
LA avg rainfall (in/mo)	4.89	3.76	2.48	0.86	0.59	0.25	0.01	0.00	0.05	0.29	1.30	2.24	16.734
LA irrigation requirements to match ET (in/mo)	-3.406	-1.837	0.4	2.976	3.622	4.434	5.072	4.96	3.788	2.562	0.616	-0.752	
LA irrigation requirements, ignoring excessive rainfall periods (in/mo)	0	0	0.4	2.976	3.622	4.434	5.072	4.96	3.788	2.562	0.616	0	28.43

TABLE A-10 Seattle Irrigation Requirements to Meet ET Deficit

	Jan	Feb	Mar	Apr	May	Jun	Jul	Aug	Sept	Oct	Nov	Dec	Annual
Seattle ETo, in/mo (reference)	0.775	0.989	1.8	2.85	3.255	4.05	4.805	3.875	2.25	1.705	1.2	0.775	
Turf grass coefficient	0.8	0.8	0.8	0.8	0.8	0.8	0.8	0.8	0.8	0.8	0.8	0.8	
SeaTac ET, in/mo (corrected for turf grass)	0.62	0.791	1.44	2.28	2.604	3.24	3.844	3.1	1.8	1.364	0.96	0.62	22.663
SeaTac avg rainfall (in/mo)	6.20	5.12	4.05	2.63	1.28	1.21	0.78	1.06	1.24	4.00	7.92	6.22	41.694
SeaTac irrigation requirements to match ET (in/mo)	-5.576	-4.327	-2.612	-0.348	1.328	2.034	3.064	2.044	0.556	-2.632	-6.964	-5.598	
SeaTac irrigation requirements, ignoring excessive rainfall periods (in/mo)	0	0	0	0	1.328	2.034	3.064	2.044	0.556	0	0	0	9.026

For a year and 100 acres, this amounts to 4.82 Mgal/yr. The indoor per capita water use and population density values were assumed to be the same for all of the medium-density, residential areas examined.

Table A-18 summarizes the monthly Los Angeles water uses for the three water demand scenarios examined in the report: conservation irrigation, toilet flushing, and conservation irrigation plus toilet flushing combined. Table A-19 shows the calculated potential water savings from the WinSLAMM model for the 5 years of rainfall data in a Los Angeles, 100-acre, medium-density, residential, study area. Values were obtained for both the roof areas alone and the total area to check the water savings values. The model calculations for water savings were averaged to obtain the annual runoff savings in both ft3 and millions of gallons.

VERIFICATION OF ORIGINAL ANALYSIS

The committee performed several levels of verification on this original analysis of water savings potential to ensure that the results are sound. The committee members performing the analysis vetted the assumptions of the analysis with

TABLE A-11 Lincoln Irrigation Requirements to Meet ET Deficit

	Jan	Feb	Mar	Apr	May	Jun	Jul	Aug	Sept	Oct	Nov	Dec	Annual
Lincoln ETo, in/mo (reference)	0.93	1.4125	3	4.5	5.58	6.3	6.2	5.27	4.5	3.72	2.1	0.93	
Turf grass coefficient	0.8	0.8	0.8	0.8	0.8	0.8	0.8	0.8	0.8	0.8	0.8	0.8	
Lincoln ET, in/mo (corrected for turf grass)	0.744	1.13	2.4	3.6	4.464	5.04	4.96	4.216	3.6	2.976	1.68	0.744	35.554
Lincoln avg rainfall (in/mo)	0.58	0.55	1.47	3.19	5.29	4.30	2.44	3.89	1.82	1.69	2.15	0.31	27.675
Lincoln irrigation requirements to match ET (in/mo)	0.1665	0.5775	0.935	0.4075	-0.826	0.7425	2.52	0.326	1.78	1.286	-0.47	0.434	
Lincoln irrigation requirements, ignoring excessive rainfall periods (in/mo)	0.1665	0.5775	0.935	0.4075	0	0.7425	2.52	0.326	1.78	1.286	0	0.434	9.175

TABLE A-12 Madison Irrigation Requirements to Meet ET Deficit

	Jan	Feb	Mar	Apr	May	Jun	Jul	Aug	Sept	Oct	Nov	Dec	Annual
Madison ETo, in/mo (reference)	0.31	0.565	1.5	3.6	4.96	5.1	5.58	4.34	3	2.17	1.2	0.31	
Turf grass coefficient	0.8	0.8	0.8	0.8	0.8	0.8	0.8	0.8	0.8	0.8	0.8	0.8	
Madison ET, in/mo (corrected for turf grass)	0.248	0.452	1.2	2.88	3.968	4.08	4.464	3.472	2.4	1.736	0.96	0.248	26.108
Madison avg rainfall (in/mo)	1.49	0.83	1.81	3.46	3.13	5.55	4.07	3.18	1.59	2.60	1.33	0.59	29.62
Madison irrigation requirements to match ET (in/mo)	-1.244	-0.378	-0.614	-0.576	0.842	-1.47	0.394	0.288	0.812	-0.86	-0.368	-0.338	
Madison irrigation requirements, ignoring excessive rainfall periods (in/mo)	0	0	0	0	0.842	0	0.394	0.288	0.812	0	0	0	2.336

TABLE A-13 Birmingham Irrigation Requirements to Meet ET Deficit

	Jan	Feb	Mar	Apr	May	Jun	Jul	Aug	Sept	Oct	Nov	Dec	Annual
Birmingham ETo, in/mo (reference)	1.24	2.26	3.3	4.5	4.96	4.8	4.96	4.65	4.2	3.72	2.1	1.55	
Turf grass coefficient	0.8	0.8	0.8	0.8	0.8	0.8	0.8	0.8	0.8	0.8	0.8	0.8	
Birmingham ET, in/mo (corrected for turf grass)	0.992	1.808	2.64	3.6	3.968	3.84	3.968	3.72	3.36	2.976	1.68	1.24	33.792
Birmingham avg rainfall (in/mo)	6.88	4.32	5.96	4.26	3.96	2.66	3.86	3.36	3.12	4.92	3.72	2.82	49.84
Birmingham irrigation requirements to match ET (in/mo)	-5.888	-2.512	-3.32	-0.66	0.008	1.18	0.108	0.36	0.24	-1.944	-2.04	-1.58	
Birmingham irrigation requirements, ignoring excessive rainfall periods (in/mo)	0	0	0	0	0.008	1.18	0.108	0.36	0.24	0	0	0	1.896

TABLE A-14 Newark Irrigation Requirements to Meet ET Deficit

	Jan	Feb	Mar	Apr	May	Jun	Jul	Aug	Sept	Oct	Nov	Dec	Annual
Newark ETo, in/mo (reference)	0.62	0.8475	2.7	4.2	5.27	5.1	5.58	4.96	4.2	3.1	2.7	1.24	
Turf grass coefficient	0.8	0.8	0.8	0.8	0.8	0.8	0.8	0.8	0.8	0.8	0.8	0.8	
Newark ET, in/mo (corrected for turf grass)	0.496	0.678	2.16	3.36	4.216	4.08	4.464	3.968	3.36	2.48	2.16	0.992	32.414
Newark avg rainfall (in/mo)	4.56	3.07	3.71	3.70	3.89	2.94	4.30	2.65	4.65	3.60	3.58	2.86	43.514
Newark irrigation requirements to match ET (in/mo)	-4.06	-2.388	-1.554	-0.336	0.328	1.14	0.16	1.32	-1.292	-1.122	-1.424	-1.872	
Newark irrigation requirements, ignoring excessive rainfall periods (in/mo)	0	0	0	0	0.328	1.14	0.16	1.32	0	0	0	0	2.948

TABLE A-15 Example Watershed Demand and Available Stormwater by Land Use in Los Angeles

		ft^3 of irrigation water/acre/mo														
	Landscaped Area (% of total land use)	Jan	Feb	March	April	May	June	July	Aug	Sept	Oct	Nov	Dec	Total Annual Irrigation Demand to Meet ET (ft^3/acre)	Total Annual Roof Runoff (ft^3/acre)	Total Annual Land Use Runoff (ft^3/acre)
Commercial	14.9	0	0	216	1,610	1,959	2,398	2,743	2,683	2,049	1,386	333	0	15,377	14,014	42,822
High density residential	46.4	0	0	674	5,013	6,101	7,468	8,543	8,354	6,380	4,315	1,038	0	47,885	11,894	30,603
Medium density residential	52.5	0	0	762	5,672	6,903	8,450	9,666	9,453	7,219	4,883	1,174	0	54,180	10,344	27,939
Low density residential	79.6	0	0	1,156	8,599	10,466	12,812	14,655	14,332	10,945	7,403	1,780	0	82,148	4,596	14,892
Industrial	24.3	0	0	353	2,625	3,195	3,911	4,474	4,375	3,341	2,260	543	0	25,078	10,074	33,534
Institutional	41.2	0	0	598	4,451	5,417	6,631	7,585	7,418	5,665	3,832	921	0	42,519	9,676	30,920

TABLE A-16 Example Irrigation Demand for Land Uses in the Los Angeles, Lincoln, and Newark (gal/day per 100 acres of land use area)

gal/day per 100 Acres of Land Use for Tank Modeling	Roof area (%)	Jan	Feb	Mar	Apr	May	Jun	Jul	Aug	Sept	Oct	Nov	Dec
Los Angeles commercial	28.1	0	0	5,323	39,605	48,202	59,009	67,499	66,009	50,412	34,096	8,198	0
Los Angeles high density residential	20.7	0	0	16,577	123,335	150,107	183,759	210,200	205,558	156,987	106,177	25,529	0
Los Angeles med. density residential	18.0	0	0	18,406	141,504	166,665	210,830	233,386	228,233	180,114	117,890	29,290	0
Los Angeles low density residential	8	0	0	28,439	211,583	257,511	315,242	360,601	352,638	269,313	182,149	43,795	0
Los Angeles industrial	20.2	0	0	8,682	64,591	78,612	96,236	110,083	107,652	82,215	55,606	13,370	0
Los Angeles institutional	19.4	0	0	14,719	109,513	133,285	163,165	186,643	182,521	139,393	94,278	22,668	0
Lincoln commercial	25.0	2,082	7,221	11,692	5,096	0	9,285	31,230	3,420	22,258	15,393	0	5,271
Lincoln high density residential	20.7	6,900	23,933	38,749	16,888	0	30,772	103,504	11,335	73,769	51,017	0	17,468
Lincoln medium density residential	18.1	9,165	34,878	51,465	23,177	0	42,231	138,707	17,944	101,241	70,784	0	23,888
Lincoln medium density residential	18.1	9,339	32,393	52,445	22,857	0	41,648	140,088	15,341	99,842	69,048	0	23,642
Lincoln low density residential	14.9	9,830	34,095	55,201	24,058	0	43,836	147,449	16,147	105,089	72,677	0	24,885
Lincoln industrial	10.2	2,275	7,892	12,777	5,569	0	10,147	34,130	3,738	24,325	16,822	0	5,760
Lincoln institutional	24	6,469	22,438	36,327	15,833	0	28,848	97,035	10,626	69,158	47,828	0	16,377
Newark commercial	28.1	0	0	0	0	4,365	15,171	2,129	17,567	0	0	0	0
Newark high density residential	20.7	0	0	0	0	13,593	47,245	6,631	54,705	0	0	0	0
Newark medium density residential	15.9	0	0	0	0	16,464	57,224	8,031	66,259	0	0	0	0
Newark low density residential	8.0	0	0	0	0	23,320	81,050	11,375	93,847	0	0	0	0
Newark industrial	20.2	0	0	0	0	7,119	24,743	3,473	28,649	0	0	0	0
Newark institutional	19.4	0	0	0	0	12,070	41,950	5,888	48,574	0	0	0	0

the entire committee. Once the analysis was completed, two committee members and one staff person reviewed the spreadsheets containing the graywater analysis and the pre- and post-processing of the stormwater model analysis in detail to check for errors. Assumptions between the two analyses were compared for consistency, and a cell-by-cell assessment was performed to check that the appropriate values and formulas were used. Following this verification, a few minor errors that were detected were discussed with the staff and committee members responsible for the analysis and subsequently corrected.

Additionally, the analysis (including Chapter 3, Appendix A, and associated spreadsheets and input files) was sent to two independent unpaid consultants who were familiar with stormwater modeling to review. They were asked to assess whether the analysis and related assumptions were reasonable and appropriate and to identify any concerns or errors in the analysis. Feedback from the independent consultants was used to strengthen the discussion of uncertainties and the appropriate use of the scenario analysis findings.

TABLE A-17 Site Characteristics and Storage Volumes for the Los Angeles Example

Parameter	Calculation
Roof area (ac per 100 ac of medium-density residential land uses, MDR)	$18\%\ of\ 100\ acres = 18\ acres$
Number of homes in 100 acres (1500 ft^2 roof)	$18\ acres * \frac{43{,}560\ ft^2}{acre} * \frac{home}{1{,}500\ ft^2} = 523\ homes$
Rain barrel storage (gallons/100 ac)	$\frac{70\ gallons}{home} * 523\ homes = 36{,}590\ gallons$
Rain barrel storage (ft^3/100 ac)	$36{,}590\ gallons * \frac{ft3}{7.4805\ gallons} = 4{,}891\ ft3$
Rain barrel storage (ft^3/ft^2 roof area)	$\frac{4{,}891\ ft3}{18\ acres * 43{,}560\ ft2/ac} = 0.006 \frac{ft^3}{ft^2} of\ roof\ area$
Water tank storage (gallons/100 ac)	$\frac{2{,}200\ gallons}{home} * 523\ homes = 1{,}149{,}984\ gallons\ gallons$
Water tank storage (ft^3/100 ac)	$1{,}149{,}984\ gallons * \frac{ft^3}{7.4805\ gallons} = 153{,}731\ ft^3$
Water tank storage (ft^3/ft^2 roof area)	$\frac{153{,}731\ ft^3}{18\ acres * 43{,}560\ ft^2/ac} = 0.20 \frac{ft^3}{ft^2} of\ roof\ area$
Landscaped area (ac per 100 ac of MDR)	$52.5\%\ of\ 100\ acres = 52.5\ acres$

TABLE A-18 Average Monthly Water Use Patterns for Los Angeles Scenario

Gallons/day/100 ac Medium-Density Residential (MDR) Area	Southwest Minimum Irrigation Requirements (gal/day)	Southwest Toilet Flushing (gal/day)	Southwest Minimum Irrigation Plus Toilet Flushing (gal/day)
Jan	0	13,200	13,200
Feb	0	13,200	13,200
Mar	18,406	13,200	31,606
Apr	141,504	13,200	154,704
May	166,665	13,200	179,865
Jun	210,830	13,200	224,030
Jul	233,386	13,200	246,586
Aug	228,233	13,200	241,433
Sept	180,114	13,200	193,314
Oct	117,890	13,200	131,090
Nov	29,290	13,200	42,490
Dec	0	13,200	13,200

TABLE A-19 WinSLAMM Calculated Water Use Savings for Los Angeles, Medium-Density, Residential Scenario Using One 2,200-gallon Water Storage Tank per Household

	Minimum Irrigation	Toilet Flushing	Minimum Irrigation Plus Toilet Flushing
% volume reduction of roof runoff	31.26	35.48	42.36
Total roof runoff (ft^3/5 yrs/100 ac MDR)	5,172,000	5,172,000	5,172,000
Volume of roof runoff used to replace domestic water use (ft^3/5 yrs/100 ac MDR)[a]	1,616,767	1,835,026	2,190,860
% volume reduction of entire MDR area	11.6	13.1	15.68
Total MDR runoff (ft^3/5 yrs/100 ac)	13,970,000	13,970,000	13,970,000
Volume of runoff used to replace potable water use (ft^3/5 yrs/100 ac)[a]	1,620,520	1,830,070	2,190,496
Average annual volume of potable water replaced by roof runoff using water tank (ft^3 per year/100 ac)	323,729	366,510	438,136
Average annual volume of potable water replaced by roof runoff using water tank (Mgal/yr/100 ac)	2.42	2.74	3.28

[a]These values should be the same and were therefore used to verify the calculations.

B

Summary of State Laws and Regulations for Graywater and Stormwater

Additional Legal and Regulatory Tables

TABLE B-1 Examples of Rooftop Runoff and Stormwater Capture Regulations in Prior Appropriation States and Some Riparian States

State	Statute/Regulation	Source
Alaska	None. State plumbing codes based on 2009 UPC, which do not include stormwater use.	http://labor.alaska.gov/lss/forms/plumbing-stats-regs.pdf
Arizona	None, although Arizona water rights laws limit surface water appropriation to water "flowing in streams, canyons, ravines or other natural channels, or in definite underground channels...and of lakes, ponds and springs on the surface." The state previously offered tax credits for rainwater harvesting systems.	ARS § 45-141
California	The state recognizes that rainwater and stormwater capture can contribute to local water supplies and, specifically, reduce reliance on the Delta. The state recommends rain gardens, cisterns, and other landscape features and practices that increase rainwater capture and create opportunities for infiltration and/or on-site storage. The Rainwater Recapture Act of 2012 exempts rooftop capture and on-site use from water rights permitting. In September 2014, the Stormwater Resources Planning Act was amended to require entities developing stormwater resource plans to identify and prioritize opportunities using existing publicly owned lands to capture and use stormwater or dry weather runoff. State plumbing code modified from 2012 UPC.	Cal. Water Code § 10571(c) (West 2014); Cal. Water Code § 10562(a); Cal. Code Regs. tit. 23, § 492.15; Cal. Code Regs. tit. 23, § 5003
Colorado	Colorado's "prior appropriation system" of water rights prevents the capture and use of rainwater falling on private property, unless senior water rights are satisfied. There are two exceptions: 1. Rural residential property owners whose water is supplied by certain wells may utilize "rooftop precipitation collection systems" for capture and use of stormwater. 2. New developments that qualify as one of a limited number of pilot projects may harvest rainwater from impervious surfaces for nonpotable uses as long as the entire amount harvested is replaced to the stream by some other source. If, after a time, the development can calculate the amount of harvested rainwater that would have been consumed by evapotranspiration from native vegetation, then the development can ask the water court for permission to not replace that amount.	Colo. Rev. Stat. § 37-90-105; Colo. Rev. Stat. § 37-92-602; Colo. Rev. Stat. § 37-60-115
Georgia	Appendix I "Rainwater Recycling Systems" of the Georgia 2009 Amendments to the 2006 International Plumbing Code allows rainwater harvesting in certain applications throughout the state. The state also provides state-level guidance on rainwater harvesting.	http://www.dca.state.ga.us/development/constructioncodes/programs/documents/GARainWaterGdlns.040209.pdf
Idaho	None (Note: The Deputy Attorney General issued a 2008 opinion that on-site rainwater harvesting is legal under the state constitution "as long as there is no injury caused to the water rights of others.")	Rassier (2008)
Kansas	No permit is required for rainwater that is used for domestic purposes or for the annual diversion and beneficial use of not more than 15 acre feet of rainwater. (Kansas Statute 82a-701(c) defines "domestic use" as the use of water by any person or by a family unit or household for household purposes, or for the watering of livestock, poultry, farm and domestic animals used in operating a farm, and for the irrigation of lands not exceeding a total of two acres in area for the growing of gardens, orchards and lawns.)	Kan. Stat. Ann. § 82a-728
Maryland	The Maryland Department of the Environment requires a detailed supply and demand analysis for systems larger than 150 gallons. Indoor use is typically limited to toilet flushing, cleaning, and laundry washing.	http://www.mde.state.md.us/programs/Water/StormwaterManagementProgram/SedimentandStormwaterHome/Documents/ESDMEP%20Guidance%20RWH.pdf

(Continued)

TABLE B-1 Continued

State	Statute/Regulation	Source
Minnesota	None, but does provide state-level guidance on rainwater harvesting, including water quality criteria for irrigation.	http://stormwater.pca.state.mn.us/index.php/Stormwater_re-use_and_rainwater_harvesting
Montana	None. State plumbing codes based on 2009 UPC, which do not include stormwater use.	http://bsd.dli.mt.gov/bc/current_codes.asp
Nebraska	None	
Nevada	None	
New Mexico	New Mexico Office of the State Engineer encourages the rooftop capture and use of rainwater at residential and commercial sites for on-site landscape irrigation and other on-site domestic uses. "The collection of water harvested in this manner should not reduce the amount of runoff that would have occurred from the site in its natural, pre-development state."	http://www.rmwea.org/reuse/NewMexico.html
North Carolina	North Carolina Department of Environment and Natural Resources encourages water reuse practices including rainwater harvesting and graywater. The state building code permits the use of cisterns to provide water for flushing toilets and outdoor irrigation.	http://www.ncleg.net/Sessions/2011/Bills/House/PDF/H609v6.pdf http://www.ncleg.net/Sessions/2009/Bills/House/PDF/H749v4.pdf
North Dakota	None	
Ohio	Allows rooftop capture for potable uses, toilet flushing, and laundry with a permit.	Ohio Administrative Code (OAC) 3701-28-12
Oklahoma	The Water for 2060 Act initiates grants for pilot programs including rainwater capture projects.	Okla. Stat. tit. 82 § 1088
Oregon	The Oregon Building Codes Division approved rainwater rooftop harvesting systems as a statewide alternative method for providing water for nonpotable uses, including irrigation, toilet flushing, and HVAC makeup water, and provides detailed design criteria applicable to all projects except irrigation uses. The collection of precipitation water from an artificial impervious surface and the use of such water do not require a water right application, permit, or certificate.	Or. Rev. Stat. § 455.060; Or. Rev. Stat. § 537.141
South Dakota	None	
Texas	The state has adopted laws that encourage municipalities to promote rainwater harvesting, outline safety and health standards for indoor and outdoor use of harvested rainwater, and prohibit homeowner associations from banning rainwater collection.	Tex. Loc. Gov't Code Ann. § 580.004 (West, 2014); Tex. Health & Safety Code Ann. § 341.042 (West, 2014); Tex. Prop. Code Ann. § 202.007 (West, 2014)
Utah	A person who has registered with the state may capture and store 2,500 gallons of rainwater in storage containers above or below ground. A person who has not registered is limited to two storage containers of not more than 100 gallons each.	Utah Code Ann. § 73-3-1.5 (West, 2014)
Virginia	None, but does provide state-level guidance regarding the use of graywater and the harvesting and use of rainwater.	http://www.vdh.state.va.us/EnvironmentalHealth/ONSITE/gmp/documents/2011/pdf/GMP_154.pdf
Washington	The Department of Ecology allows onsite capture of rooftop rainwater without a water right. The Washington Plumbing Code, based on the 2012 UPC, allows for rainwater capture with storage up to 360 gallons without a permit when used for drip or subsurface irrigation. A permit is also not required for rainwater catchment at single family dwellings where all system components are located on the exterior of the building. The state requires that counties reduce by at least 10% the rate they charge for storm or surface water sewer systems for any new or remodeled commercial building that utilizes a permissive rainwater harvesting system.	Wash. Rev. Code § 36.94.140(3); Water Resources Program Policy Regarding Collection of Rainwater for Beneficial Use, Wash. Dept. of Ecology (Oct. 9, 2009); Wash. Admin. Code 51.56.1700
Wyoming	None	

TABLE B-2 Summary of State Laws Regulating Reuse of Graywater at Lower Treatment Standard Than Reclaimed Wastewater[a]

State	Law	Source
Arizona	Arizona water quality statute allows graywater reuse with a permit issued by Department of Environmental Quality (except in certain circumstances where reuse would interfere with water right). Department of Environmental Quality regulations governing direct reuse of reclaimed water include tiered graywater permit criteria (see Table B-3).	Ariz. Rev. Stat. § 49-201 et seq. (LexisNexis, 2014); Ariz. Admin. Code § 18-9-701 et seq.
California	The California Health and Safety Code directs the Building Standards Commission to adopt plumbing code standards for the construction, installation, and alteration of graywater systems for indoor and outdoor uses. The 2013 edition of the California Plumbing Code authorizes graywater systems in residential and nonresidential buildings subject to permits and code standards. The code authorizes single fixture systems without a permit as long as they follow specific best management practices.	Cal. Health & Safety Code § 17922.12, § 18941.8 (West, 2014); Cal. Code Regs. tit. 24, part 5, 1602.0 (2013)
Colorado	Colorado statute signed into law in 2013 directs the Water Quality Control Commission to make rules describing requirements, prohibitions, and standards for the use of graywater for nondrinking purposes, to encourage the use of graywater, and to protect public health and water quality. The regulations restrict graywater use to areas where the local government has adopted an ordinance or resolution approving use of graywater. Graywater ordinances will not alone create new water use rights, and that use of graywater shall be allowed only in accordance with the terms and conditions of the decrees, contracts, and well permits applicable to the use of the source water rights or source water and any return flows therefrom.	Colo. Rev. Stat. § 25-8-205, § 30-11-107, § 37-90-102
Florida	Florida Plumbing Code allows graywater use for toilet flushing (disinfection required). Public health statutes governing on-site sewage disposal systems empower the Department of Health to approve the installation of individual graywater disposal systems where blackwater is treated by central sewage system. Additionally, water conservation statute "urges" public-owned and investor-owned water and sewerage systems to reduce connection fees and regular service charges for customers who utilize water-saving graywater systems.	2010 Florida Plumbing Code, Appendix C (adopted by reference at Fla. Admin. Code § 61G-20-1.001); Fla. Admin. Code Ann. r.64E-6-011 et seq.; Fla. Stat. § 373.619
Georgia	Health code authorizes private residential direct reuse of graywater for outdoor landscape irrigation under a set of criteria outlined in the statute (and county boards of health are directed to adopt the criteria by regulation). Natural resources code requires GA state agencies to provide rules and regulations to encourage the use of graywater in lieu of potable water and exempts graywater from outdoor watering restrictions. The Georgia Amendments to the 2006 IPC allow graywater fixtures to discharge to an approved graywater system for flushing of toilets or subsurface irrigation.	Ga. Code Ann. § 31-3-5.2; Ga. Code Ann. § 12-5-4, 12-5-7; Ga. Comp. R. & Regs. 110-11-1.19-1.21
Hawaii	Pursuant to water pollution statutes, the health department may authorize any county to implement a graywater recycling program for irrigating lawns and gardens, but all use of graywater must conform to the state plumbing code. Department of Health wastewater disposal regulations affirm that graywater may be used for subsurface irrigation. State-specific amendments to 2006 Uniform Plumbing Code (UPC), with state-specific amendments to the graywater chapter authorizing use of graywater for subsurface irrigation.	Haw. Rev. Stat. § 342D-7; Haw. Code R. § 11-62-31.1; Haw. Code R. § 3-183-13 (adopts UPC by reference with amendments)
Idaho	Graywater is defined as untreated wastewater from bathtubs, showers, and bathroom wash basins. Laundry water is not included in the definition of graywater. Residential graywater reuse systems are permitted for subsurface irrigation only and require a permit. The Idaho State Plumbing Code is modeled after the 2009 UPC; Idaho amendments clarify that graywater fixtures up to, but not including, exterior irrigation tanks must be inspected by the authority issuing building permits, while Department of Environmental Quality has jurisdiction to inspect and approve the installation of the exterior irrigation system tank and all piping up to the point of disposal in accordance with Individual/Subsurface Sewage Disposal Rules.	Idaho Admin. Code r. 07.02.06.011; Idaho Admin. Code r. 58.01.03
Indiana	The 2012 Indiana Plumbing Code, modeled from the 2009 International Plumbing Code (IPC), allows for the reuse of graywater for indoor and outdoor uses under permit.	675 Indiana Admin. Code 16-1.4

(Continued)

TABLE B-2 Continued

State	Law	Source
Kansas	In 2014, the Kansas Department of Health and Environment issued guidance for permitting subsurface residential graywater irrigation systems under a variance to KAR 28-5-2 through 7. Local authorities may adopt or prohibit such systems within their jurisdictions.	See http://www.kdheks.gov/nps/lepp/download/Graywater_System_Specification_FINAL.pdf.
Maine	Maine has adopted the 2009 UPC but "does not adopt 'Part I, Graywater Systems,' in its entirety." Maine's subsurface wastewater disposal regulations allow for "limited" graywater systems with no more than 1,000-gallon storage capacity.	02-395-004 Code R. § 1; 10-144-241 Code R. § 1-13
Massachusetts	On-site sewage treatment and disposal regulations issued by the Department of Environmental Protection set out requirements for graywater systems in residential, commercial, and public facilities. The Uniform State Plumbing Code governs all plumbing systems, including nonpotable water supply lines. Under the plumbing code, water recycling systems—including graywater systems—can be installed with special permission.	310 Mass. Code Regs. 15.000 et seq.; 248 Mass. Code Regs. 10.00 et seq.
Montana	Environment code requires the Board of Environmental Review to establish rules allowing diversion of graywater from wastewater treatment systems. The graywater rules allow graywater use for toilet flushing without a permit from the environmental quality department, but a permit is required for irrigation. Montana has also adopted the 2009 UPC. Graywater may be land-applied at approved sites without vector or pathogen reduction only if it will not pollute state waters.	Mont. Code Ann. § 75-5-326 et seq.; Mont. Admin. R. 17.36; Mont. Admin. R. 24.31.301 (adopts UPC); Mont. Admin. R 17.50.810
Nevada	Nevada has adopted the UPC. Sewage disposal regulations allow single-family dwellings to use graywater systems for subsurface irrigation.	Nev. Rev. Stat. § 444.350; Nev. Admin. Code § 444.7616, 444.7825, 444.837, 444.8732
New Hampshire	Under New Hampshire's plumbing code, based on the 2009 UPC, graywater fixtures are not required to discharge to the sanitary drainage system where such fixtures discharge to an approved graywater system for flushing of water closets and urinals or for subsurface landscape irrigation. However, New Hampshire has not as of 2014 approved any uses beyond disposal.	N.H. Admin. Rules, Bcr § 304.01
New Mexico	Graywater rules, located in the liquid waste disposal regulations, set out requirements for graywater use for irrigation and toilet flushing. New Mexico adopted the 2009 UPC with significant amendments to Chapter 16. New Mexico water quality statutes allow use of small quantities of graywater for residential irrigation without a permit. Surface drip irrigation is allowed.	N.M. Stat. Ann. § 4-6-4; N.M. Code R. § 20.7.3. 809-10; N.M. Code R. § 14.8.2.27. See also http://www.ose.state.nm.us/water-info/conservation/pdf-manuals/NewMexGWGuide.pdf
New York	Residential buildings are required to connect all plumbing systems to the sanitary drainage system. Appendix C of the 2010 New York Plumbing Code, based on the IPC, provides standards for use of graywater for toilet flushing and subsurface irrigation in nonresidential buildings only.	N.Y. Comp. R & Regs tit. 19, § 1222.1
North Carolina	The 2012 NC plumbing code is adapted from the IPC and allows installation of graywater systems for toilet flushing. Public health statute states that graywater systems shall be regulated by the health department under rules promulgated by the environmental management commission to encourage and promote the safe and beneficial use of graywater. In 2008, North Carolina passed House Bill 2499, a drought bill that allows graywater reuse by watering with buckets in case of a drought year, which will sunset when graywater rules are promulgated.	N.C. Gen. Stat. § 130A-335; N.C. Gen. Stat. § 143-350, 355.5
Ohio	Ohio health and safety statute directs the board of health to prescribe standards for regulation of graywater recycling systems. According to the Department of Health, the final draft of proposed rules (effective January 2015) will be adopted into the administrative code at § 3701-29-17 (http://www.odh.ohio.gov/rules/drafts/3701-29.aspx). The proposed rules permit graywater reuse for irrigation only. Ohio Plumbing Code allows segregation of graywater from wastewater streams when the graywater discharges to a graywater recycling system approved by Ohio Environmental Protection Agency (EPA) in accordance with on-site disposal regulations.	Ohio Rev. Code Ann. § 3718.02; Ohio Admin Code 4101:3-3-01
Oklahoma	Oklahoma passed a law in 2012 that allows graywater reuse without a permit when applying less than 250 gallons per day of untreated private residential graywater for gardening, composting, or landscape irrigation of the resident if a set of simple best management practices (BMPs) is followed that restrict surface ponding, spray irrigation, and off-site discharge. Surface drip irrigation is allowed.	Okla. Stat. 27A, § 2-6-101 et seq.

TABLE B-2 Continued

State	Law	Source
Oregon	Public health statute declares it the public policy of the state to encourage reuse of graywater for beneficial uses. The statute prohibits any person from installing or operating a graywater system without a permit from the Department of Environmental Quality. Rules were approved by the Environmental Quality Commission in 2011 that prescribe tiered requirements for the permitting of graywater reuse systems.	Or. Rev. Stat. § 454.607; Or. Rev. Stat. § 444.610; Or. Admin. R. 340-053-0050 to 0110
South Dakota	On-site wastewater disposal regulations establish basic requirements for graywater reuse for toilet flushing or irrigation.	S.D. Admin. R. 74:53:01:38
Texas	The Texas Health and Safety Code and the Texas Water Code require the Commission on Environmental Quality to adopt rules implementing minimum standards for the use and reuse of graywater for a variety of uses; the Health Code specifies that regulations may not require a permit for the domestic use of less than 400 gallons of graywater each day if the graywater originates from a private residence and is used for gardening, composting, or landscaping (observing certain simple BMPs). Environmental quality regulations—or the graywater code—provide criteria for domestic, industrial, commercial, institutional, and irrigation uses.	Tex. Health & Safety Code Ann. § 341.039, § 366.012; Tex. Water Code § 26.0311; 30 Tex. Admin. Code § 210.81 et seq.; 30 Tex. Admin. Code § 285.80 et seq.
Utah	In 2013, Utah adopted water quality regulations governing graywater systems that apply to the construction, installation, modification, and repair of graywater systems for subsurface landscape irrigation for single-family residences. It is unlawful to construct, install, or modify a graywater system in a building or on a lot without first obtaining a permit from the local health department.	Utah Code Ann. § 15A-3-313; Utah Admin. Code R317-401-1 et seq.
Virginia	Regulations issued in 1999[b] require health and building code permits prior to installation of a graywater system. Graywater systems must be inspected by the State Health Department prior to operation. Virginia uses an amended version of the IPC that governs graywater systems for toilet flushing, irrigation, and other nonpotable applications. Utilization of harvested graywater is expressly exempted from requirements of the Water Reclamation and Reuse regulations.	Va. Code Ann. § 32.1-248.2; 13 Va. Admin. Code § 5-63-320; 13 Va. Admin. Code § 5-63-210; 9 Va. Admin. Code § 25-740-10 et seq.
Washington	State amendments to the 2009 UPC allow graywater to be used in lieu of potable water for indoor nonpotable uses and subsurface irrigation where permitted by the Department of Health rules and apply UPC standards to graywater systems. Department of Health regulation "Greywater Reuse for Subsurface Irrigation" sets out a comprehensive three-tiered permitting system.	Wa. Admin. Code § 51-56-1600; Wa. Admin. Code § 246-274-001 et seq.
Wyoming	As of July 2014, Wyoming's graywater permitting system was undergoing a transition from permit-by-rule approach to a more comprehensive regulatory approach. However, the status of this effort is unclear.	

[a]This table summarizes generally applicable state statutes and regulations addressing graywater reuse. The summary does not include state constitutional provisions, court decisions, guidance documents, or local ordinances that may be relevant to the legality of graywater reuse in a particular location. This table does not include state laws establishing green infrastructure tax credits or grant programs. Inconsistencies among two or more areas of a state's code may affect the legality of graywater reuse. Table 6-3 describes the plain language of statutes and regulations addressing graywater reuse (retrieved from LexisNexis and Westlaw using keyword searches, June 2014) and is not intended to determine the legality of graywater reuse in light of ambiguous or contradictory provisions.

[b]https://www.vdh.virginia.gov/EnvironmentalHealth/ONSITE/regulations/FormsDocs/documents/2010/pdfs/Graywater%20Use%20guidelines%20by%20VDH_feb99.pdf.

TABLE B-3 Arizona Tiered Regulations for Graywater Irrigation

Permit Type	Requirements for Direct Reuse of Graywater
Type 1 Reclaimed Water General Permit for Graywater Graywater irrigation systems with a flow of <400 gallons per day	No prior approval or notice to the agency is necessary if the following 13 best practice measures (BMPs) are followed: Human contact with graywater and soil irrigation by graywater is avoided. Graywater originating from the residence is used and contained within the property boundary for household gardening, composting, lawn watering, or landscape irrigation. Surface application of graywater is not used for irrigation of food plants, except for citrus and nut trees. Graywater does not contain hazardous chemicals derived from activities such as cleaning car parts, washing greasy or oily rags, or disposing of waste solutions from home photo labs. The applications of graywater are managed to minimize standing water on the surface. The graywater system is constructed so that if blockage, plugging, or backup of the system occurs, graywater can be directed into the sewage collection system, or on-site wastewater treatment or disposal system, as applicable. The graywater system may include filtration to reduce plugging and extend system lifetime. Any graywater storage tank is covered to restrict access. The system is sited outside of a floodway. The system is operated to maintain a minimum vertical separation distance of at least 5 feet from the point of graywater application to the top of the seasonally high groundwater. If using an on-site wastewater treatment facility for blackwater treatment and disposal, the use of a graywater system does not change the design, capacity, or reserve area requirements for the on-site waste system. Any pressure piping used in a graywater system that may be susceptible to cross connection with a potable water system clearly indicates that the piping does not carry potable water. Graywater applied by surface irrigation does not contain water used to wash diapers or similarly soiled or infectious garments unless the graywater is disinfected before irrigation. Surface irrigation by graywater is only by flood or drip irrigation. Graywater cannot be used for other purposes besides subsurface irrigation or drip irrigation.
Type 3 General Permit for Graywater Graywater irrigation systems with a flow between 400 - 3,000 gallons per day	A notice of intent to operate a graywater irrigation system must be submitted to the department 90 days before construction beings. The system must meet the setback and soil absorption rates under the on-site wastewater treatment facility requirements for shallow trenches. The depth of the graywater dispersal trenches should be designed for appropriate irrigation use but not more than 5 feet below the finished grade of the soil. The void space of the aggregate fill in the dispersal trench must allow enough capacity to contain 2 days of graywater at design flow. The department has the authority to review design plans and details to accept a graywater irrigation system different from the typical system provided sufficient performance and protection are provided.
On-site disposal systems with flow >3,000 gallons per day	For large systems, the department handles permits on case-by-case basis.

SOURCE: Ariz. Admin Code § 18-9-711, 719. Ariz. Rev. Stat. § 49-242, 243.

C

Water Science and Technology Board

D

Biographical Sketches of Committee Members and Staff

RICHARD G. LUTHY (NAE), *Chair*, is the Silas H. Palmer Professor in the Department of Civil and Environmental Engineering, and Senior Fellow at the Woods Institute for the Environment at Stanford University. His area of teaching and research is environmental engineering and water quality with application to water reuse and management of contaminated sediments. He is the director of the National Science Foundation's Engineering Research Center for re-inventing the nation's urban water infrastructure (renuwit.org). The Center is a collaboration among four universities that promotes more sustainable solutions to urban water challenges. His work includes study of persistent and bio-accumulative contaminants and emerging contaminants. Dr. Luthy is a past chair of the the National Academies of Science, Engineering, and Medicine's Water Science and Technology Board, and he has served on various Academies committees. He is a former president of the Association of Environmental Engineering and Science Professors. He is a registered professional engineer, a board-certified environmental engineer, water environment foundation fellow, and a member of the National Academy of Engineering. He received a B.S. in chemical engineering from the University of California, Berkeley, a M.S. degree in ocean engineering from the University of Hawai'i at Manoa, and an M.S. and Ph.D. degrees in environmental engineering from the University of California, Berkeley.

RICHARD W. ATWATER is the executive director of the Southern California Water Committee (SCWC), a nonprofit public education partnership focused on solving the water problems of southern California. Mr. Atwater has more than 35 years of experience in water resources management and development in the western United States and has pioneered many award-winning water projects and implemented numerous innovative water resource management programs. Prior to joining the SCWC, he served as the CEO and general manager of the Inland Empire Utilities Agency, which provides wholesale water and wastewater utility services to more than 850,000 customers. Throughout his career, Mr. Atwater has accumulated extensive public agency management experience in directing the development of some of the largest water projects in the United States. This includes the water recycling program for the West and Central Basin Municipal Water Districts in California, which at the time were the largest in the country. He is the recipient of the Conservation Service Award, the highest citizen award for resources management, and has participated in policy formulation workshops and expert panels for the National Academy of Sciences, Western Governors Association, Western Water States Council, and the National Water Research Institute. Mr. Atwater received his B.S. degree in geology and environmental science from Stanford University and his M.P.L. degree in urban and regional planning from the University of Southern California.

GLENN T. DAIGGER (NAE) is a professor of engineering practice in the Department of Civil and Environmental Engineering at the University of Michigan. He is also president and founder of One Water Solutions LLC. Previously, he was the senior vice president with CH2M HILL in Englewood, Colorado. He served as chief wastewater process engineer and was responsible for wastewater process engineering on both municipal and industrial wastewater treatment projects on a firm-wide basis. Dr. Daigger was the first technical fellow for the firm, an honor that recognizes the leadership he provides for CH2M HILL and for the profession in development and implementation of new wastewater treatment technology. He was also the chief technology officer for the firm's Civil Infrastructure Client Group, which includes the water, transportation, and operations businesses. From 1994 to 1996, Dr. Daigger served as professor and chair of the Department of Environmental Systems Engineering at Clemson University. Dr. Daigger is a registered professional engineer in the states of Indiana and Arizona and a board-certified environmental engineer. He also has Academies experience, having recently served as chair of the Committee to Review

EPA's Economic Analysis of Final Water Quality Standards for Nutrients for Lakes and Flowing Waters in Florida. Dr. Daigger received his B.S.C.E. degree, his M.S.C.E. degree, and his Ph.D. degree, all in environmental engineering, from Purdue University.

JÖRG DREWES is chair professor for urban water systems engineering at Technical University of Munich, and he is a professor emeritus and research professor at the Colorado School of Mines. He brings extensive knowledge and experience with graywater systems in Germany. His research interests focus on water and wastewater treatment engineering and potable and nonpotable water reuse. In particular, he focuses on technologies leading to indirect potable reuse (soil-aquifer treatment versus microfiltration/reverse osmosis); beneficial reuse of produced water during natural gas exploration; desalination and concentrate volume minimization; state-of-the-art characterization of natural and effluent organic matter; fate and transport of emerging contaminants (such as endocrine disrupting compounds, pharmaceutical residues and household chemicals) in natural and engineered systems as well as the rejection mechanisms of organic micropollutants in high-pressure membranes. Dr. Drewes received his B.S., M.S., and Ph.D. degrees in environmental engineering from the Technical University of Berlin.

BENJAMIN H. GRUMBLES is Secretary of the Environment for the state of Maryland. Previously, he served as president of the U.S. Water Alliance, where he worked to unite people and policies for water sustainability throughout the country. Prior to joining the Water Alliance, Mr. Grumbles led Arizona's Department of Environmental Quality working on air quality and climate change, energy policy and waste management, water efficiency, and wastewater recycling. Regional priorities in this effort included protecting the Grand Canyon, Colorado River, and Arizona-Mexico border environment. Mr. Grumbles served as Assistant Administrator for Water at the U.S. Environmental Protection Agency (EPA) from 2004 through 2008. He launched the EPA's water efficiency labeling program, WaterSense, and initiatives on green infrastructure, water and climate change, and pharmaceuticals. He carried out and defended the nation's clean water, drinking water, ocean and coastal, and wetlands laws and worked on great waterbody collaborations from coast to coast. Mr. Grumbles is currently a member of the Academies' Water Science and Technology Board. He received his B.A. degree from Wake Forest University, a master's degree in environmental law from George Washington University Law School, and a J.D. degree from Emory University Law School.

ARPAD HORVATH is a professor of civil and environmental engineering at the University of California (UC), Berkeley. He heads the Energy, Civil Infrastructure and Climate Graduate Program and is the director of UC Berkeley's Consortium on Green Design and Manufacturing and of UC Berkeley's Engineering and Business for Sustainability Certificate Program. His research focuses on developing models for life-cycle environmental and economic assessment of products, processes, and services, particularly of civil infrastructure systems. He has worked the environmental implications of transportation systems, buildings, construction, water and wastewater systems, and various service industries. Dr. Horvath is a member of the Environmental Engineering Committee of the EPA's Science Advisory Board, as well as the EPA's Scientific and Technological Achievement Awards Committee. He is associate editor of the *Journal of Infrastructure Systems* and is on the editorial boards of *Environmental Science & Technology, Environmental Research Letters*, and the *Journal of Industrial Ecology*. Dr. Horvath was conference chair of the 6th International Conference on Industrial Ecology in 2011. He is a recipient of the American Society of Civil Engineers' Walter L. Huber Civil Engineering Research Prize, the Laudise Prize "for outstanding achievements in industrial ecology by a young scientist or engineer" of the International Society for Industrial Ecology, and the Excellence in Review Award from *Environmental Science & Technology*. Three of his co-authored papers have been named among the top three papers in *Environmental Science & Technology* in 2008, 2011, and 2012. He received a Dipl. Eng. (M.S.) degree in civil engineering from the Technical University of Budapest and M.S. and Ph.D. degrees in civil and environmental engineering from Carnegie Mellon University.

ROBERT E. PITT is the Emeritus Cudworth Professor of Urban Water Systems in the School of Engineering at the University of Alabama, Tuscaloosa. His major area of interest is in stormwater management, especially the integration of drainage and water quality objectives associated with green infrastructure and combined sewers, development of stormwater treatment systems at critical source areas, system modeling or urban water systems, and the beneficial uses of stormwater. His research has also examined stormwater effects on groundwater. He is a member of the American Society of Civil Engineers and the American Water Resources Association and has served as a member of the Academies Committee on Reducing Stormwater Discharge Contributions to Water Pollution and the Groundwater Recharge Committee. He received his B.S. degree in engineering science from Humboldt State University, his M.S. degree in civil engineering from San Jose State University, and his Ph.D. degree in civil and environmental engineering from the University of Wisconsin.

MARCUS M. QUIGLEY is a founding partner and chief executive officer of OptiRTC, Inc., a company formed by Geosyntec, focused on intelligent control and decision support solutions for stormwater management. Previously, Mr. Quigley served as principal civil and environmental engineer at Geosyntec Consultants, where he worked in the areas of surface water hydrology, hydraulics, water quality, and stormwater and erosion and sediment control permitting and management. He is recognized as a national technical leader in stormwater best management practice (BMP) design, research and development, modeling, data analysis, and field data acquisition. He has been the lead designer for a number of conventional and low impact development (LID) controls systems, and has directed groundbreaking monitoring work to demonstrate the effectiveness of LID. Mr. Quigley regularly conducts and directs complex surface water quantity and quality modeling efforts, and during the past 10 years he has provided technical leadership and project management for the International Stormwater Best Management Practices Database project. He received his B.S. degree in environmental engineering from the University of Notre Dame and his M.S. degree in civil and environmental engineering from Oregon State University.

ROBERT S. RAUCHER is a founding partner and principal at Stratus Consulting/Abt Associates. He specializes in economics, risk management, strategic planning, and regulatory policy analysis related to water utilities, water resources, and environmental quality. He is a noted expert on water resources management, benefit-cost analysis and water-related valuation issues, regulatory policy, and climate change-related vulnerability assessment and adaptation strategies for water resource management. Dr. Raucher has been involved in desalination and water reuse planning and implementation issues, and he is actively engaged in research assessing reliability values associated with water supply portfolio diversification through desalination and other "new" water options that offer drought-resistant yields. Dr. Raucher is an active member of the water supply and wastewater community, serving on numerous expert panels and committees, including three workgroups for the National Drinking Water Advisory Council, and he is a member of the WateReuse Foundation's Research Advisory Committee. He received a B.A. degree in economics and anthropology, an M.S. degree in econometrics, and a Ph.D. degree in natural resource economics and public finance.

SYBIL SHARVELLE is an assistant professor in the Department of Civil and Environmental Engineering at Colorado State University (CSU). Her research interests focus on wastewater and graywater reuse; biological process engineering; conversion of waste to energy; and integrated urban water management. Dr. Sharvelle is also a member of the CSU Sustainable Urban Water Research Working Group, where she works to address new and innovative infrastructure design concepts in water management through a multidisciplinary approach both in the United States and abroad. She received her B.S. and M.S. degrees in civil engineering from the University of Colorado and while pursuing her M.S. degree, she optimized the nitrification and denitrification steps in a biological processor for treatment of wastewater highly concentrated with ammonia. Dr. Sharvelle also received a Ph.D. degree from Purdue University, where her research involved design and optimization of a biotrickling filter for simultaneous treatment of graywater and waste gas. A major component of this research effort was to examine the fate of surfactants in the biotrickling filter.

CLAIRE WELTY is a professor of civil and environmental engineering and director of the Center for Urban Environmental Research and Education at the University of Maryland, Baltimore County. Her research is focused on developing an end-to-end system of field-deployed sensors and fully coupled groundwater-surface water mathematical models to quantify and predict the urban hydrologic cycle and coupled biogeochemical cycles from neighborhood to regional scales. Her goal is to be able to assimilate sensor data into hydrologic and water quality models in near-real time for predicting flow paths, fluxes, and stores of water and chemicals on land surfaces and in the subsurface. Dr. Welty has Academies experience, having served on a number of committees and as chair of the Water Science and Technology Board. She received her B.A. degree in environmental sciences from the University of Virginia, her M.S. degree in environmental engineering from the George Washington University, and her Ph.D. degree in civil engineering from the Massachusetts Institute of Technology.

MARYLYNN V. YATES is a professor of environmental microbiology at the University of California, Riverside. Dr. Yates conducts research in the area of water and wastewater microbiology. Her research focuses on assessing the potential for the contamination of water by human pathogenic microorganisms. As the intentional use of reclaimed water and biosolids (which may contain pathogenic microorganisms) increases, it is necessary to understand the potential impacts of these practices on public health. Dr. Yates has

Academies experience, currently serving as a member of the Water Science and Technology Board. She received her B.S. in nursing from the University of Wisconsin, her M.S. degree in chemistry from the New Mexico Institute of Mining and Technology, and her Ph.D. degree in microbiology and immunology from the University of Arizona.

Staff

STEPHANIE E. JOHNSON, study director, is a senior program officer with the Water Science and Technology Board. Since joining the National Academies of Science, Engineering, and Medicine in 2002, she has worked on a wide range of water-related studies, on topics such as desalination, wastewater reuse, contaminant source remediation, coal and uranium mining, coastal risk reduction, and ecosystem restoration. She has served as study director for 15 committees, including all 6 Committees on Independent Scientific Review of Everglades Restoration Progress. Dr. Johnson received her B.A. degree in chemistry and geology from Vanderbilt University and her M.S. and Ph.D. degrees in environmental sciences from the University of Virginia.

MICHAEL J. STOEVER is a research associate with the Water Science and Technology Board. He has worked on a number of studies including Desalination: A National Perspective, the Water Implications of Biofuels Production in the United States, and the Committee on Louisiana Coastal Protection and Restoration. He has also worked on National Research Council studies on the review of Everglades restoration progress, the effect of water withdrawals on the St. Johns River, and Chesapeake Bay restoration. Mr. Stoever received his B.A. degree in political science from Stockton University in Pomona, New Jersey, and is currently working toward his M.S. degree in environmental sciences and policy from Johns Hopkins University.